The Earliest Wheeled Transport

From the Atlantic Coast to the Caspian Sea

The Earliest
Wheeled Transport

From the Atlantic Coast to the Caspian Sea

Stuart Piggott

Cornell University Press
Ithaca, New York

Quis? Quid? Ubi? Quibus auxiliis? Cur? Quomodo? Quando?

What was the crime? Who did it? When was it done and where?
How done and with what motive? Who in the deed did share?

Eighteenth-century criminological maxim by
Joachim Georg Darjes of Frankfurt-am-Oder

Maps and line drawings by H. A. Shelley

First published 1983 by Cornell University Press

International Standard Book Number 0-8014-1604-3
Library of Congress Catalog Card Number 82-73810

Printed and bound in Hungary

Contents

Foreword

One single invention of outstanding importance to all human societies was the wheeled vehicle which first appeared at the end of the fourth millennium BC. In this book I have attempted a study of the relevant archaeological evidence, considered in terms of the adoption and adaptation of this vehicle over a specific period of time and a clearly defined geographical area. This covers Europe and western Russia, running from the Atlantic to the Caspian Sea, and the three thousand years from the time of the original invention to the establishment of the Roman Empire.

My interest in prehistoric wheeled vehicles was largely developed as a result of Cyril Fox's work on the La Tène chariot remains in the Llyn Cerrig Bach find, published in 1946, at a time when I was trying to interpret the references to chariotry in the Sanskrit *Rig Veda* as part of a study of prehistoric India. His enthusiastic support and co-operation encouraged me to go further, as did Gordon Childe, concerning himself at that time with the origins of wheeled transport in the Old World. A third distinguished archaeologist no longer with us was Terence Powell, that perceptive student of ancient chariotry and horsemanship, who travelled with me in more than one part of Europe and contributed greatly to my understanding of later European prehistory. In the preparation of the book in its present form over the last few years I owe an outstanding debt of gratitude to Mrs Mary Aitken Littauer, who has so generously shared with me her unrivalled knowledge and sound judgment of the realities of animals and vehicles in antiquity.

So far as the differing nature of the evidence allows (with, for instance, the paucity of representational art north of the Alps), it is hoped that this study may form a complementary survey to that of Mrs Littauer and T. Crouwel (1979a) for the ancient Near East. With a few exceptions it takes account of published and unpublished material known and accessible to me to June 1981. In a wide survey such as this accidental omissions are not easy to avoid, and for these I offer my apologies: they would have been more numerous had it not been for the willing co-operation of fellow scholars thoughout Europe. The failure, however, of not a few individuals and institutions to reply to enquiries has led to gaps in information which I have been unable to fill.

In a synthesis necessarily based on the work of others, my debt to numberless colleagues here and abroad is obvious. In this country I am particularly grateful to Proressor Warwick Bray, who guided me in the Mesoamerican sources; to Mr David and Dr Francesca Ridgway for their critical comments on the draft of much of Chapter 5 and to Professor A. L. F. Rivet for a similar salutary reading of the last sections of Chapter 6. Mr. Richard Langhorne kindly allowed me to use his unpublished MA dissertation on *British Iron Age Warfare: the texts and the archaeology,* and for help in various ways I am indebted to Professor John Coles, Dr Dominique Collon, Mr Heinrich Härke, Professor Martyn Jope, Dr Carol McNeil, Professor Vincent Megaw, Dr Andrew Sherratt and Dr Ian Stead. On the Continent I would especially thank Professor A. Arribas, Dr J. A. Bakker, Dr A. Cahen-Delhaye, Professor W. Dehn, Dr S. Esayan, Dr A. Haffner, Dr A. Häusler, Professor H-J. Hundt, Dr H-E. Joachim, Professor W. Kimmig, Dr E. E. Kuzmina, Dr A. E. Lanting and Professor D. van der Waals, and in the United States Professor M. Gimbutas and Professor S. Milisauskas. The work would not have been possible without the resources and the courtesy and help of the staff of the Ashmolean Museum Library in Oxford, and that of the Society of Antiquaries of London and its Librarian, Mr John Hopkins. Mrs Mary Figgis transformed an idiosyncratic manuscript into faultless typescript, and to her and to the staff of Thames and Hudson the book and its author owe much. I would also like to thank Mrs Elizabeth Fowler for preparing the index.

The late Dame Ngaio Marsh, aided by Professor John Crook, was very helpful in tracking down the source of the Latin tag, quoted in her detective fiction, and here used as an epigraph, as appropriate to an archaeological, as much to a criminological, investigation.

STUART PIGGOTT

1 Introductory: history and the technological background

The history of the study

An interest in the wheeled vehicles of antiquity arose early in antiquarian studies, primarily as a result of the repeated references to them in the ancient texts which became increasingly available to scholars from the Renaissance onward. But before that western Christendom had been made vividly aware of a particular type of vehicle, the oriental chariot, by its repeated appearance in its fundamental historical document, the Old Testament. Here one had to read no further than Genesis 41 to encounter Joseph, honoured by a place in Pharaoh's second chariot *(super currum suum secundum)*, and four chapters later bidden to take his people's wagons out of Egypt *(tollant plaustra de terra Aegypti)*; henceforward the drama of Exodus proceeds with the annihilation of the Egyptian chariot forces in the Red Sea. The chariot becomes a potent symbol, human and divine, and the world of the second and earlier first-millennium Levant was seen as it was in fact, one in which St Jerome's *currus* and *plaustrum* had important parts to play in warfare or in agriculture. To the Middle Ages, easily transposing the contemporary world into that of remote Biblical antiquity, and incurious of the history of technology, a past or a people without wheeled transport would be difficult to visualize: later conjectures on the inventor of the wheel usually looked to Tubal-cain. When at the dawn of early modern times the Americas presented to Europe a people who did not include wheeled transport in their technology, the fact does not appear to have aroused comment and, had it done so,

there were alternative explanations ready to hand, such as the inevitable cultural degeneration of peoples who had wandered so far from Ararat, or even their exclusion from the ranks of true human beings.

With the Renaissance and the discovery of the classical past of Greece and Rome both the literary and increasingly the iconographical evidence for a wide variety of wheeled vehicles in antiquity became apparent. Some of the Latin tradition had, of course, been transmitted by the Church and its lexical richness in vehicle terms appreciated and commented on by encyclopaedists such as Isidore of Seville, but now texts and scholia revealed more, and at first hand. And not only for the civilized Mediterranean world, but for the barbarians, as Celtic chariotry became known from such writers as Caesar, Tacitus, or Pomponius Mela, and with the availability of Greek texts, Herodotus on Scythians and others, including the Persians, and above all the chariots of Homer. By the beginning of the seventeenth century the textual evidence available for wheeled transport in the Greek and Roman world was virtually as we have it today, and commentaries on it soon began to appear, of which the most remarkable was the encyclopaedic study, *De vehiculari veterum*, by John Scheffer in 1671. Scheffer was himself a remarkable man, called from his home town of Strasburg to the prestigious chair of Politics and Rhetoric in the University of Uppsala in 1648, where he was to make an international reputation as a classical scholar, an historian of Sweden and a writer on the Lapps. In 1634 he published a treatise on ancient naval warfare, the *De militia navali veterum*, and followed this with his study of wheeled transport. This is basically lexical, dealing with the various named vehicles in Latin and Greek, but solidly practical in its discussion of the constructional features such as draught-pole, axles and wheels, as well as draught-animals, yokes and harness and a discussion of alternative transport by pack animals and sledges. He distinguishes disc and composite disc wheels from those with spokes, illustrating a composite disc with internal dowels which, together with another drawing of a slide-car, must represent contemporary Swedish farm vehicles. He saw the earliest vehicles as ox-drawn four-wheeled wagons, with the horse as a later domesticate associated with two-wheeled carts and chariots; a view with which one would largely agree today. In England comment on Caesar's description of British chariot warfare was made throughout the seventeenth century, with Clement Edmonds in his *Observations upon Caesar's Commentaries* of 1604 noting that as chariots were not encountered by the Roman armies on the Continent, but frequently referred to in oriental contexts, their British presence might be taken to support Geoffrey of Monmouth, and 'to prove the Britaines descent from Troie in Asia'. Aylett Sammes in his *Britannia Antiqua Illustrata* (1676) devoted an entire chapter to 'The Customs of the BRITAINS in their Wars, and Manner of Fighting', and has a long discussion of the terminology of chariots—*covinnus, essedum* and so on—with a spirited picture of a battle. He again thought of British chariotry in terms of the Iliad and the Greeks, so that it 'will be thought to proceed either immediately from that Nation, or else from the *Phoenicians*'. Ancient British chariotry (from the texts) became a part of the romantic im-

1 age of the past, from time to time eliciting antiquarian discussion such as Samuel Pegge's 'Observations on the chariots of the antient *Britains*' (Pegge 1795), which

pointed out that Caesar's description of running forward along the chariot-pole demanded vehicles 'formed very low in the fore part', unlike known classical types, and small draught-ponies.

Only twenty years later the first archaeological evidence for prehistoric wheeled vehicles in Europe was encountered and recognized, and appropriately enough it was in the form of two British chariot burials, excavated in 1815–17 at Arras in Yorkshire. The Rev. E. W. Stillingfleet's notes of his operations are scrappy and were not published for many years, but the iron hoop-tyres, bronze and iron horse-bits and harness fittings, and indeed the skeletons of a pair of horses in one grave, brought prehistoric British vehicles out of the classical texts and into archaeology (Stead 1979). Similarly, from the early nineteenth century, Egyptian chariotry became known from the publication of the monuments on which it was depicted, and in 1836 the Italian Egyptologist Ippolito Rosellini published a completely preserved chariot found in grave-plundering in Thebes. This famous vehicle, now in Florence, he thought, on the grounds of its birch-bark bindings, to be a trophy of war from an engagement between Egypt and Scythian invaders in 611 BC but it is of middle second-millennium date, and the non-Egyptian woods point rather to north Syrian sources used by Egyptian craftsmen (Rosellini 1836). In 1817 the Bavarian Royal Inspector of Carriage-Building, J. C. Ginzrot, published the first part of his massive treatise on ancient wheeled transport (Ginzrot 1817), a work which was to have an unfortunate influence in later discussions, as he too readily imported into antiquity his professional knowledge of nineteenth-century wagon-building design and practice.

In Europe the rapid development of archaeology in the nineteenth century soon led to the discovery and record of primary evidence for wheeled vehicles from excavations of varying degrees of competence, or

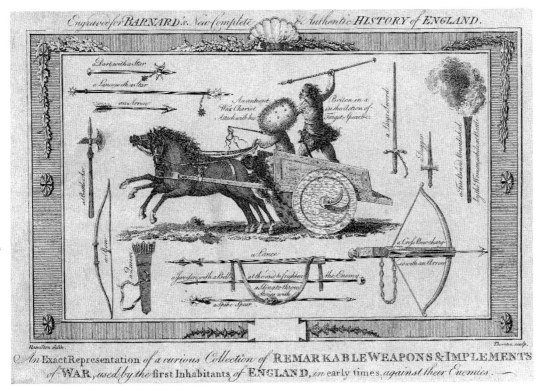

1 An 'Ancient British' chariot, from Edward Barnard, 'New, Comprehensive and Complete History of England', 1783

the recognition of chance finds. The first of the now famous series of third-millennium wooden disc wheels from the peat bogs of the Netherlands was found in 1838 and published ten years later (Van der Waals 1964), and somewhere around 1830 what is still the only tripartite disc wheel in Britain was found in peat at Blair Drummond in Perthshire. It was not published until 1873, when it was described as an ancient wooden shield, and nearly a century elapsed before its true character was put on record (Piggott 1957). On the European continent the Baron de Bonstetten was excavating Hallstatt wagon graves in Switzerland in 1848 (Drack 1958), and in the 1860s and 1870s the wheels from the north Italian terremare sites of the second millennium BC at Mercurago and Castione de Marchese were recovered. By now widespread digging of the La Tène cemeteries of Champagne and the Marne

was revealing the French counterparts of Stillingfleet's early Yorkshire discoveries of chariot graves. The excavations were disastrously numerous and at minimal standards of execution and publication, and it has been estimated that not less than 150 chariot burials must have been dug into, mainly apparently robbed in antiquity, and of no more than three or four of the intact graves have we reasonably adequate records (Joffroy & Bretz-Mahler 1959; Stead 1965). By the 1870s two interpretative essays on the evidence so far available were published by Fourdringier (1876) and Mazard (1877), and in Germany the Dörth (Waldgallschied) Early La Tène chariot grave was found in 1850, that of Waldalgesheim in 1869, while the Early Urnfields wheeled cauldron of Milavec in Czechoslovakia was published in 1882, and the Late La Tène carriages of Dejbjerg in Denmark were found in

11

1881–83. The publication of Joseph Déchelette's fourth volume of his *Manuel d' Archéologie* in 1914 appropriately submitted the Hallstatt and La Tène material to magisterial analysis, and marked the end of an epoch.

The interest in evolutionary sequences and typologies characteristic of late nineteenth-century anthropology inevitably led to theoretical discussions on 'the origin of the wheel', as those of Tylor and especially Haddon (1898). Haddon used the archaeological evidence from Europe such as it then was (basically the wheels from the Italian terremare), and he and others were particularly interested in the recent and contemporary types of sleds, slide-cars and block-wheeled vehicles in the British Isles and the Iberian peninsula, as preserving archaic transport modes, and a generation later, as we shall see, this line was to be taken up again in England and Scandinavia. In the meantime, a post-war summary of knowledge was provided by the articles on 'Rad' and 'Wagen' in Ebert's *Reallexikon der Vorgeschichte* (1929). In England Leonard Woolley's discoveries at Ur seem to have aroused interest in disc and tripartite disc wheels, or so it seemed to Cyril Fox when he, like Haddon, turned to review the relationship of sleds, carts and wagons (Fox 1931), a profitable ethnographic approach to be developed for North Europe and Scandinavia in a brilliant study by Gösta Berg (1935). By this time too new discoveries of wheels of late prehistoric date had become available from La Tène itself (Vouga 1923) and in England, Glastonbury (Bulleid & Gray 1917), and of native type in Roman contexts from the Saalburg (T. Jacobi 1897), and Newstead and Bar Hill in Scotland (Curle 1911; Macdonald & Park 1906). Two important contributions to the knowledge of wagon graves of Hallstatt C and D were made by Paret (1935) in publishing the Bad Cannstatt grave, and in Czechoslovakia by Dvořák (1938), excavating and publishing those of Hradenín: both assembled and discussed

comparative material. An outstanding work of synthesis and interpretation, the thesis of which was to have repercussions up to today, was that of the Commandant Lefebvre des Noëttes on the harnessing and tractive powers of horses from antiquity to recent times with its significant sub-title 'Contribution à l'histoire de l'esclavage' (des Noëttes 1931).

Further developments had to await the consolidation and reorientation of professional archaeology in the various countries of Europe after the Second World War, resulting both in new finds derived from excavations, which were on the whole becoming conducted and published at standards of acceptable reliability, and in synthesis within a comprehensive framework of prehistory. A new beginning was made by Gordon Childe in two papers (Childe 1951, 1954a) in which he reviewed the Old World evidence for the beginnings of wheeled transport and found he could accommodate it within a diffusionist model, with priority accorded to Mesopotamia and the pictographs in the Protoliterate script on clay tablets from Level IV of Uruk (Warka). The problems presented are basic, and largely dependent not only on the accident of survival and discovery but on an accurate chronological framework. They will inevitably be discussed at length at a later stage in this book, and the Near Eastern evidence has now been put on a new and modern basis by the masterly work of Littauer and Crouwel (1979a). Immediately after the war Fox had published a new and stimulating study of the 'Celtic' chariot as a result of the wartime discovery of a votive deposit of metalwork at Llyn Cerrig Bach in Anglesey (Fox 1946), and rescue excavations in connection with hydro-electric schemes in the Caucasus led to the discovery of spectacular waterlogged wagon burials of the second millennium BC at Trialeti (Kuftin 1941), and later at Lchashen on Lake Sevan (Mnatsakanyan 1957; 1960 a, b.; Piggott 1968). The publication of the third-millennium wagon model from Budakalász in the 1950s showed the potentialities of East and Central Europe for new information, fully demonstrated by Boná's survey of the wagon and wheel models up to the later second millennium BC (Boná 1960). The Dutch bog-finds of massive single-piece disc wheels were made the occasion of a magisterial study, employing a systematic programme of radiocarbon dating, by Van der Waals (1964), and a few years later the writer attempted a general synthesis of the evidence for the earliest wheeled transport in Europe and Asia, incorporating the Caucasian finds and the new third-millennium BC south Russian vehicle graves then becoming available (Piggott 1968). In the later prehistory of vehicle technology the study of Hallstatt and La Tène vehicle graves, to which the way had been pointed by Paret's work of 1935, was extended by the discovery and publication of new and sometimes spectacular finds such as Vix (Joffroy 1954) and the Hohmichele (Riek & Hundt 1962), as well as comparative studies of the material (Schiek 1954; Drack 1958; Joffroy 1958; Joffroy & Bretz-Mahler 1959; Stead 1965a), and new primary material was continuing to appear, from Ireland (Lucas 1972) to the Urals (Gening 1977). An outstanding popular summary of the use of wheeled vehicles from prehistory to the nineteenth century, with excellent documentation and bibliography, is that of Lázló Tarr (1969).

The evidence for the manufacture and use of wheeled vehicles in prehistoric Europe is now comparatively full for a period over at least three and a half millennia, despite gaps in our knowledge which may be geographical—the paucity of early material west of the Rhine for instance—or chronological. As an initial innovation, and in its subsequent development and adoption, it raises classic questions of monogenesis or polygenesis, diffusion and independent invention, acceptance or rejection for economic

or social reasons; beyond Europe the same problems concern us on a larger scale in ancient Asia. As a problem of world prehistory wheeled transport presents us with a phenomenon which in its restricted distribution alone calls for a fuller examination of its earliest manifestations, and in the history of technology it represents a complex of inventions and adaptations in response to an equally intricate series of changing social circumstances. The prehistoric European evidence, closely linked with that from the early literate Near Eastern civilizations at more than one stage, deserves as full an examination as the surviving material permits over its whole chronological range, from the early stone-using agricultural communities in which it first appears, up to its final developments at the eve of, and into, the incorporation of much of Europe into the Roman Empire. In previous works of synthesis there has been a tendency to consider either the beginnings of European wheeled transport, or its final specialized forms such as Celtic chariotry, and although the intervening phases may not be so comparatively well documented, technological continuity in workshop practice is perceptible throughout. The study now attempted is offered as a contribution to the history of technology, so far as possible, not as an antiquarian end in itself, but to the degree that evidence from non-literate prehistory permits, in terms of the social and economic conditions inferred from the evidence of contemporary material culture as a whole. Failure to achieve this end in literate antiquity has with justice been deplored by ancient historians: 'the history of technology and economic history have tended to go their separate ways, so that students of each are often astonishingly unaware of the work of the other' (Finley 1959, 120). The divorce between technology and social circumstances may be thought to be less apparent in prehistory, where the latter are necessarily ultimately inferred from knowledge of the former, but the warning is apposite nevertheless, as is a recognition of the danger of circular arguments in such a context.

Europe, Asia and the New World

Attention has already been drawn to the fact that the early use of wheeled transport is a phenomenon limited to a part only of the prehistoric world, unlike the use of boats, rafts and other forms of water transport, or indeed, on land, the utilization of various forms of sledge, slide-car or travois. Wheeled vehicles make their first appearance on present showing around the fourth millennium BC within an area from the Rhine to the Tigris; they do not appear in India until the third, and in both Egypt and China the second millennium BC, and are unknown in South East Asia, Africa south of the Sahara, Australasia, Polynesia and the Americas before the historical periods of outside contacts. The problem posed by this curiously restricted distribution will be discussed more fully below in the specific areas of Europe and Asia involved, but the wider issue must be considered here. Geographers, notably Carl Sauer, in the 1950s, have pointed to basic distinctions in the agricultural practices of the prehistoric world, especially the contrast between the planters and gardeners on the one hand, and the sowers of seed, husbandmen and herdsmen, on the other (Sauer 1972). Basically a distinction between methods of plant cultivation, a differentiation in types of domesticated animals associated with these two groups of agricultural traditions also exists, partly but not wholly itself linked to the natural zoological distribution of potentially domesticatable animals. In the Old World the classic area of planting economies was South East Asia, with dog, pig, fowl, duck and goose as household, not herd animals, and this complex extends to China and Japan, tropical Africa and the Pacific. The New World presents a complementary and independent complex of early planting economies in Mesoamerica—what Sauer called vegetative planters south of the Isthmus of Panama, seed-planters to the north—with the domestication in the Andes of the guinea pig, llama and alpaca, the turkey further north. On the other hand, to return to the Old World, the earliest agricultural economics of Western Asia and Eastern Europe typify Sauer's second complex, with seed agriculture (basically cereals) and a range of domesticated herd animals—sheep, goat, cattle—as well as pigs and dogs. To the herd animals was later to be added the horse.

The critical factor which emerges here is that of non-human tractive power, provided above all by domestic cattle and, especially where speed was concerned, by the horse, both available as domesticates among the cereal cultivators from at least the sixth millennium BC as regards cattle, the fourth for the horse, and among no other agricultural groups in early antiquity. As we shall see, other factors are involved in explaining the appearance of wheeled vehicles in this specific context in the fourth millennium BC, notably the traction plough, but the main reason for the absence of such transport elsewhere (the New World, for instance) seems reasonably explicable. The stimulus to the invention of a wheeled vehicle, whatever the more imponderable economic factors which might make it an acceptable technological instrument to a society, must have been the availability of motive power beyond human capacity. But before examining the Old World problem in more detail, the situation in the Americas deserves further comment, since it is not quite as simple as the foregoing brief summary might suggest, and has important general implications. Here the writer is deeply indebted to Professor Warwick Bray for guidance.

The unexpected factor in Mesoamerican archaeology is that, while there is no evi-

2 Mesoamerican (Aztec) pottery wheeled toy from Mexico

dence of wheeled transport in the full-size, practical sense from the archaeological, pictorial or historical evidence before or at the time of the Spanish Conquest, there is a comparatively large number of pottery toys of animals on wheels (Ekholm 1946; Haberland 1965; de Borhegyi 1970). These are spread throughout the Mesoamerican area of high culture, from the Panneo region near the Texas border down to Salvador, and in central, west and east Mexico, with a time range from Early Classic (*c.* AD 300–600) to the Conquest. The original figurines are perforated through their usually stumpy legs to take a wooden axle carrying perforated clay disc wheels, precisely in the manner of the Mesopotamian or East European vehicle models. As in the Old World, the question of distinguishing detached wheel models from spindle whorls or other perforated discs has been raised, but the presence of three figurines and twelve wheels in a closed find at Tres Zapotes seems convincing (Ekholm 1946, 223): we shall later return to the Old World examples. The wide distribution in time and space justifies the assumption that 'the idea of wheeled toys was definitely established in the cultures of Mesoamerica. It was passed from one group to another and from one generation to another for a considerable period of time'. It follows that 'there was therefore some knowledge of the principle of the wheel in pre-Conquest Mexico ... we have toys or miniatures involving an important mechanical principle not put to practical use in the culture' since 'it seems quite clear and certain that wheeled vehicles were not actually used as a means of transport anywhere in the New World up to the time of the first European contact'. We return to the point made earlier on, the natural circumstances of terrain and available domesticates for draught, and to the phenomenon already touched on, and to be taken up again in the instance of prehistoric Europe, the acceptance or rejection of a technological innovation by a society in terms of its fulfilling, or its failing to fulfil social and economic needs. The wheeled toys can be seen as 'a prime example of an invention or idea the practical value of which was not realized, or, if it were realized, was not useful because of the nature of the existing culture'; with no adequate draught-animals 'the lack of such animals would be a prime deterrent to an application of the wheel to transport. The generally rugged nature of the Mexican terrain and the thick forests of the coastal lowlands would also stand in the way'. And with 'a tendency towards conservatism in the American Indian cultures, especially in the field of material culture' the traditional transport of pack-animals and the back-pack on human beings might be thought to be in no need of alteration or augmentation (quotations from Ekholm 1946). The domesticated American camelids, the llama and the smaller alpaca, could not supply adequate draught, the former having a tractive power of not much more than a dog. It is, however, a hardy pack-animal, capable of carrying a load of 45 kg for an average 25–30 km a day in high altitudes.

15

Both are difficult to breed and the related vicuña remains undomesticated (Cranstone 1969).

To return to the Old World, and to prehistoric Europe, we have two main themes to consider before turning to a detailed examination of the individual finds which constitute our evidence, one technological and the other cultural. The technology is that of wheeled vehicles themselves, their basic requirements first in raw materials and secondly in the supporting technology of the tools required for their fabrication. Even the simplest vehicles demand a complex of skills, and in the more elaborate an interdependence of craftsmanship of a high order. The cultural background, already touched on in the instance of the unique Mesoamerican phenomenon, concerns not only natural resources in the form of potential domesticates for draught purposes—also in effect raw material—but the social and economic circumstances in which the inception and early development of wheeled vehicles, themselves labour-intensive in manufacture and maintenance with their appropriate draught teams, could be found acceptable rather than, as it seems in Mesoamerica, irrelevant to a society whose transport needs had been satisfied by alternative means. Here too we shall have to consider not only the relationship of such contemporary or antecedent uses of animal draught as for the traction plough or for various forms of wheel-less sledges and slide-cars, but questions of chronological priorities between various cultural provinces in the adoption of the new technology. As Childe saw when he made his classic studies in 1951–54, this could be a test case for a diffusionist model as against one invoking independent invention in more than one place in the Old World, just as the principle of wheeled transport was certainly discovered, as we have seen, wholly independently in the New World, though not exploited in practice.

Materials, technology and tools

Until the Industrial Revolution and the advent of railways and motorcars, all wheeled transport was fundamentally of wooden construction even if metal fittings came to play a part from later prehistory onwards. From the point of view of archaeological survival it is unfortunate that we are dealing with a material so subject to decay, and still under modern laboratory conditions so difficult to conserve in the waterlogged state in which its survival is most likely in the climatic conditions of temperate Europe. Nevertheless, a surprising number of actual vehicles or their recognizable parts, usually wheels, has survived and constitutes our primary evidence, supplemented by models or linear representations in representational art styles, however schematic. From the first millennium BC onwards the metal fittings just referred to may give valuable information or indeed constitute the only evidence of a vehicle.

The wheels are, of course, of prime importance, and constitute the technological innovation from which all else followed, by concentrating all friction on to the bearing surfaces of pairs of vertically rotating discs. All sliding mechanisms of transport seek to minimize the friction of moving a load horizontally over a land surface: a sledge on narrow or plank-edge runners is more efficient than on a pair of logs, and the slide-car or travois by tilting the structure from the horizontal transfers the friction to the lower ends only of the component pair of poles. At a time when simplistic evolutionary schemes were popular, an inevitable 'origin of the wheel' was found in rollers used under a sledge, the centres thinned to form an axle, the untrimmed ends the prototype of disc wheels. Haddon (1898, 170, fig 23) produced just such a sequence, from a garden roller to a thinned-out log with disc ends, said to have been seen in Portugal, but ignored the practical fact that such a log

16

would inevitably break in the middle and the point that in both the weight was taken by an inserted iron rod clearly shown in the drawings. Similarly, solid disc wheels were frequently envisaged as cut transversely from a tree stem like thick slices of salami, ignoring the inherent proneness to fracture such a disc would have, and the virtual necessity of using a large metal saw of modern type, unknown in early antiquity. While a relation between rollers and wheels is at least conceptually probable there is no evidence for any direct link, nor does it seem likely that a generalized concept of rotary motion applicable in infinite contexts, from spindle-whorl and potter's wheel to transport, is appropriate to ancient modes of thinking which may be non-logical. There is certainly no archaeological evidence that these three uses of a rotating disc in antiquity were related applications of a single technological concept.

Archaeology supports the long-standing assumption that the earliest wheels would be of solid disc form, single-piece or composite. It is here that the nature and structural growth of tree-stems must be considered, growing in annual rings and so in cross-section presenting concentric growth-circles from the central heartwood to the exterior sapwood, cambium, bast and bark. In addition, the longitudinal 'grain' of the wood is embodied in a series of medullary pith rays which provide natural radial cleavage lines. To cut a cross-section of a tree-trunk is to ignore its radial structure, and is technologically difficult to achieve without the use of a saw, which must be of suitably springy metal and therefore out of the question among societies with a stone and flint edge-tool technology, within which the earliest disc wheels, often of large size, are attested. The earliest saws, from Early Dynastic Sumer onwards, are small, draw-cut, cabinet-makers' tools (Piggott 1968, 269 with refs) with the exception of the Late Minoan series of very large copper or

bronze blades of up to 170 cm long, which are without parallel elsewhere (H. B. Wells 1974). In consequence, to cut a transverse section of a tree-stem up to a metre in diameter, a size approximated to by many prehistoric disc wheels, strains technological capacity beyond possibility and, furthermore, no such cross-cut wheel has been recorded. The wood structure just described, with its growth-rings and medullary rays, would make a thin disc of any size liable to fracture under strain when used on an axle and, in fact, all surviving disc wheels are cut from the plank in a manner shortly to be discussed. Note must be taken here, though, of a find, establishing something which might be thought to constitute an exception, from the fifth-century BC tombs of Pazyryk in the Altai. Here roughly made trolleys—in effect, as we shall see, slide-cars with wheels, were evidently made *ad hoc* to transport heavy materials such as the tree-trunk coffins and stones to build the burial cairns. The 'wheels' were cylinders of a solid trunk of larch (*Larix* sp.) 30–47 cm diameter, 35–40 cm long, with a large central perforation 12–16 cm diameter, often showing signs of wear, mounted on a fixed axle and secured with wooden linchpins (Rudenko 1970, 187). It seems likely that some of the four-wheeled ox-drawn vehicles depicted in the Syunik rock carvings in Russian Armenia and shown with extremely small 'wheels' are vehicles of 'trolley' type: they constitute over 40 per cent of the classifiable vehicle representations which have been published (Karakhanian & Safian 1970; Littauer 1977a) but, of course, the actual type and construction of the wheels is unknown. An exceptional modern instance of trunk-cut rough disc wheels is recorded among the Hudson Bay fur-traders in 1801—'small carts, the wheels of which are each of one solid plate, sawed off the ends of trees whose diameter is three feet.' As Fox said, commenting on this record, the wheels presumably 'worked successfully for a time at least', but the circumstances are those of improvization with modern saws available, and two years later the settlers were making 'real wheels' (Fox 1931, 198).

The earliest sawn wood planks in prehistoric Europe that have been identified and published seem to be those of the main burial chambers of the Hohmichele barrow in south Germany, of Hallstatt D (sixth century BC) date, which had every appearance of having been sawn with a large timber-yard saw: they were 6.6 m long, about 0.35 m wide and 5 cm thick (Riek & Hundt 1962, 44). Normally, as in ancient shipbuilding in northern waters, planks were split from the tree-trunk (though the Bevaix Gallo-Roman boat in Switzerland had sawn planks: Arnold 1978). The splitting or riving process has been described in the instance of prehistoric woodworking in Britain.

'This method of reducing the log relied upon the exploitation of its natural planes of weakness. Most woods will split along the medullary rays: judiciously placed blows will split the wood along the ray and by further wedging the log can be riven along its full length. The first split is to halve the timber, and the process is then repeated to quarter and subdivide into radial segments of the desired size. Partition along rays minimises the risk of decay from damp/fungus/bacteria penetration through ruptured cell walls, and shrinkage in drying' (Coles *et al.* 1978, 27).

Once the tree has been felled and trimmed the radial splitting process is not difficult. Wedge-section planks obtained by splitting were used to build a full-size replica of the Ladby Viking ship by Danish boy scouts—'it really sounds silly when a little boy stands there with a little woodman's wedge in his hand and says he is going to split the trunk which is beside him, when it is half as thick as he is high... it is more a question of knack than of strength.' A small wedge is driven in to start the crack, and this is followed up with further and larger wedges, so that 'it takes only a few hours to split a trunk about 5 m long' (Hjorth quoted in Fenwick 1978, 188). The fullest and latest study for such woodworking practice in prehistoric Europe is from the Somerset Fens in south-west England, where abundant planking of the fourth to second millennium BC has been described and summarized (Coles *et al.* 1978 with refs). Some planks were wedge-sectioned, some adzed flat, and others had been split 'on the chord' to yield a square-cut and thicker plank, by 'halving or quartering and then twice cutting at right angles to the halving split, or once parallel with the quartering split'. Similarly split planking from Storrs Moss in Lancashire has radiocarbon dates of *c.* 3460 bc. (Powell *et al.* 1971) This technique is important to us because it is from massive planks of this type that the large single-piece disc wheels, dating from the third millennium BC at least and found in north-west Europe, were made, as were the component planks of the equally early tripartite wheels. Unfortunately, it is not in all published instances that the wood grain has been examined with a view to determining the technique of manufacture, though it can on occasion be inferred. By and large we seem to be dealing with planks split on the chord.

From the point of view of raw material— the parent tree-stem from which the original *3* plank was split before further working—it is clear that the diameter of a final one-piece disc wheel must be somewhat less than the width of plank from which it is cut; if split on the chord, the chord will be proportionate to and, even with a halved tree-trunk, slightly less than the diameter of the tree. Similarly, with wheels having massive hubs or naves worked from the solid, the overall thickness of the plank must be not less than the total nave length. Some twenty single-piece disc wheels from the Netherlands, north Germany and Denmark have diameters ranging from about 60 to 90 cm and averaging 80 cm, implying planks of something over this width which, allowing for trimming, including removal of bark, waney sapwood etc., demand trees of a metre or so in diameter (and if radial planks were used, twice this). Thicknesses, where a nave is worked on both faces, are around 20 cm or rather more: the centre planks of the very massive Caucasian tripartite disc wheels would have been at least twice this thickness. The Irish Late Bronze Age circular wooden shields and moulds for leather versions, all around half a metre in diameter, were worked from planks in precisely the same manner (Coles 1962). Similarly, the Irish late prehistoric cauldrons of alder or poplar, hollowed from blocks of wood at least 60 cm across, testify to the same competence in craftsmanship (Coles *et al*. 1978). The large disc wheels of third-millennium northern Europe would then call for tree-stems (normally oak but occasionally alder) of at least a metre in diameter.

The relationship of large wood artifacts in prehistory to natural timber resources has recently been considered, very appositely to our enquiry related to land transport, in the instance of water craft of the logboat ('dug-out canoe') type, hollowed from halved tree-trunks (McGrail 1978). As with the wheels, oak is almost inevitably used (and occasionally alder). Here too dimensions are often much of the same order, an original parent stem of something over a metre in diameter. In natural stands of European deciduous woodland, when the woodman's convention of measuring the girth at 'breast height' of 5 feet (1.52 m) is used, the bole, trunk or stem diameter of mature oak ranges from 1.2 to 1.8 m, while free-standing oaks can reach twice this figure. As a rule-of-thumb, a free-standing tree is reckoned to add 1 inch (2.5 cm) of circumference for each year of growth, a forest tree half this rate, implying annual tree-ring widths averaging 4.0 or 2.0 m respectively (Rackham 1976, 27), so that using this approximate scale for the forest trees with which we would be concerned in antiquity, an oak 1.0

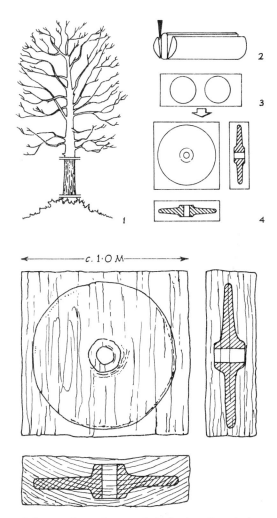

3 Stages of manufacture of single-piece disc wheel

m in diameter would have an age of some two and a half centuries and represent a mature tree. In large-size, over-mature oaks the trees may develop a rot known as 'brittle heart', and it has been noted that 'almost all modern oaks greater than *c*. 2 m in diameter are palpably hollow' (McGrail 1978, 121). It looks therefore as though 1.0–1.5 m diameter oaks, from 200–300 years old at most, would represent the optimum source for sound timber for large woodworking needs, including logboats and disc wheels, and indeed that the recorded wheel diameters themselves reflect this. So too do the known sizes of timbers used in the construction of prehistoric buildings, domestic or ceremonial, where the post-core has been measured within the packing of its post-hole. The Linear Pottery long-houses of the fifth millennium BC were framed on uprights from 0.4 to 0.5 diameter or somewhat greater, and in British long barrows and ceremonial monuments of the fourth and third millennia even larger wooden uprights were used. Posts of around 0.9 m in diameter were used freely in the Fussells Lodge long barrow and the timber circles at Arminghall and Woodhenge, and at Waylands Smithy long barrow the two split halves of a tree-trunk, D-shaped in section, had an original diameter of 1.2 m (Piggott 1968, 269). Clearly the felling of mature timber of some size was well within the capabilities of the stone-using agriculturists of Europe at this time, even if an oak trunk 0.6 to 0.9 m diameter and some 3.75 m long (as used in Fussells Lodge long barrow) would weigh when freshly thrown over 2,358 kg. Mature trees of the type required—oaks of two hundred or more years old—would also be relatively scarce compared to younger growth, and the Swedish botanist Knud Jessen (1948, 177) reckoned that in stands of natural deciduous woodland oaks with stems of 25 to 50 cm diameter would occur at about 28 to the hectare; those of 50 to 75 cm, 31; of 75 to 100 cm, 19, and those of 1.0 m and above no more than 7 to the hectare. Nor need the provision of timber and smaller wood for making even the earliest known European wheeled vehicles have depended on haphazard search in the primeval 'Wildwood', as the evidence so elegantly inferred for conscious woodland management by the third-millennium Neolithic communities of south-west England shows (Rackham 1977). Here in particular the demonstration of coppicing and pollarding to obtain small poles is apposite to the need, in later prehistoric spoked wheels with single-piece bent felloes, for such raw material.

The felling of trees of a size necessary for plank wheels of the type under discussion must have presented a problem, but one which was clearly solved. Stone axes are certainly effective for cutting down small trees of up to about 30 cm diameter, but for those three or more times this size, fire-setting was perhaps employed (Coles 1973, 20–21). At all events, when thrown, the trunk would have been worked green by wedge-splitting, axes, adzes and chisels or gouges of stone, flint or metal according to the technological status of the communities involved, and with the skills of craftsmen like the logboat-builders of the Society Islands, who delighted Sir Joseph Banks in 1768–69 when he saw them work 'without Iron and with amazing dexterity: they hollow with their stone axes as fast at least as our Carpenters could do and dubb tho' slowly with prodigious nicety'—to 'dub' is an obsolete verb meaning to smooth with an adze (quoted by McGrail 1978, 36). The use of adzes, essential to all traditional woodworking especially in the absence of large saws or of planes, demands special consideration. An axe-blade or axe-head is hafted with the cutting edge parallel with the shaft, the adze at right angles, and strictly speaking and with modern tools, the axe-blade is symmetrical in section, the adze asymmetrical with the cutting edge flat on the inner and bevelled on the outer edge. Using this criterion alone, one can in the European Neolithic cultures make a distinction between 'an "axe province" extending from the Alps to the Iberian peninsula and an "adze province" from the Alps to the Balkan peninsula' on the typology of stone implements assignable to either class (Childe 1950, 159). Literally applied, this would severely limit the areas within which adzed woodwork could be expected with the stone-using peoples among whom the earliest wheeled transport in Europe is attested, but as Childe realized, and as ethnographic evidence shows, 'in practice of course any celt could be used as an adze if mounted on a knee-shaft' (1950, 157), and this was demonstrated by Van der Waals in the instance of the wooden adze-haft found under the log trackway at Nieuw-Dordrecht, itself adjacent to a disc wheel and of third-millennium BC date, and the contemporary and similar haft found with symmetrical-section stone 'axes' at Stedten in Saxony (Van der Waals 1964, 56). Stone or flint 'axe-heads' could then in many instances also serve as adze-blades: the adzed planks of Neolithic Somerset are firmly within the 'axe province', for instance.

The question of unrecognized adze-blades may be taken a stage further in connection with disc-wheel technology. When discussing the massive tripartite disc wheels of the Caucasus, with internal dowels held in tubular mortices, the writer drew attention to the relation of such techniques to adequate copper or bronze narrow chisels and above all gouges with a hollow blade, in the absence of augers in early antiquity, and instanced closed finds of associated metal tools including these (Piggott 1968, 305). From the rich tombs of the Maikop group in the western Caucasus comes a large series of arsenic-bronze tools, Maikop itself (Hančar 1937, fig. 30, pl. XLIII) having the well-known shaft-hole adze and axe-adze, as well as a shaft-hole axe and small 'flat axes', one with visible remains of transverse binding on its corroded surface. The association of shaft-hole axes, gouges and similar small thin blades occurs at Novosobodnaya, Vozdvizhevskoy and Makhoshevkaya (Popova 1963, pls IX, XXI), and the reasonable conclusion is that the flat blades are those of adzes fastened to a knee-haft in the Egyptian manner (Petrie 1917, pl. LV; Goodman 1964, 18). In Central Europe the Staré Zamky hoard of the third millennium BC in Moravia consists of a shaft-hole axe and gouge of Caucasian type, an awl and a 'flat axe', again presumably an adze-blade, and one suspects that in the contemporary hoard from Ezero in Bulgaria, of the two axes the smaller is an adze-blade, found with

narrow chisels and a knife (Benešova 1956; Georgiev & Merpert 1966, fig. 2). All finds would constitute basic carpenters' kits of the essential tools—axe, adze, gouge or chisel—appropriate to all woodworking including wheeled vehicles. Indeed, we may go a stage further, and consider that since adzes as such have never been identified in the rich equipment of the European full Bronze Age from the early second millennium BC onwards, it can only mean that bronze 'axe-blades' were as capable of being cross-hafted as those of stone, and of functioning as adzes. With the axe as the primary tool for shaping the components of a wooden vehicle, finishing, dressing and smoothing can follow with the adze, and holes and mortices cut with chisels and gouges. All jointing was morticed, though wood pegs or treenails could also be used: in one of the second-millennium BC covered wagons from Lchashen in Russian Armenia it was estimated that 12,000 mortices had been cut in its 70 component parts, 600 alone in the bent yew framework of the arched tilt. Glue may also be employed, especially with bark or bast bindings on light bent-wood vehicles of chariot type, as it was in the manufacture of composite bows, and the practical needs of adequate vessels of pottery or metal for obtaining strong fish or animal glue has been commented upon by Rausing (1967, 146, 155).

Vehicles and their wheels; typology and technology

With these preliminaries, we may now turn to the vehicles themselves. Ambiguity and confusion among archaeologists has sometimes arisen in the use of the simple series of technical terms employed by wagon-builders and wheelwrights in pre-industrial technological contexts (Sturt 1923; Jenkins 1961, 1962; Arnold 1974); further difficulties may be presented in the international literature by a failure to obtain precise equi-

Glossary

	French	German	Italian	Russian
Axle Axle-bed Axle-arm	Essieu Corps de fusée	Achse Achschenkel	Asse, Assale	Os'
Bit	Mors	Gebiss, Trense	Morso (equino)	Udila
Body, box	Caisse	Kasten, Bock	Letto, cassone	Kuzov
Bridle	Bride	Zaumzeug	Briglia	Uzda
Canon (of bit)	Canon, tige			
Carriage	Voiture	Prunkwagen	Carozza	Kolyaska
Cart	Charette	Karren	Carro, caretto	Dvukolka, arba
Chariot	Char	Streitwagen, Rennwagen	Cocchio, biga	Kolesnitsa
Cheek-piece (of bit)	Branche	Knebel	Montante	Psalien
Draught-pole, Pole	Timon	Deichsel	Timone	Dyshlo
Felloe	Jante	Felge	Gavello	Obod
Gauge (of wheels)	Voie	Spurbreite	Carregiata	Koleya
Harness	Harnais	Geschirr	Bardatura, finimenti	Sbruya, Upryazh'
Linchpin	Esse, clavette	Achsnagel	Acciarino	Cheka
Mouth-piece (of bit)	Embouchure	Mundstück	Imboccatura	Udila
Nave, hub	Moyeu	Nabe	Mozzo	Stupitza
Nave-band	Frette de moyeu	Nabenring		
Perch	Perche	Langbaum	Timone (?)	
Rein(s)	Guide(s)	Leine(n)	Redine, Briglia	Povod'ya, vozhzhi
Shafts	Brancards	Gabeldeichsel	Stanghe	Ogloblya
Sledge	Traineau	Schlitten	Slitta, traino	Sani
Slide-car	Travois(?)	Stangenscheife	Tréggia (?)	Poloz'ya
Spoke	Rais, rayon	Speiche	Raggio, razza	Spitza
Terret, rein-ring	Clef, passe-guide	Führungsring	Anello per le redini	Zga
Tilt	Bâche	Plane	Tenda, tendone	Verkh, naves
Traces	Traits	Stränge	Tirelle	Postromka
Tyre	Bandage	Reifen	Cerchione	Shina
Undercarriage	Chassis	Unterwagen	Telaio	
Vehicle	Véhicule	Wagen	Veizolo	Povozka
Wagon	Chariot	Ackerwagen	Carro (a 4 ruote)	Telega

	French	*German*	*Italian*	*Russian*
Wheel	Roue	Rad	Ruota	Koleso
Disc wheel	Roue pleine	Scheibenrad	Ruota a disco, ruota piena	Massivno Koleso
Spoked wheel	Roue à rais	Speichenrad	Ruota a raggi	Koleso c' spitzani
Cross-bar wheel		Strebenrad		
Whipple-tree Swingle-tree	Palonnier	Ortscheit	Bilancino	Vaga
Yoke	Joug	Joch	Giogo	Yarmo
Neck-yoke	Joug d'encolure	Nackenjoch		
Horn yoke	Joug de cornes	Hornjoch		
Yoke-saddle	Fourchon d'encolure	Jochsattel		

valents in translation. An attempt to remedy this situation is set out in a table for the main European languages. The vehicles themselves may basically be divided into two groups, *wagons* having four wheels, and *carts* having two, but in English usage both terms carry with them an implication of heavy-duty vehicles, especially those used in agriculture or commercial transport: when raised in domestic status and, used for parade, pleasure, ceremonial or ostentation, the wagon became a *carriage* which, in its eighteenth and nineteenth-century apotheosis, proliferated innumerable fashionable variants, but in European prehistory its undifferentiated use might be thought applicable to the elaborate vehicles buried in Hallstatt graves, or to those of Dejbjerg, rather than the more bucolic 'wagon'. As to the two-wheeled vehicles, the term *chariot* has traditionally been used for light vehicles for military or ceremonial use; in the late prehistoric Celtic world Stead has for long campaigned for the usage 'cart burial' rather than the customary 'chariot burial' where a two-wheeled vehicle has been buried with the dead, on the grounds that warrior status is not always implied by the accompanying grave-goods (Stead 1965a, 1979). But the burials are demonstrably those of the upper classes in a stratified society, and if one sees the chariot playing the part of a status sym-

bol equally to be used in battle as a *war chariot* and in peaceful circumstances as a vehicle of ceremonial, parade or demonstration of one's hierarchical position in society, surely *chariot* in this wider sense is permissible usage, avoiding the incongruously plebian and utilitarian sense of 'cart'. It is amusing to note that Lewis and Short's Latin dictionary of a century ago glossed the vehicle terms of *covinnus* and *cisium* as a 'tilbury' and a 'cabriolet' respectively, the former used of Celtic chariots by Roman observers, and itself of Celtic etymology, but both here equated with Victorian pleasure carriages.

In construction a simple wagon or cart has a *frame* or *undercarriage* supporting a *body* or *box* which may be provided by some sort of arched cover, awning or *tilt*. The *axle* or *axle-tree* embodying the axle-bed and the arms on which the wheels turned or to which they were fixed, carried the essential pair of *wheels*, constituting with them a unit. Wheels are necessarily the most distinctive features of the vehicles under consideration, and survive in greater numbers than other recognizable parts of carts or wagons, while their types and methods of construction form a basis of important technological distinctions. The primary classification which, as we shall see, also has important chronological and cultural implications is that between *disc* or *block* wheels, and *spoked* wheels; both groups can be subdivided and there is a specialized form of rare occurrence in surviving European finds, the *cross-bar* wheel. All wheels which rotate on the axle need a tubular hub or nave to maintain a relative narrow rotating disc in a constantly vertical position; when fixed to an axle which rotates with the wheel, the nave may be minimized or absent as stability is obtained by the rigid axle-wheel joint. Disc wheels may be single-piece, cut from a single plank, or composite, normally *tripartite,* and made from three or more planks held together with *dowels* held in *mortices,* and both single-piece and tripartite disc wheels may have either *integral naves* worked from the solid plank (which has to be of an original thickness to allow for this) or *inserted* as a separate tubular member, thus enabling a thinner plank to be used from the beginning. A more detailed glossary will be found in Littauer and Crouwel 1979a.

The tripartite disc wheel, a standardized form occurring, as we shall see, over a wide geographical area of Europe and Asia at the chronological outset of our knowledge of wheeled transport in the late fourth or early third millennium BC, demands special discussion. Owing to the conservatism of the wheelwright's craft in non-industrialized societies, such wheels were still being manufactured in the traditional manner in Ireland in the 1950s, when the process was recorded in detail (Lucas 1952, 1953). The wheels were small (0.62 m in diameter) and fixed to the axle, but otherwise conformed to a large series of prehistoric examples. The original wood plank was cut so as to be twice as long and half as wide as the eventual diameter of the wheel, and was halved transversely, and one of the halves itself bisected laterally, thus producing three planks, two half the width of the third. These were then jointed into a square, the narrower planks flanking the broader, by two pairs of dowels in internal tubular mortices. This was then cut into a disc and the slightly projecting nave worked from the solid, the modern tools

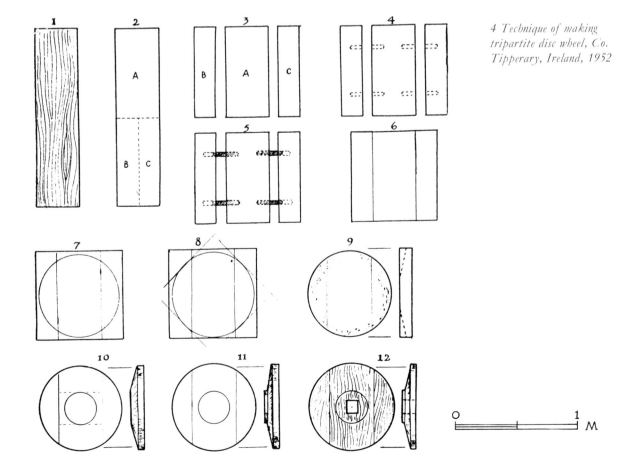

4 *Technique of making tripartite disc wheel, Co. Tipperary, Ireland, 1952*

employed being saw, adze, chisel and auger. A square mortice was cut in the centre to hold the wheel on an axle which turned with it. All the features of this modern Irish wheel appear in antiquity (though the square mortice on a turning axle is only known in third-millennium Switzerland at present) and, most significantly, the 2:1 proportion of the centre plank to its flanking members, enabling a tripartite disc wheel to be made from a chord-split plank from a tree of half the diameter necessary to produce a single-piece disc of equivalent diameter. When one realizes that the Dutch single-piece disc wheels were made from such planks, originally with a content of three cubic metres of oak weighing 322 kg, one appreciates the advantage of the tripartite disc at the very beginning of the wheel-making process.

There are a number of typological variants within the tripartite disc class in Europe, whether with integral or inserted naves, in which lunate openings of varying dimen-sions are cut in the inner edges of the flanking members (and from the solid in some single-piece discs). The cutting of internal tubular mortices, known from the Caucasus and south Russia to Scotland and Ireland in prehistory, of course demands metal tools and could hardly have been achieved with stone or flint chisels, and the method of jointing otherwise employed was by cutting dovetail mortices on one or both faces of the wheels holding transverse exterior battens, as in the third-millennium Zürich wheels or those of later date elsewhere. A sophisticated variant was to cut the mortices as arcs of circles so that the batten could then be sprung into the groove and thus achieve greater tension, as at Mercurago in the second or Buchau in south Germany in the early first millennium BC. The inserted naves were, in fact, hollowed lengths of small tree-stems, slightly tapered in some instances to enable them to be hammered home into the perforated disc and held by friction, perhaps aided by glue. It is worth while making the

5 Types of prehistoric disc wheels:
1, single-piece with integral nave;
2, single-piece disc with inserted nave;
3, tripartite disc with external battens
and turning axle; 4, tripartite disc
with integral nave and internal
dowels; 5, tripartite disc with inserted
nave, curved battens and lunate
openings

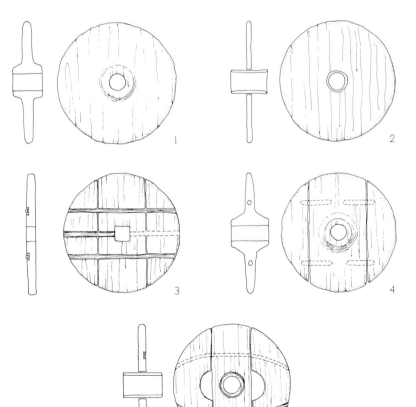

point that simple and tripartite disc wheels continued to be made in Europe (as elsewhere in the Old World) from their Neolithic inception throughout prehistory and up to the present day, side-by-side with every spoke-wheeled variant. Typology is therefore no guide to date, which can only be obtained by direct radiocarbon assay or an archaeological context: the undated tripartite disc from Blair Drummond in Scotland could equally be of the eighteenth century AD or BC. The surviving disc-wheeled vehicles in, for instance, Anatolia, Sardinia or the Iberian Peninsula, and their only recently extinct counterparts in northern countries, are of great value in understanding the prehistoric evidence (cf. Berg 1935; Koşay 1951; Galhano 1973). Taking the prehistoric disc-wheeled vehicles as a whole, where botanical identifications have been made we can see certain conscious choices made from the Northern European deciduous forests, particularly for the wheels themselves which usually alone survive, with oak (*Quercus* sp.) used in the Caucasian tripartite wheels at Lchashen and Trialeti XXIX, ten of the Dutch single-piece wheels and the Ezinge Iron Age tripartites, three of the Danish single-piece discs and some of the tripartite wheels from the late prehistoric Rappendam votive find. Alder (*Alnus glutinosa*) was used in a pair of Dutch single-piece wheels and that from Nonnebo in Denmark, the four single-piece inserted nave wheels from Glum, and tripartites from Dystrup and Rappendam in Denmark and Doogarymore in Ireland. The Glum naves were of birch (*Betula* sp.) and the dowels of Trialeti XXIX were of box (*Buxus supervirens*) and at Doogarymore of yew (*Taxus baccata*). The Trialeti V tripartite wheels were exceptionally of pine (*Pinus* sp.). In Switzerland and north Italy one moves into a more southerly forest zone, with the Zürich tripartites of beech (*Fagus sylvatica*) planks and ash (*Fraxinus excelsior*) dowels, and the Castione de Marchese and

Mercurago tripartite discs were of walnut (*Juglans regia*), the latter with dowels of larch (*Larix decidua*). The Scottish Blair Drummond wheel was of ash (*Fraxinus*) for both planks and dowels, and lime (*Tilia* sp.) occurred in the Rappendam group. Little is known of the remainder of the vehicles except at Lchashen where bodies and axles were of elm (*Ulmus* sp.) and yokes of elm or oak, and yew (*Taxus*) used for the arched tilts.

Surviving vehicles and contemporary representations show that the disc-wheeled vehicles of early antiquity were heavy constructions of carpentered timber, the wagons necessarily of rectangular plan, the carts (of which less structural detail is known) either square or rectangular or, exceptionally in the Caucasus, of the triangular or A-shaped type surviving today in, for instance, the Anatolian and Sardinian disc-wheeled carts, and all, except the few Swiss and possibly north Italian finds, with the wheels rotating on the axle, secured by a vertical peg or *linchpin*. All involve draught by a pair of animals, normally cattle or castrated oxen, as with a plough harnessed to a central *draught-pole*, and such paired draught, in later prehistory supplemented by the alternative of horse-traction, continued throughout antiquity. Oxen are invariably harnessed to a *yoke*, and yoke-harnessing continued with equid traction until the introduction or adoption of the rigid horse-collar and shafts from late Roman times into the Middle Ages in Western Europe, perhaps owing to contacts with the Orient, as shaft-and-collar harnessing was known in China from the Han dynasty in the fourth century BC (Needham & Lu 1960; des Noëttes 1931, 83–84, 120–24; Lynn White 1962, 61). The draught-pole in a cart would be fixed, any vertical play owing to slight discrepancies in the height of the draught-animals at the flexible junction of pole and yoke being accommodated by tilting the vehicle, but in a four-wheeled wagon the horizontal position of the body must be maintained by pivoting the draught-pole so as to allow for vertical movement (as we see in the complete Lchashen wagons). Lateral movement, whereby the axle of the front pair of wheels is pivoted, as in modern vehicles from the Middle Ages to today, is only dubiously and exceptionally attested in prehistoric or classical antiquity, for the problem applies to vehicles with any type of wheel, disc or spoked. This long-standing question is further discussed in detail in Chapter 5, but for the present we may assume that ancient wagons normally had both front and back axle-tree immovably fixed to the bed of the vehicle, with consequent manœuvrability restricted to a large turning circle even with wheels rotating on their axles: with fixed wheels the restriction is even greater, though such wagons survive until today in, for instance, China. The two-wheeled cart with freely turning wheels is an infinitely more flexible vehicle.

The alternative to the disc wheel, that with *spokes* radiating from the nave to a circumferential *felloe,* a new device dating from hardly before the beginning of the second millennium BC in Western Asia or Europe, takes us into not merely a new typology, but a novel technology. As Childe pointed out many years ago, 'though naturally an expression of the wheel idea, the spoked wheel was a new invention rather than a modification of the tripartite disc' (1954b, 214). The novelty comprises not only the wheels, but the vehicles they were designed to serve, and the animal traction they employed. The disc-wheeled wagons and carts, drawn by oxen, were slow, heavy, structures primarily designed for conveying heavy or bulky goods; they were, to borrow engineering terminology apt in the present circumstances, compression structures. The first spoked-wheel vehicles on the other hand were drawn by small horses, and were two-wheeled, fast, light contrivances or chariots to be used for carrying men in the pursuit of warfare, parade and ceremonial, and in technology were largely tension structures (cf. Gordon 1978, 145). They were the products of new social demands, and drew on new skills and the alternative forest resources needed for bent-wood construction.

The spoked wheel, as we have just seen, has three main components, the nave (invariably freely rotating on the axle and secured by a linchpin) into which are morticed spokes of varying numbers (from the basically necessary four), at their outer end again morticed into an encircling felloe that may be bound with a *tyre* which, in metal-using contexts, may be bronze or iron, or the rim of the felloe may be given an armature of close-set large-headed studs or nails. Heat-bent wood may be used (in exceptional surviving Egyptian examples) to form intricate nave-and-spoke composite units, and frequently for the felloes, either a single complete hoop (characteristic of prehistoric Europe in the later Celtic Iron Age) or in two or more arcs (as also in Egypt and China). As an alternative to the *single-piece bent-wood* felloe in ancient Europe, various forms of *segmental* felloes cut from the plank were also devised for wagon and cart or chariot wheels, and what became the standard modern European form, a felloe in which the segments are dowelled one to another, each carrying two spokes, is known in a north German find from Barnstorf (Oldenburg) of the second millennium BC (Hayen 1978b) as well as from later prehistoric European examples. The 'dishing' of spoked wheels, giving them a shallow conical rather than a discoidal cross-section, though practised in China from the second, and perhaps the fourth century BC (Lu *et al.* 1959) does not appear in Western Europe until the Middle Ages, the earliest known examples being apparently those of the fourteenth century from Boringholm Castle in Denmark (Witt 1969).

History and the technological background

The new necessities for material suitable for light, and frequently bent-wood structures made demands on woodland resources of a different order from those of the heavy disc-wheeled vehicles of earlier antiquity. In traditional English usage, from the Middle Ages onwards, we turn from *timber* to *wood*: 'timber, medieval Latin *meremium*, is big stuff suitable for making planks, beams and gate-posts; wood (Latin *boscus*) is poles, brushwood and similar small stuff suitable for light construction' (Rackham 1976, 23). In view of the evidence for deliberate woodland management from the early fourth millennium BC in the Somerset Fens, already commented on, it is no anachronism to think in these terms in a prehistoric context, and the Somerset evidence is much of it related to the practice of coppicing and pollarding to obtain light wood for the construction of the fen causeways there. Available wood identifications are unfortunately few, and confined to a handful of British and North European finds of late prehistoric date or native work within the Roman period. Oak (*Quercus* sp.) was used for naves at Glastonbury and probably Dejbjerg (an alternative identification is ash (*Fraxinus*)); ash at Glastonbury and the Saalburg; birch at Holme Pierrepont and elm *(Ulmus)*, the wood favoured by all recent wagon and carriage-builders (cf. Sturt 1923; Jenkins 1961; Adams 1837) at Newstead and Bar Hill. In this context, the question of the use of a new woodworking technology, that of lathe-turning, arises. Spokes were of ash at the Saalburg, oak in both the La Tène wheels and at Holme Pierrepont, willow (*Salix* sp.) at Newstead and Bar Hill and hornbeam *(Carpinus betulus)* at Dejbjerg. Ash was used for the single-piece bent felloes in one of the La Tène wheels and those of Dejbjerg, Newstead and Bar Hill, and for the segmental felloe at Holme Pierrepont, and the bent felloe of the second La Tène wheel was of elm. In Southern Europe, the cross-bar wheel from Mercurago in north Italy was of

walnut (*Juglans regia*). Indirect and marginal evidence from the Mediterranean is contained in the references to chariots and their wheels in the Mycenaean Linear B inventories—'one old chariot of elm-wood', and elm, willow and cypress (*Cupressus* sp.) for wheels (Ventris & Chadwick 1959, 361–75). The bent-wood felloe referred to in Iliad IV 282–86 was of poplar (*Populus* sp.). The Egyptian evidence for imported woods for chariot-building is discussed by Littauer and Crouwel (1979a, 81), and was originally commented on by Rosellini in the instance of the Florence chariot, where elm, ash, willow and birch bark or bast was used; elm was also used in wheels from the tombs of Tutankhamen and Amenophis III (Western 1973). Similar imports (including birch-bark) were used in the composite bows of the Egyptian New Kingdom, products of another bent-wood technology (McLeod 1970). The single-piece bent felloes are the most interesting from a technological view point, since their construction calls for certain constraints in the proportions of the wood employed, as Kossack has pointed out. 'Practical experience in a modern workshop, using mechanical devices, shows that for a wheel of 80–100 cm diameter the maximum thickness of felloe obtainable is 6.5 cm. With north European elm the proportion of thickness to radius reached 1:10, if modern methods are set aside' (Kossack 1971, 146). In terms of the woodland resources already referred to, an ash felloe for a 1.0 m diameter wheel, allowing for an overlapping scarf joint, would have required a straight-grained stem about 6 or 7 cm in diameter well over 3.0 m in length.

The technique of heat-bending wood in recent non-industrial societies deserves fuller study. Berg (1936, 121) commented on this in the context of bent-felloe wheels, drawing attention to these in Central Asia: modern light carts in Asia Minor preserve this form of felloe, now bent commercially by steam-heating, and the type may have

135
136

28

been introduced with the Turkish conquests of the Middle Ages. A recent study in a different context is that of Andrews, describing the traditional Turkic and Mongol wood-framed felt tents, the type usually misnamed a 'yurt', a word strictly denoting a group of these dwellings for which the individual name is *alacheikh* (Littauer *in litt* 1978). Andrews (1973, 1978) has shown that the earliest evidence for such tents is in the eleventh century AD, but the absolute antiquity of the type is of less moment than the traditional technique of its construction which embodies a 'roof-wheel' framing the central smokehole, 2.0 m in diameter, made of two or three bent segments, morticed externally to receive the similarly bent roof struts. (The wood used is said to be probably maple (*Acer* sp.) or 'tamarind', but presumably tamarisk (*Tamarix* sp.) would be more likely in Iran than *Tamarindus indica*). The wood, cut to shape, was soaked overnight and heated in a vertical open-topped chimney-like structure over a hearth for half an hour, and then bent manually and by leverage to the required curvature on a pegged bench; once bent it was held in shape between stakes in the ground for five or six days, when it was ready for constructional use. Such observations indicate the type of simple technology behind the single-piece felloes of prehistory, and moisture in green wood would aid the process.

We may conveniently conclude this introductory chapter with a reminder of the nature of the primary evidence for prehistoric European wheeled transport we are about to consider in detail. In the first place, obviously the evidence of actual finds of wooden vehicles or their recognizable parts, notably wheels, is by far the most important, and here we are dependent on soil conditions inimical to decay, especially the anaerobic conditions of waterlogging in stray or votive finds, or in graves where vehicles were deliberately buried. The prac-

tice of vehicle burial in graves is a recurrent feature in ancient Europe and Asia among varied societies in which wheeled vehicles became objects of prestige, demonstrative of status, and provides us with information of outstanding value. High standards of excavation and recording techniques play a large part in the recovery of such evidence, especially in the detection and recognition of wood remains which may be no more than soil stains. When, from the second millennium BC, metal fittings of bronze or iron were used on wooden vehicles, these alone may survive, so that many late prehistoric vehicle burials virtually depend on their recognition as such on the presence of iron hoop-tyres and metal fittings which alone can survive to attest the burning of a vehicle in the funeral pyre of a cremation burial. In burials again the provision of the grave with features such as wheel-slots dug in the floor may give a clue to a vanished vehicle. The next class of evidence is that provided by models, normally of baked clay or metal, which fortunately for our purpose were sporadically but not infrequently made in prehistoric Europe, as in the ancient Near East, as funerary or votive offerings, toys, or other, to us unknown, purposes. Ambiguity may unfortunately arise when miniature clay wheels alone survive, for while some are clearly models of actual wheel types, others can have a confusing morphological similarity to such objects as spindle-whorls. Finally, where some sort of representational art style, however schematic, is current among prehistoric communities using wheeled transport, vehicles may be depicted, and while rock carvings, as in the Caucasus, north Italy or Scandinavia, afford our richest sources, incised drawings on pottery or other artifacts can on occasion provide information of great value. In sum, the evidence is far from negligible, and must now be surveyed within its chronological and cultural sequence.

2 The European evidence to the end of the third Millennium

Constraints and limitations

The geographical area comprised in our survey is that of the European continent in its widest sense, from the Atlantic coasts on the west to the Ural Mountains, the River Volga, the Caspian Sea and the Caucasus on the east. While the western boundaries are obvious, the eastern limits allow of an element of choice. Coles and Harding in their recent survey of the European Bronze Age set their eastern bounds on the River Don for good and sufficient reasons (1979, 113), deliberately excluding the Caucasus: for us, its exceptionally rich vehicle finds of the second millennium equally demand its inclusion, and similarly the Volga and Urals constitute important areas for the specialized study under consideration. Northwards, southern Scandinavia and the adjacent areas of the North European plain form a natural and cultural boundary, while to the south, so far as the period dealt with in this chapter is concerned, the north Italian lakes and the Maritime Alps offer tentative evidence at the end of the millennium. The Aegean and southern Balkans provide no evidence until a later date.

Over this vast area there is naturally a wide diversity of terrain—mountain, plain, marsh, uplands—as well as plant cover, significant in terms of the provision of timber for vehicle-building, in addition to providing the natural elements in a landscape which, however modified by man, might exercise constraint on the use of wheeled transport. Studies of recent pre-industrial transport in Europe have tended to stress a dichotomy between the use of the heavy

four-wheeled wagon with its limited manœuvrability on the plains, as against the adaptable two-wheeled cart in broken hilly country. This over-simplification, often illustrated by very schematic maps all apparently deriving from that of Deffontaines (1938; cf. Jenkins 1961, fig. 10), need not hold for early antiquity, but may contain an element of practical reality at least in certain limited circumstances. For the third millennium we must envisage a European landscape basically forested but already severely modified by human activity, within the east areas of steppe grasslands themselves bordering upon mixed forest-steppe to the north and desert to the south and east. Though direct and detailed evidence from pollen analysis or other palaeobotanical sources seems lacking, there is a general assumption that the steppe of south Russia, with its westward extension on to the Hungarian Plain, existed as a natural area of open grassland or prairie in post-glacial neothermal times, and is not a man-made phenomenon arising from artificial deforestation. Movement by any form of transport over such terrain is relatively uninhibited, but in naturally wooded areas of Europe with only partial clearance for plough or pasture, we may have to consider recognized trackway systems by as early as the third or fourth millennium (cf. Bakker 1976).

The diversity of terrain and subsoil conditions can be an important factor in causing an unequal distribution-pattern of survival for wooden artifacts such as vehicles. Temperate Europe does not provide the conditions of aridity which have led to the survival of chariots in Egyptian tombs, nor the abnormal permafrost preserving the vehicle from Pazyryk in the Altai, and wood has its best chances of survival in the anaerobic waterlogged conditions of peat bogs, as in north-west Europe (*cf.* Coles *et al.* 1978) or the sites on the margins of lakes, as in Switzerland and Italy. The survival of evidence of this direct (and most important) kind is therefore of a different order than, for instance, that for pottery, or stone and metal tools and weapons, and further constraint is exercised when one turns to human factors affecting the archaeological record. These include the great rarity of representational art, however schematic, which might depict vehicles, outside certain restricted areas with rock carvings such as those in the southern Alpine regions, south Scandinavia or Russian Armenia. In three-dimensional form, the making of model vehicles for whatever purpose—toys or cult objects—is capriciously distributed in time and space among European prehistoric communities, as is the funerary practice of depositing a vehicle as a whole or in part (its wheels) in the grave. Childe's view that such burials reflected the burial rites of third-millennium Sumer or Elam and that there was an 'intimate association of wheels and royalty' can hardly be maintained, nor that all wheeled vehicles demanded 'an adequate supply of metal for carpenters' tools' (Childe 1951, 177, 193), but in the latter context, although we have simple disc and even tripartite disc wheels among north-west European communities with a stone tool equipment, the availability of metal tools in the third and earlier second millennium could introduce technical constraints, as in the cutting of internal mortices in thick tripartite discs. Certain archaeological phenomena, such as the burial of putative draught-animals in pairs, may suggest paired traction of a vehicle rather than a plough, especially as in certain instances a burial vehicle could have existed, undetectable or at least undetected by the

excavators. And this brings us to a final point, applicable throughout archaeology but especially relevant here. It must be frankly admitted that standards of excavation, not only in the past but at the present time, are extremely unequal as between one area of Europe and another, and that the quality of their publication is similarly variable. The recognition of fugitive substances such as wood, which may be detectable as no more than a change in soil colour and texture, calls for skills in observation and record not everywhere regarded as normal and necessary. So too with publication; prompt, fully documented and illustrated accounts are not the commonplace they should be, and the employment of ancillary scientific techniques such as radiocarbon dating is far from systematic. The factors enumerated in this section do not invalidate the evidence we have at our disposal, but they should not be forgotten in our assessment.

Cultures and chronology

It is not the purpose of this book, nor is it within its author's competence, to present a prehistory of Europe from the fourth millennium, within which the adoption and development of wheeled transport plays an incidental part. Our primary concern is to examine an episode in the history of technology so far as it can be inferred from archaeological evidence in pre-literate conditions, and to this end the material must be ordered in terms of cultural groups in their relative and absolute chronological relationships. The status, and indeed the existence of what have conventionally been termed by prehistorians archaeological cultures, inferred from artifacts, has been called in question by certain schools of thought, but for the present purpose the general pattern accepted in current European prehistory is followed. Nevertheless, it must be borne in mind that in so doing we are to a large extent ignorant of what any given archaeo-logical 'culture' may mean in terms of human societies, let alone ethnic groups or linguistic affiliations. Many 'cultures' have been defined from no more than a regional pottery style or a shared burial rite, and recent European political history has resulted in what may be relatively homo-geneous units changing their names as they are crossed by political or linguistic boun-daries of modern formation.

Since we are dealing with a technological innovation appearing in identical or closely related forms among societies constituting distinct and separate archaeological cul-tures, and widely distributed over Europe, we must necessarily consider possible means of transmission, or alternatively of indi-genous development among the disparate societies themselves. It is here that we come up against the question of diffusion, espe-cially from more complex or 'civilized' societies to those less complex, or 'barbar-ian', a thesis which, in the instance of prehis-toric Europe, Gordon Childe pursued over decades of devoted study. We now have a wide spectrum of variant approaches to this problem, which does not seem to have been so great a matter of concern to archaeolo-gists in the Near Eastern or classical fields as to prehistorians concerned with Europe as a whole. On the one hand we have what, to pursue the metaphor, might be called the in-fra-red position, where 'folk movements' and the 'invasion hypothesis', rejected or at most very warily invoked as acceptable models by most modern prehistorians, are enthusiastically endorsed. Of direct concern to our present enquiry is the 'Kurgan cul-ture' concept of Marija Gimbutas (e.g. 1970, 1974, 1978) which must be examined in detail later in this chapter as wheeled trans-port and the use of the domesticated horse for traction are essential components of the thesis, which abounds in 'invasions', 'thrusts' and 'waves' of peoples, with maps upon which the arrows of movements ever

thicken and blacken. The other extreme, or ultraviolet position on the spectrum, may be represented by Colin Renfrew (1973, 15–19), wholly rejecting the 'diffusionist view' as a result of the application of the calibrated radiocarbon dates, shortly to be discussed here. They invalidate previous accepted models of an east-west diffusion of megalithic chambered tombs and so 'if the procedures and assumptions used to build up this single, if important, piece of prehis-tory are wrong, then they are wrong for the rest of prehistory as well, at all times and places'. Diffusion of any kind is a totally un-acceptable model and we must look to other explanatory theories.

Between such extremes it is reasonable to seek a *via media* in which we are concerned with 'diffusion' as with the modes of transmission of innovations or inventions, and their acceptance or rejection by this community or that, as technological incre-ments regarded as expressions of social needs, may be varied and inconstant as between individual situations. They cer-tainly do not inevitably demand transmis-sion or imposition by any social, ethnic or linguistic group, always supposing that such entities can be regarded as valid inferences from non-literate archaeological evidence. But whatever the mechanisms involved in the transmission may be, they cannot be dis-cussed except in the framework of compara-tive and better, absolute chronology. Only with this can we order our material and assess priorities in the primary adoption and employment of wheeled vehicles by prehis-toric European communities hitherto using alternative means of land transport. Above all, it is basic to any consideration of a cen-tral problem, that of the validity of the long-held assumption that wheeled transport in prehistoric Europe afforded a classic demonstration of well-documented diffu-sion from the ancient Near East, as Childe believed. There is now agreement among archaeologists that the only reasonably satisfactory chronological framework up at least to the early first millennium BC (when historically computed chronologies may profitably be used) is that provided by C14 or radiocarbon dating.

The theory and practice of this means of isotopic dating are now too well known to demand restatement here, but one or two points merit reiteration, especially for those readers more accustomed to historically computed chronologies. A radiocarbon date does not represent a fixed and exact moment in time, but the expression of a statistical probability within a margin of error (ineradicable by the nature of the process) which is formally stated: that of the Zürich tripartite disc wheels later to be described is 2340 + 60 BC. 'Error terms (equivalent to +1 standard deviation) do not define a bracket within which the true date must fall; there is still 1 chance in 3 that the true date may be outside these limits, or 1 chance in 20 that it may be outside twice these limits' (CBA 1971). It therefore follows that a single radiocarbon date, though valuable and preferable to none at all, is of less validity than a series from the same context with replicated readings. It should further be remembered that the date obtained is that of the sample submitted to assay, and that its archaeological significance depends on such factors of excavation technique as the observation and record of precise context and stratigraphy. And since wood or charcoal samples may come from trees up to two or more centuries in age, the original position of the sample as between heartwood and sapwood could introduce another minor variable, as could the reuse of old timbers, for instance in beams of buildings. Finally, it has now been demonstrated that radiocarbon 'years' before the present are not the equivalent of calendrical or solar years, but diverge from them increasingly as the time interval between past and present increases. The precise form of the curve which can be plotted by calibrating the radiocarbon readings

agaínts the dendrological annual tree-ring count, is still not determined in certain details even if its main outline is firm, and some archaeologists have chosen to adopt 'calibrated' radiocarbon dates *in toto*. A convention has been established whereby normal uncalibrated dates are expressed in years 'bc', and the calibrated dates, in common with those arrived at by historical means, as 'BC': the Zürich date just quoted would appear as 2340 + 60bc or *c.* 3030 BC. In this book the practice of using uncalibrated radiocarbon dates has been followed, enabling like to be compared with like in non-literate and so prehistoric Europe, where very large numbers of dates have now provided a consistent and reliable time-scale. Where calibrated dates, giving a higher antiquity, have to be considered is in contexts where comparison is needed between radiocarbon and historical chronologies, notably in the chronological relationship between the earliest wheeled vehicles in the Near East and those of Europe, which will be considered at the end of this chapter. The correction factors necessary to convert dates 'bc' to dates 'BC', which will be used, are those of R. M. Clark 1975, Table 8, and the conventional half-life of 5568 is employed.

Preconditions of wheeled transport

The economic background of prehistoric Europe, since the inception of the earliest food-producing communities in the southeast and the Mediterranean coast from the sixth millennium BC, was one of developing agricultural settlement and forest clearance across the continental land-mass to the north and west, and eastward into the south Russian steppe and forest-steppe zones. Convenient modern surveys in terms of radiocarbon dates are given in Phillips 1975 for the west, and Tringham 1971 for the east. The latter, ending her account of the earlier agriculturalists of this region at a nominal, calibrated date of 3000 BC (in uncalibrated terms about 2300 bc) has moved to a point when copper metallurgy was replacing a stone-technology for edge-tools. 'It is not to be expected' she writes 'that the year 3000 BC will be taken seriously as an actual year in which all the settlements and cultural phenomena discussed in this chapter suddenly ceased and disappeared. But it is interesting to note that throughout Europe, especially eastern and central, in the first half of the third millennium BC (according to radiocarbon dates) profound environmental and cultural changes took place' (Tringham 1971, 205). The environmental changes are best documented in Northern Europe, owing to the unequal development of paleobotanical and palynological research among the countries of modern Europe, and are marked by the transition from the post-glacial vegetation (and inferentially climatic) zones of Atlantic to Sub-Boreal. This change, marked particularly by a decline in the tree population (especially elm) and an increase in open grassland, has frequently been attributed to man's deliberate interference with the natural plant cover by forest clearance for agricultural purposes, not only for cereal cultivation but for the increased pasturage for herd animals—cattle, sheep and goats, especially the former. The archaeological evidence of this period, that of the later third millennium bc, certainly suggests discontinuities, changes of emphasis in agricultural practice and the siting of settlements (large numbers of the long-established 'tell' settlements of the Balkans are abandoned, for instance); the increasing exploitation of non-ferrous metallurgy and other technological developments, of which the first appearance of wheeled transport is our direct concern.

The later third millennium is therefore the critical period for the beginning of our detailed enquiry, and certain general considerations of the technology which immediately antedated the adoption of wheeled transport, and made its use possible and

acceptable, now become of considerable relevance. They relate to concepts already touched on in their wider aspects in connection with Sauer's thesis of the dual nature of Old World types of crop cultivation and animal management, and have recently been made the subject of an important and stimulating study (Sherratt 1981). Briefly, and in so far as it relates to our present purpose, Sherratt is concerned with ovicaprids, equids and cattle, and particularly the last in their intimate association with plant domestication and cereal cultivation in their function as traction animals for a plough, but sees this as part of a larger system of exploitation which moves beyond the primary use of flocks and herds as a direct source of protein in the human diet in the form of meat. What he suggests naming, rather unhappily, the Secondary Products Revolution in prehistoric animal management, includes the use of milk products from cows, ewes, goats and mares; wool from ovicaprids; and traction by cattle and horses. The evidence suggests that this stage of utilization was not primary to the first known agricultural communities (in Europe from about the sixth millennium and earlier in the Near East), but begins to emerge in Europe around the turn of the fourth and third millennia. The adoption by prehistoric societies of these related technological innovations could be seen as an example of Joseph Needham's concept of the 'packaged transmissions' of innovations from one cultural milieu to another (Needham 1970, 61). From our present point of view the use of cattle for draught is the important factor, for the horse, though domesticated in south Russia by the fourth millennium bc, has a later and more specialized function in vehicle traction to be discussed in the next chapter. With cattle, the practice of castrating bulls to produce docile oxen plays an important part, and their employment as draught animals in plough agriculture replacing earlier traditions of hoe or digging-stick cultivation assumed (though not, in fact, directly documented) for primary Neolithic agriculture.

Castration can be inferred from bone modifications detectable in osteological studies and has been documented for the Central European Linear Pottery culture of the fifth millennium (Bökönyi 1974, 116) and the Polish Funnel-Beaker (TRB) culture of the third millennium bc (Bakker *et al.* 1969, 224 n.; Murray 1970, 37, 61). Further instances are likely to be recorded as such detailed studies by palaeozoologists develop and become widespread. Draught can, of course, be effected by cows or heifers, and mixed draught, sometimes by strangely assorted animals, can be presumed in later prehistory in the manner familiar in the Near East and elsewhere today. The evidence for oxen in Linear Pottery contexts naturally raises the question of the use to which they were put at this early date: the possible function of the 'shoe-last celt' as a stone ploughshare has more than once been suggested (cf. Jankuhn 1969, 27, with refs).

The evidence for the use of a traction-plough can take the form of pictorial representations, actual surviving implements as a whole or in part (e.g. plough-shares) or in soil traces of plough marks or furrows, on ancient land-surfaces. Technically all the implements of earlier prehistory are of the light type without mould-board known as an *ard,* and this term should be employed in their description. In the Near East the pictographs of the Late Uruk period and contemporary proto-Elamite scripts, conventionally dated to the late fourth millennium BC, show a type of ard of distinctive form with double beam, and this form continues in many later depictions, but is not known in Europe. For Europe the fundamental study is that of Glob (1951) with subsequent radiocarbon dates (Lerche 1968, Shramko 1971) which show that the surviving North European and south Russian finds of ards are not earlier than about the middle of the second millennium. The stone ard-shares of

Shetland and Orkney are again of this date, if not slightly earlier (Fenton 1964), and an early second millennium ard was found in the Lago di Ledro settlement in north Italy (Barfield 1971, 72, pl. 34). This Southern European find leads us to the rock carvings around Monte Bego in the Maritime Alps, notably the Vallée des Merveilles with over 150 scenes of ploughing with paired oxen (Bicknell 1913; Glob 1951, 91–98; de Lumley *et al.* 1976). Barfield (1971, 63) has made a case for dating these, on the grounds of the accompanying engravings of metal daggers and halberds, as contemporary with the later third-millennium Remedello culture of northern Italy, and they are certainly not later than the early second millennium. They would then be contemporary with similar ard and oxen representations among the Val Camonica rock carvings lying in the Italian Alps 300 km to the north-east, of Anati's Phase III A and on a contemporary stele, Bagnolo II (Malegno), (Anati 1975). Here, as we shall see, on one rock composition of this phase, Cemmo 2, such a ploughing scene is accompanied by a representation of an ox-drawn wagon. The remaining European rock carvings depicting ploughing, in south Scandinavia, cannot be dated before the second millennium bc (Coles & Harding 1979, 317). Finally, but of decisive significance, are the traces of ard-ploughing in the form of scoring or 'furrow' marks preserved on buried land-surfaces, usually at right angles indicative of cross-ploughing. Of these the earliest known are on the old surface beneath the TRB long barrow, Sarnowo no. 8, in Poland. The barrow itself is of the Wiorek phase of the TRB culture, of the earlier third millennium bc, and occupation on the old surface has a radiocarbon date of 3620 ± 60 bc (GrN-5035) (Bakker *et al.* 1969, 224; Gabalowna 1970). There are four comparable examples in Denmark of the Middle Neolithic, early to middle third millennium in date, and in England under the South

Street long barrow at Avebury dated to 2810 ± 130 bc (BM-356) (Ashbee *et al.* 1979). Thenceforward such ard-marks are known in Britain and north-west Europe into and beyond the second millennium bc.

The cumulative evidence for the use of the ox-drawn ard for cultivation in Europe from the third millennium is therefore widespread and convincing, and from plough-traction the use of draught oxen for other purposes existed as a potential adjunct to manpower in other allied circumstances, including that of wheeled vehicles when they became available to early agricultural communities. But alternative land transport for which draught-animals could be employed existed in the form of the wheelless vehicles coming under the general heading of sledges, slide-cars and travois. These have direct relationship to certain wheeled vehicle forms, and have long played an important part in hypothetical reconstructions of vehicle evolution. The rather scanty archaeological evidence for such means of land transport must now be reviewed.

Sledges and their congeners

Sledges, as devices for minimizing friction for loads dragged over the ground, achieve this aim by limiting the bearing surface to a pair of relatively narrow horizontal runners. The variants within the *slide-car* type, surviving in agricultural use in the British Isles and Scandinavia until modern times, side-by-side with sledges, further reduce friction, at the expense of a horizontal loading surface, by taking the form of a pair of poles or shafts joined by cross-bars, and a small platform tilted at an angle between the ground and the point of harnessing to the draught-animal or animals so that only the far ends of the poles slide in contact with the ground. A simple form of such a contrivance was in use among North American Indians and named a *travois* by the French Canadian colonists. The classic study of

sledges and slide-cars in relation to wheeled vehicles in recent Northern Europe is that of Gösta Berg (1935); detailed studies of the comparable evidence from Great Britain and Ireland include those of Fox (1931), Thompson (1958), Jenkins (1962) and Fenton (1973).

Sledges are usually associated with movement over snow and ice, but can equally well be used on hard ground such as dry sandy or gravel soils or over grassland and steppe country. The earliest archaeological evidence does, in fact, come from near-Arctic Northern Europe in the form of runners from composite built-up sledges among early hunter-fisher communities, one at least, that from Heinola in Finland, nearly 2.50 m long and of pinewood, belonging to the early Boreal climatic phase, contemporary with the early Maglemose Mesolithic cultures about 7000 BC. Dog traction has been assumed for these early northern sledges, as the dog was an Early Mesolithic domesticate: even earlier domestication has been claimed, among the late glacial Palaeolithic mammoth-hunters of the Ukraine, but awaits detailed zoological confirmation (G. Clark 1975, 230; Klein 1973, 56). In temperate Europe a sledge-runner of length equal to that of Heinola comes from the Gorbunovo peat-bog site in the Urals and has been assigned to the late third—early second millennium. Other fragments seem to indicate a built-up sledge (Berg 1935, 39). From the same site came a simple worked log which appears to have been the primitive type of single 'guide-runner' in Berg's classification (Gimbutas 1958; Berg 1935, pl. 2), and the same description could apply to the birch-wood object from Barrow 5 at Pazyryk in the Altai, of the fifth century BC (Rudenko 1970, 192). What appear to have been the remains of a wooden sledge were found in a Bell Beaker grave (site XII) at Dorchester-on-Thames in Oxfordshire with a likely date around 1900 bc (Atkinson 1960, 114; Lanting & Van der Waals 1972).

From surviving finds we can turn to a few representations. Two model sledges have been published from Tripolye culture contexts from southern Russia, dating from the early or mid third millennium bc. One of these shows a sledge on log runners, the other on edge-on planks, and it is likely that other examples exist, as Kravets only published material in the archaeological collections of Lvov (Kravets 1951; Hančar 1956, pl. II). The remaining European sledge representations are among rock carvings or rock paintings. In the Val Fontanalba area of the Monte Bego carvings (already referred to in the context of their ploughing scenes, and of late third—early second millennium bc date) are at least three depictions of paired oxen seen in plan view, drawing a sub-rectangular or trapezoid frame with cross-bars, with the narrower end at the yoke, and as long as, or longer than, the draught-animals. No wheels are shown, and at first the assumption was that harrows, or more likely sledges, were represented until Kothe argued for slide-cars, sloping backwards from the attachment to the horn-yokes of the oxen (Kothe 1953; Tarr 1969, 14). That this is not necessarily the case is, however, implied by the second series of carvings, in the Syunik area of Russian Armenia and probably in the main of the mid-second millennium (Karakhanian & Safian 1970; Littauer 1977a). Among some fifty representations of ox-drawn vehicles, at least four show a trapezoid or A-shaped object with cross-bars, very much in the manner of the Val Fontanalba outlines, the pointed end similarly joined to the ox-yoke. A couple of additional examples show similar frames with a pair of wheels at their mid point (in fact, A-frame carts), and in two more the same triangular frame has four small wheels even though in one instance the pointed end of the frame joins the yoke. Four wheels demand horizontality, and it seems preferable to regard both series as representing, within the bounds of a sche-

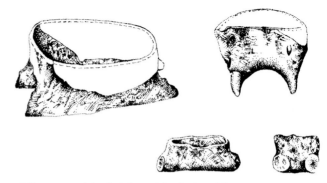

6 Pottery models of sledges with plank and log runners, Tripolye culture, 4th millennium bc, Lvov region, USSR

matic plan view, horizontal sledges rather than sloping slide-cars. Boat-shaped or triangular sledges have good counterparts in recent European ethnography (Berg 1935, 20–26, pls II, X). There remain for consideration the rock paintings of the Peñalsordo (Badajoz) region of Spain (Breuil 1933, 64–65, fig. 20; Almagro 1966, 194, fig. 79). These are undated, but since they include both spoked-wheeled and cross-bar-wheeled vehicles are unlikely to be earlier than the second half of the second

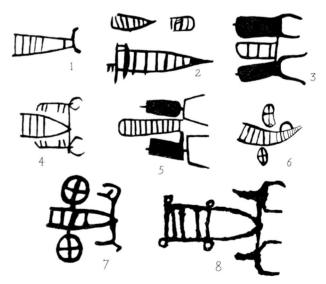

7 Representations of sledges and sledge-vehicles, 3rd to 1st millennium bc. 1, 4, 7, 8, Syunik, Caucasus; 2, 6, Peñalsordo, Spain; 3, 5, Val Fontanalba, Ligurian Alps

millennium or even later. They include sub-rectangular and triangular objects very similar to those of Val Fontanalba (no draught-animals are shown) and the reasonable presumption is that we are once again dealing with sledges of trapezoid or triangular plan.

Note must be taken here of the well-known 'cart tracks' on the exposed limestone plateaux of the islands of Malta and Gozo. These pairs of ancient ruts cut deeply into the rock have an average width between them of about 1.40 m, and observation and practical experiments combined to show that such ruts could not have been made with wheeled vehicles or sledges, but only by some form of slide-car. Dating is difficult to establish but the most plausible conclusion is that they are related to Bronze Age inhabitants of the islands about 1400 BC (Evans 1971, 202–4). They appear to have no counterparts in the Mediterranean or elsewhere.

The evidence from the ancient Near East, well known and of great importance in any consideration of relative chronology between Europe and the Orient, has recently been discussed by Littauer and Crouwel (1979a, 10, 25) and needs little comment here, though it is returned to in the final chronological considerations at the end of the chapter. Briefly, sledges are schematically represented among the pictographic signs on the inscribed tablets of Uruk level IV a in southern Mesopotamia,

8 with runners upturned at one end, 'as well as similar sledges raised over what may be either two (captive) rollers or four disk wheels... wheels seem already more likely here than rollers'. Both the sledges and the 'sledge cars' have a roofed or canopied structure upon them, and a contemporary stone plaque shows a person or a god conveyed in such a covered sledge, drawn by one or a pair of bovids. The conventional date of these representations is at the end of the fourth millennium BC. Remains of an actual 'sledge-throne' were found in the

8 *Pictographs in Protoliterate script, 4th millennium* BC, *from Uruk, Iraq.* A, *sledge symbols;* B, *sledge-on-wheels vehicle symbols*

tomb of Queen Pu-Abi at Ur (PG 800), which had been drawn by bovids: it is of Early Dynastic III date, around the middle of the third millennium BC.

Finally the scanty evidence for ox-yokes from earlier European prehistory—which could be used for any form of traction, of plough, sledge or wagon—should be noted. Recent examples fall into two groups of recognizably different form, horn-yokes or head-yokes and shoulder or withers-yokes according to their place and function in the draught harness. A withers-yoke from Petersfehn in north Germany has been vaguely assigned to an early second millennium date on palynological grounds and a horn-yoke from Vinelz in Switzerland comes from a lakeside settlement of this or earlier date. The remaining finds are un-dated, or in contexts of the Early Iron Age (Gandert 1964; Fenton 1972). In the same context of unspecified ox-draught come the burial of animals in pairs, already referred to, the copper models of a pair of oxen from a hoard from Bytyń and the yoked beasts on a pottery handle from Krężnica Jara, both *12* of the Polish TRB culture and further discussed below (Piggott 1968, 307). As Littauer and Crouwel point out (1979a, 14), the traditional manner of controlling bovids is by a line to a nose-ring, or cord passing through the nasal septum. The latter is admirably shown on pottery models from Hittite Anatolia of the fifteenth or sixteenth *9* century BC, as the pair from Büyükale IV at Boghazköy (Macqueen 1975, pl. 22) or the head from Tokat (Arts Council 1964, no. 176).

The scattered evidence does then suggest that sledges of some form, and less certainly slide-cars, were in use for land transport in prehistoric Europe from Spain to the Urals and from the Maritime Alps to Finland, as well as in ancient Mesopotamia. The use of rollers under a sledge to minimize friction has always been assumed, and modern experiments at Stonehenge showed how their employment reduced the tractive man-power by 56 per cent, but also necessitated an extra team constantly laying the rollers ahead of the advancing sledge (Atkinson 1960, 114–15). The theoretical intermediary between such traction and the wagon or cart is that of 'captive' rollers held between pegs or blocks, thus leading to a turning axle with fixed wheels, but this is nowhere archaeologically attested. The horizontal sledge rather than the inclined slide-car, with paired traction yoked to a central draught-pole, seems the more likely proto-type for the wagon, and this is indeed sug-gested by the Uruk pictographs. It must be remembered that the A-frame cart in prehis-tory is attested only in the Caucasus in the second millennium BC, and such evidence as we have is consistently that of a rectangular body (Littauer & Crouwel 1979a, 8–10). The various evolutionary diagrams of sled-ge-slide-car-cart-wagon relationships, such as that of Haudricourt (1948, fig. 14), fre-quently reproduced in its original or modi-fied form, must in our present state of knowledge be regarded as misleading and purely hypothetical.

The earliest European wheeled vehicles: the Funnel Beaker (TRB) culture

We are now in a position to return to the phase of European prehistory already touched on, that of inferred social, econo-mic and technological changes in the third

9 Painted pottery vessels in the form of oxen with nose-ropes, 15th century BC, *from Boghazköy, Turkey*

millennium bc. It is convenient to start with south-east Europe and very briefly review its archaeology in broad outline by the second half of the fourth millennium, when communities of stone-using Neolithic peo-ples, some with long ancestries stretching back to the sixth millennium bc at least and others of more recent emergence from Mesolithic hunting and gathering societies, were beginning to acquire and exploit non-ferrous metallurgy in copper and gold. The Karanovo VI–Azmak phase of the deeply stratified 'tell' settlements of the Bulgarian plains, around 3600–3000 bc, is characteris-tic of this region as is its counterpart the Gumelnitsa culture across the Danube in Romania; eastward of this the Cucuteni-Tri-polye settlements extend to the River Dnieper, and the still rather imperfectly determined cultures antecedent to that of the Pit Graves, such as that represented at Sredny Stog, Mikhailovka, Evminka and Dereivka (Tringham 1971; Sulimirski 1970,

133–36). These last are of particular importance in the present context as in addition to sharing with the remainder of the cultures mentioned the normal herd complement of cattle, sheep, goats and pigs, they also afford the first evidence, in their steppe environment, of the domestication of the horse for food, and soon apparently for the secondary uses of riding or traction (Telegin 1971; Kuzmina 1971; Bökönyi 1974, 238; Lichardus 1980). The evidence from these sites, which have radiocarbon dates from *c.* 3565 to 2850 bc, is discussed more fully later; chronologically they are followed by the Pit Grave culture with a span of *c.* 2500–1800 bc (Berger & Protsch 1973; Protsch & Berger 1973; Bökönyi *et al.* 1973; Coles & Harding 1979, 124, 153; Telegin 1977). To the south, across the Black Sea in eastern Turkey and Transcaucasia, and on the borders of the ancient Near East, the Kura-Araxes or Early Transcausian culture, with dates from *c.* 2900–2300 bc, must have been emerging, to run roughly concurrently with the end of Sredny Stog and the earlier phases of Pit Graves (Piggott 1968, 276–78, Burney & Lang 1971, 43–85).

North and west of the Carpathian Ring the adoption of Neolithic agricultural economies had led to the genesis of the widespread Linear Pottery culture, eventually to spread from the River Seine to the Vistula and Dneister, with radiocarbon dates spanning virtually the whole of the fifth millennium bc. Towards the end of this period derivatives are contemporary with the emergence of the Lengyel culture on the Pannonian plain west of the Danube and other regional Late Neolithic cultures in Czechoslovakia: Late Lengyel dates range from *c.* 3500 to *c.* 2850 bc (Tringham 1971, 191). Finally, from perhaps as early as *c.* 3500 bc (Bakker *et al.* 1969) and certainly from *c.* 3000 bc, the Funnel Beaker (TRB) culture as a Neolithic stone-using agricultural economy based on local Mesolithic hunting and fishing traditions, comes to occupy the North European plain from the Netherlands to the River Vistula, taking within its ambit south Scandinavia and probably, in a westernmost variant form, much of southern and eastern Britain (cf. Madsen 1979).

It is in these disparate and widely distributed cultures of the third millennium bc that the first archaeological evidence for the use of wheeled transport in Europe appears, at dates beginning around 2500 bc, and thenceforward is taken up in slightly later contexts not yet enumerated but before *c.* 2000 bc. The earliest evidence comes from the TRB culture in south Poland and the Late Copper Age in Hungary, the Kura-Araxes culture in Transcaucasia, the Pit Graves of south Russia, and the Corded Ware cultures and their counterparts in the late third millennium bc in north-west Europe. Earlier evidence has been claimed, and must be dealt with briefly now, if only as we shall see, to dismiss it. The writer accepted (with reservations) a model wheel as from a Gumelnitsa (Karanovo VI) context at Bikovo in Bulgaria, and a cord-ornamented example from Gremnos-Magula in Thessaly: of these the Bikovo model was in fact, from the later levels of the site, contemporary with Ezero, (*c.* 2300–2500 bc) and commented on below; the Gremnos-Magula object must be a spindle-whorl (Piggott 1968, 302; A. E. Lanting *in litt.*; C. Renfrew *in litt.*). 'Wheel models' of pottery are notoriously the least reliable evidence, owing to their morphological similarity to spindle-whorls, particularly when they occur in isolation or in archaeological and chronological contexts where no corroborative evidence of wheeled transport exists: the Gumelnitsa IIIa example from Tangiru in Romania, claimed as a wheel model by Bichir (1964), is a case in point. This was published and illustrated by the excavator as a spindle-whorl, and should be accepted as such (Berciu 1961, fig. 241.8). In general, 'wheel models' should be treated with caution unless they show marked naves, and are thus differentiated from the normal run of spindle-whorls in their contemporary context, or if (as in the Carpathian Earlier Bronze Age) model vehicles of which they could have formed a part are of frequent occurrence.

As the evidence suggests that the earliest employment or adoption of wheeled transport in prehistoric Europe is a phenomenon which cuts across any boundaries inferred by the conventional pattern of different archaeological cultures, the only logical method of treatment is a chronological one, based on or within the framework of radiocarbon dating. In such a sequence the earliest instance at present available is within the TRB (Funnel Beaker) culture already referred to, from the settlement site of Bronocice, 45 km north-east of Krakow in southern Poland (Milisauskas & Kruk 1977, 1978; Kruk & Milisauskas 1981), where a pot with incised decoration has the repeated motif of a schematically rendered four-wheeled vehicle. The TRB culture is fortunately very well dated by a large series of radiocarbon determinations, including a long series from Bronocice itself (Bakker *et al.* 1969; Tauber 1972; Kruk & Milisauskas 1977, 1981; Bakker 1979) and the stratigraphy and pottery typology of the site enables it to be divided into five phases, the wagon-ornamented vessel belonging to Phase III, with dates between 2740 ± 85 bc (DIC–718) and 2570 ± 60 bc (DIC–363) giving a round figure of 2750–2550 bc.

The small shouldered pottery cup, 10.5 cm high, is incomplete but its panelled decoration when whole would have had four

10 Pottery cup with incised wagon representations, TRB culture, late 4th millennium bc, *from Bronocice, Poland*

repeated wagon representations, separated by blocks of chequer-board and herringbone motifs, of which two survive intact *10, 11* and one in part. They are virtually identical and show in plan view an almost square body with its side members slightly projecting at the back and a wheel represented by a small circle at each corner. The draught-pole bifurcates to make a Y-junction with the front of the body, and at its far end the yoke is indicated by an open V; in one of the two complete examples the line of the pole

11 Detail of decoration on TRB cup, Bronocice

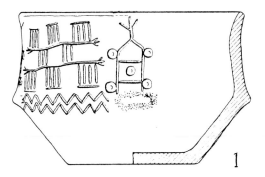

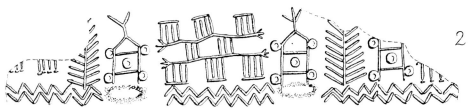

12 Copper models of paired oxen, 3rd millennium BC, *TRB culture, Bytyń, Poland*

runs through the arms of the V. Centrally between the two cross members of the wagon body is a circle equal in diameter to those representing the wheels. No draught animals are shown. We have then convincing evidence for the use of wagons with presumptively disc wheels in the Polish TRB culture just before the middle of the third millennium bc.

The earliest date for the first phase of the Polish TRB culture is that of *c.* 3620 at Sarnowo, already referred to in connection with the evidence for plough-agriculture at that site, and the Bronocice series starts at *c.* 3110 bc. In Denmark the earliest date is that from Ringkloster, *c.* 3350 bc (Tauber 1972). The latest date at Bronocice is *c.* 2250, and the final phase of TRB (Middle Neolithic V) in Denmark ends *c.* 2200–2150 bc (Malmros & Tauber 1975), much the same date as its

counterparts (Late Havelte) in the Netherlands (Bakker 1979). These consistent dates show the TRB culture beginning around the middle of the fourth millennium bc and ending just short of the end of the third, developing in its mature and later phases a number of regional variants distinguished geographically by characteristic pottery styles, and by a diversity of burial rites including in some areas megalithic chambered tombs. It had an agricultural economy which, as we have seen, included plough agriculture and in which cattle played an important part (about 60 per cent of the animal bones at Ćmielów; 42 per cent at Bronocice) and was basically stone-using, mining flint for axe-blades and other tools. Indigenous copper working is reported from Ćmielów (dates between *c.* 2825 and 2665 bc: Bakker *et al.* 1969) and Gródek Nadbużny (*c.* 3100 bc), and some copper objects are best interpreted as imports from the south-east, including such hoards as that from Bygholm in Denmark, Riesebusch in north Germany and Bytyń in Poland, all including flat copper blades that could have been used as axes or adzes for woodworking, as we saw in Chapter 1. (Jażdżewski 1965, 85; Piggott 1968, 307, pl. XXV; Kowalczyk 1970.) The whole question of copper finds in TRB contexts has been discussed recently by Bakker (1979, 127–31). The pair of model oxen from the Bytyń hoard, already noted, are *12* also significant here, as are the modelled yoked beasts (not certainly oxen) on a pottery handle from Krężnica Jara (Piggott 1968, 307, pl. XXIV; Kowalczyk 1970, pl. VII), which should be contemporary with the Bronocice wagon depictions.

It is here that we must take cognizance of another group of depictions of yoked oxen and, apparently vehicles of some kind, those at Lohne (Züschen), some 30 km south-east of Kassel in north Hesse. These were rejected as vehicle representations by Van der Waals (1964, 63) and following him, by the writer (1968, 308), but very rightly J. A.

Bakker (*in litt.* 1979) has urged a reconsideration. The representations are incised (or 'pecked') upon the side-slabs and a detached stone of a well-known stone chambered tomb, and take the form of a number of very schematized figures of oxen, reduced to a single line for the body and a U-shaped pair of horns, some yoked in pairs to a central draught-pole which terminates in a transverse bar or D-shaped object with 'wheels' shown as no more than a pair of cup-marks (Uenze 1956, 1958; Powell 1960; Dehn 1980). This last feature was one of the points of objection raised by Van der Waals: 'we can hardly believe that an important element as the wheel should have been rendered by a minuscule cup-mark'. In addition he was influenced by chronological considerations arising from the dating of the tomb by Schrickel 'to a horizon in between younger Rössen—TRB A/B—younger Chassey A and Michelsberg—older Chassey B', a position which would date from the beginning of the fourth millennium bc. Schrickel also saw Lohne and similar tombs in Germany as derivatives from the megalithic monuments of western and northern France, and compared the engravings to the earliest carvings of the Monte Bego group, which as we saw have no wheels depicted. Both objections can be commented upon, especially that based on chronology.

The Lohne tomb was found and excavated in 1894, and the engravings and the grave-goods demand new and full publication. For the former, the 1894 drawings are *13* more informative than recent photographs (cf. Uenze 1956, pls. 34, 35; 1958, fig. 4) and in default of new illustrations must be used with caution. A D-shaped 'vehicle' appears to be shown twice on the wall-slab 'b1' and on a stone illustrated by Powell (1960, fig. 2a), and a bar terminating in cup-marks four times on slab 'b2'. However, if these are to be interpreted in terms of actual vehicles, it should be pointed out that of the many representations of four-wheeled vehicles among the second millennium BC rock carvings at Syunik in Russian Armenia (referred to above in connection with sledges and further described in Chapter 3) a large proportion have their wheels represented as proportionately no larger than the Lohne 'cup-marks', and imply a type of trolley with very small wheels in use in the Caucasus at that time. It is at least plausible that its distribution may have been far wider in antiquity, and if so, the Lohne representations could be of such vehicles. On the whole, and pending a critical publication of Lohne in the detail it deserves, the engravings may be accepted as ox-drawn vehicles of some sort.

The affinities and dating of the tomb itself raise controversial issues which will be dealt with as succinctly as possible. It was a slab-lined narrow trench-grave with at the northeast end a 'port-hole' stone, a slab with a large circular central perforation, a well-known type of restricted tomb entrance found in many stone chambered tombs. There were no signs of roofing-slabs when

13 Carvings of vehicles and oxen on stone of burial chamber, probable early 3rd millennium BC, *at Lohne (Züschen), Germany*

it was first found. The grave-goods included pottery from Early Neolithic to Late Corded Beakers, and Uenze pointed out that a second series of engraved herring-bone or zigzag patterns on the upper part of the stones might date from the latter period, when the tomb was already partly filled with earlier burial deposits. Among the pottery from the tomb were collared flasks, a well-known TRB type of vessel. Lohne has close local parallels in similar trench-graves, partly slab-lined as at Altendorf or boulder-walled enclosures as at Calden. All form part of Fischer's Hercynian Foothills group (U. Fischer 1937) and have a counterpart in the timber-and-boulder grave of Stein in the Netherlands, as Modderman pointed out in his excavation report on this site, which contained a collared flask (Modderman 1964a). Collared flasks, as has recently been demonstrated, are not exclusively early in the TBR sequence, but continue in use, especially in peripheral regions, throughout the third millennium—*c.*2150 bc at Angelslo in the Netherlands, for instance (Bakker & Van der Waals 1973), and are probably mainly in the Middle Neolithic around 2600–2200 bc. Modderman dated Stein to 'the 26th century' bc on comparative grounds, fearing contamination of the charcoal from the site from the underlying Linear Pottery settlement, thus rendering a radiocarbon date uncertain, but a subsequently published date of 2830 ± 60 bc (GrN–4831) might, in fact, be appropriate. In the Hercynian group of tombs, the Thuringian sites such as Nordhausen and Niederbösa are of the Walternienburg-Bernberg culture, dated at Aspenstedt to 2610 ± 100 bc (H-210/271) (Ashbee 1970, 38 with refs). In earlier assessments of Lohne, conducted within the then current model of megalithic tomb studies, much was made of 'western influences' and the French Seine-Oise-Marne group of morphologically analogous tombs, and of the significance of the 'port-hole' entrance slab. So far as the latter fea-

ture is concerned, while present in the French tombs, it is equally a feature recurring in stone tombs in, for instance, Iberia, Britain, south Scandinavia, the Caucasus, Bulgaria and India with widely disparate cultural contacts and dates. Independent and convergent developments in tomb architecture are here a preferable model to that of diffusion, nor is there any need to seek connections in the simplicities of depiction by stone tools on rock faces. Lohne and its congeners might more plausibly be seen as locally variant forms of tomb, some using stone slabs and others not, which within TRB and allied contexts might go back to those recently published by Madsen (1979) from early TRB long barrows in Denmark. The vehicle representations would then be as local as those on the Bronocice cup.

The Late Copper Age in the Carpathian Basin

Chronologically the next cultural area within which there is evidence for wheeled transport in the third millennium bc is that centred on what has been defined as the Hungarian Copper Age in the Carpathian Basin. This has been the subject of intensive research by Ida Bognar-Kutzian, culminating in a series of recent studies (1972, 1973 a, b) establishing a sequence in parallel with adjacent cultural areas (cf. Tringham 1971, 189, fig. 41). The Early Copper Age (ECA) or Tiszapolgár phase emerges from Late Neolithic roots such as Lengyel (with dates for its late phases of 3410 ± 160 (Bln-231) at Kmehlen and 2860 ± 200 bc (M-1847) at Zlotniki) and the final phases of Gumelnitsa at a time when the tell settlements characteristic of this culture were coming to an end. In general in the Copper Age tell settlements are abandoned and cemeteries separate from settlements appear. The Bronocice radiocarbon dates can be used here, as Bognar-Kutzian equates the latter half of ECA with the Lublin-Volynian phases of the Polgar culture in Poland, dated at site C at Bronocice at 2740±240 bc (DIC-364). ECA might then be dated *c.* 3000–2650 bc.

The Middle Copper Age (MCA) or Bod-rogkeresztur phase is a development of ECA, with shaft-hole copper axes and axe-adzes. Useful correlations for absolute dating are with TRB 'at its zenith', which at Bronocice would presumably be Phase III, dated as we saw to *c.* 2750–2500, and the Usatovo-Gorodsk group in the Ukraine, with dates at Mayaki of 2450±100 bc (Bln–629) and 2390±65 bc (Le–645) (Zbenovich 1973; Coles & Harding 1979, 117). MCA also appears to be contemporary with the earlier Pit Graves in south Russia, which as we shall see run from *c.* 2500 bc. A range of *c.* 2650–2400 bc for MCA would then be reasonable.

The Late Copper Age (LCA) or Baden (in Austria) or Pécel (in Hungary) phase does not appear to be a development from MCA in the same manner as that phase relates to ECA, but shows an impoverishment in the metal component and other signs of discontinuity. Its dating is difficult but critical to us, for it is here that we have our first wheeled vehicle evidence. The writer and others put up a chronology ten years ago with the phase beginning *c.* 2700 bc (Piggott 1968, 304), which appeared to be supported by Polish correlations (Bakker *et al.* 1969, 220) and by a 'late Baden' date of 2565±80 bc (Bln–476) from Oszentivan VIII. However, at Bronocice, Milisauskas and Kruk (1978) assign Phase IV of the site, with dates from *c.* 2500, to the early (Boleraz) phase of Baden, and Phase V (*c.* 2350–2000) to the 'classic' or mature Baden culture; there are also dates for this phase of 2220±120 bc (Bln–351) at Hissar IIa, and 2160±160 bc (KN–145) at Pivnica in Yugoslavia. The high date for the beginning of LCA now seems difficult to sustain, and as early Baden is equated by Bognar-Kutzian with the C1 phase of the Jevišovice stratification in Moravia, one might use the date for the follow-

14 Pottery model probably of a vehicle with pair of oxen, Boleraz culture, early 3rd millennium BC, *from Radošina, Slovakia*

ing B phase of 2310±70 bc (GrN–4065) at Homolka, which would be in agreement with Bronocice. The best approximate absolute date for LCA on present evidence would then be *c.* 2500–2200 bc.

Within this chronological framework, the earliest piece of evidence to be considered comes from a site of the Boleraz phase of the Baden culture at Radošina, 30 km north-north-west of Nitra in Slovakia (Nemejco-vá-Pavúková 1973, 300, fig. 3). It is the front part of a pottery model of a rectangular trough-shaped object, 10.5 cm wide and

14

the same surviving length, with an arched bar across the complete end and three lines of punctuated ornament along its upper edges. Projecting in front are two modelled protomes of cattle, half length with stumpy forelegs. The bottom is flat, with no signs of attachments for wheels. Despite this, the ox protomes seem clearly to denote paired draught, and the model can hardly be that of a sledge. The numerous pottery wagon models of the second millennium from the Carpathians have pierced lugs taking wooden axles with pottery disc wheels and some wooden framework conceivably held the Radošina model on wheels. On at least two of these models a pair of probably bovid protomes occur (Bichir 1964, 69).

With the next two models we are unambiguously dealing with four-wheeled transport in a Late Baden (Pécel) context. Both are square-mouthed pottery cups on wheels, the first being the well-known model from Grave 177 of the large cemetery of Budakalász, north of Budapest (Childe 1954a, 14; Banner 1956, pl. 120; Boná 1960). This is rectangular, with a broken loop-handle springing from the base and originally joining the rim, and the flat base is carried on four modelled disc wheels with their naves indicated in relief. The sides splay outwards and its corners rise in peaks, in the manner of many of the second millennium models just referred to, and its maximum height is 8.5 cm. The sides have incised zigzag decoration, and on the flat base incised lines appear to depict a plank floor within a square frame. No axles are represented. The second model also comes from a grave, in a contemporary Baden/Pécel cemetery at Szigetszentmárton, 32 km south of Budapest. This again is a square-mouthed cup, with complete handle and four disc wheels of roughly similar proportions and 8 cm high. The sides have incised zigzag and panelled ornament, and the base, proportionately smaller than in the Budakalász model, is almost wholly occupied by two massively

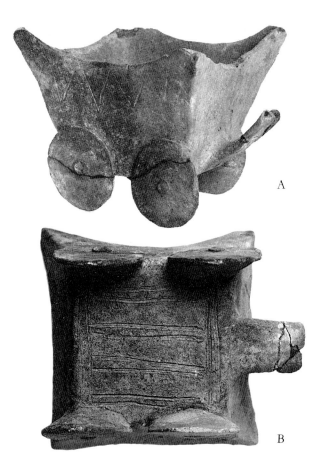

A

B

15 Pottery cup in the form of a wagon, Baden/Pécel culture, early 3rd millennium BC, *from Budakalász, Hungary.* A, *cup;* B, *detail of base*

15

16 Pottery cup in the form of a wagon, Baden/Pécel culture, early 3rd millennium BC, *from Szigetszentmárton, Hungary*

16

17 Burial of paired cattle under excavation, 3rd millennium BC, *Baden/Pécel culture, Alsónémedi, Hungary*

modelled axles joining the pairs of wheels (Kalicz 1976). In both these instances, even if we are strictly dealing with symbolically wheeled cups rather than actual model wagons (a point stressed by Boná and others), their formal resemblance to the later models, the apparently structural prototype for the decoration on the base of the Budakalász example, and the realistic wheels and axles themselves, convincingly demonstrate the existence of actual four-wheeled vehicles in Baden contexts at the time of their manufacture in the second half of the third millennium bc. We are, of course, ignorant (as with Bronocice) of any structural details, and especially the nature of the disc wheels: in small schematic renderings these could be single-piece or tripartite. Nor can we be sure of the tools employed, stone or metal, and there is nothing in the rather impoverished copper inventory of the Baden culture to give us a clue.

In an earlier study the writer drew attention to the possibility that the burial of bovids in pairs in the Baden culture implied not only paired ox-draught but in certain instances the burial of a vehicle undetectable or undetected by excavation (Piggott 1968, 307). This had in the first place been suggested by Banner (1956, 207, 222) in the instances of Grave 3 at Budakalász and Graves 3 and 28 in the contemporary cemetery of Alsónémedi (cf. Behrens 1963, 1964, 19–23). *17* At Budakalász it was further pointed out that, in common with some other roughly contemporary German and Polish graves, there was space enough for such a vanished vehicle between the paired bovids and the human burials in the same grave-pits. Objections raised against Banner's interpretation by Boná (1960, 106) were not then discussed, but must be taken into consideration here. His first point was that a decayed wooden vehicle would at least leave identifiable soil-stains, and that nothing of the kind was perceptible at either site. Here the writer can only say that his own field experience has shown that in many soils wood can disappear completely without trace, and apposite instances in the present context are provided by the probable vanished wooden vehicles in Hallstatt C graves (Chapter 5) and by recent excavations to the highest technical standards of La Tène chariot burials (such as those of Cahen-Delhaye in the Ardennes), where metal fittings and the grave-form itself provide the only evidence of an otherwise completely vanished wooden vehicle (Chapter 6). Bóna's second objection is perhaps more cogent when he pointed out that the bovid burials were not of oxen, but of cows and calves—two calves in Budakalász 3, a cow and a calf in each of Graves 3 and 28 at Alsonemedi. The only suggestion that can be made is that the paired animals were symbolic of paired bovid draught, and that rather than slaughter good oxen, more expendable members of the herd were used for funerary purposes, perhaps themselves diseased or otherwise unsuitable. Bökönyi (1968, 51) noted the frequent presence of old and lame horses in Early Iron Age and Avar graves, and suggested this implied avoidance of killing of good stock for funerary purposes. His later suggestion (1974, 290) that the lame animals were curious shamans' magical steeds can hardly apply to the majority of instances, even if a Magyar burial of man and horse, both hideously deformed (Tóth 1976), seems a convincing case. Apart from the three Baden culture graves just discussed, the writer further instanced similar paired bovid burials in Poland and central Germany, collected and illustrated by Behrens, as indicative of paired ox-draught, all of the second half of the third millennium bc, and suggested vanished or undetected vehicles in graves at Plotha in Germany, and at Zdrojowka and Parchatka in Poland (Behrens 1964, figs 10,

27, 36, 39). These instances are all inconclusive, but in the attested context of later third-millennium wheeled transport, are at least suggestive of what new excavations may reveal.

There remain for mention a dozen or so pottery wheel models in contexts which date from the late third millennium bc and before the appearance of Bell Beaker pottery in the north-west of the Carpathian region. As has already been said, such models are not the most satisfactory evidence for wheeled transport when they appear in isolation, but the status of those which have marked naves, and are dissimilar from the contemporary run of spindle-whorls, has been defended by Bóna (1960, 98) and Bichir (1964, 12) in listing examples of the second millennium and earlier. We may on this argument cautiously accept the finds in question, beginning with those in a late Baden settlement at Vel'ká Lomnica near Poprad in Slovakia (Novotny 1972) and from Staré Zamky (Jevišovice) in Moravia, the latter from the B stratum of the site, equivalent to the Řivnač phase, with a date at Homolka of 2310 ± 70 bc (GrN–4065). In the Vučedol-Zók complex, which in variant local forms follows stratigraphically upon Late Baden (Kalicz 1968, 62) and is contemporary with Řivnač in the northern Carpathian area, Bóna (1960, 91) lists half-a-dozen finds and Kalicz (1968, 70, 75, 83) adds a couple more. Others, broadly contemporary, come from Transylvanian sites; at Czikzsögöd, Horodiştea and Darabani, from Gorodsk/Usatovo contexts: the phase has dates of *c.* 2450 and 2390 bc at Mayaki (Zbenovich 1973; Coles & Harding 1979, 117, 153). In Bulgaria they occur at Ezero, *c.* 2500–2300 bc, and contemporary Bikovo (Georgiev & Merpert 1966). If these wheel models are accepted as such, and there seems a reasonable presumption for so doing, they afford interesting evidence for a knowledge of wheeled transport, in the late third millennium bc, among several diverse cultural contexts over a wide area of Central and Eastern Europe, none with evidence of more than a sparse and limited copper metallurgy.

The evidence from North, and West Europe

It will be convenient at this stage to discuss the evidence for the first wheeled transport in the areas on the western fringes of the TRB culture, notably in the Netherlands and Denmark, and that from Switzerland and north Italy, reserving the south Russian and Transcausian areas for a later section leading us to the boundaries of the ancient Near East. The nature of the evidence in north-west Europe changes from linear representations and models to actual surviving wheels with direct radiocarbon dates, but in the Italian Alps rock carvings alone serve as evidence in the late third millennium.

The dozen Dutch finds made up to the early 1960s were the subject of an exhaustive and far-reaching study by Van der Waals (1964), on which all subsequent thinking has been based, and comprised single finds and in three instances pairs of wheels. A subsequent pair from Ubbena was found in 1966 (Vogel & Waterbolk 1972); one pair from Midlaren was unfinished, with naves roughed out and unperforated, and is also exceptional in being made of alder, the remainder being of oak. They are all closely similar, and are massive single-piece discs worked from planks split from the parent trunk in such a manner as to avoid the heartwood, and their diameter ranges from 0.55 to 0.90 m. They have bifacial integral naves worked from the solid, averaging in overall length *c.* 20 cm, with central smoothly circular perforations 6 to 8 cm in diameter. The largest and best preserved, that from De Eese (found in 1960 whereas many of the others are old finds) showed carefully adzed surfaces, and must have been worked from a

18

18 Single-piece wooden wheel with integral nave, mid 3rd millennium BC, *from De Eese, Netherlands*

plank originally not less than a metre square and 25 cm thick. To anticipate, all must have been worked with stone or flint tools. All were stray finds in the peat except that from Nieuw-Dordrecht, which was found a few metres from and at the same level as a wooden trackway of corduroy construction over the bog, and dating from the end of the third millennium bc (cf. Piggott 1965, pl. IXb). The range of eight radiocarbon dates obtained from the wheels themselves is *c.* 2235–2010 bc, of the trackway *c.* 2140 bc.

The archaeological context of these wheels, which in four instances at least suggest two-wheeled carts rather than wagons, is fortunately well established. The sequence of the third millennium in the Netherlands has been exhaustively and admirably worked out in connection with a remarkably full series of radiocarbon dates, and Van der Waals demonstrated (1964, 51–54) that the dates obtained for the wheels could be set in parallel with four cultural groups or pottery styles: the Funnel Beaker (TRB) culture; the local variant of the widely dispersed Corded Ware/Battle Axe culture, in the Netherlands that of the Protruding Foot Beakers (PFB); a hybrid pottery style between this and that of the Bell Beakers; and finally that of the Bell Beakers themselves, again with an extremely wide distribution. The dates of the wheels coincided with those of the 'hybrid' PFB-Bell Beakers, and so implied that they were to be attributed to the 'very end or after the TRB culture, in a period of a developed stage of the PF Beaker Culture, when the hybrid PF Beaker–Bell Beaker groups were already in existence' (1964, 61). Association with the Corded Ware culture was archaeologically strengthened by the finding under the Nieuw-Dordrecht trackway of the wooden haft for a stone adze-blade, closely paralleled by one from a Corded Ware grave at Stedten in Saxony. Since 1964 one or two modifications to the general scheme have been made, notably the recognition that the TRB culture, represented in the Netherlands by the Late Havelte phase, extends chronologically to *c.* 2150, and that the 'hybrid' beakers of the All Over Ornamented type are likely to be directly ancestral to the developed Bell Beakers, which would then be the products of an ultimate Corded Ware tradition in that region (Bakker & Van der Waals 1973; Bakker 1979; Lanting *et al.* 1973). Correlations between the Dutch Corded Ware-Battle Axe culture and its Danish counterpart, the Single Grave culture, has been given added precision by a series of radiocarbon dates (Malmros & Tauber 1975; cf. Bakker 1979, fig. 73). On chronological grounds then Van der Waals' argument of 1964 still holds good: the wheels are contemporary with the developed Corded Ware tradition of the Netherlands and to this we can now add, possibly with a final phase of TRB in the same area. The archaeological implications of Corded Ware connections are more conveniently discussed after consideration of the Danish and Swiss evidence.

The third-millennium Danish finds comprise a pair of wheels from Kideris, a single half-wheel fragment from Bjerregaard Mose, both in the Ringkøbing district of Jutland and one from Pilkmose, Vejle (Rostholm 1977). All finds represent wheels of precisely the Dutch type, single-piece discs with massive integral naves with circular perforation, and all are of oak, from planks split to avoid the heartwood. The Kideris wheels are 73.5 and 78 cm in diameter, the Bjerregaard fragment again from a 78 cm diameter wheel. The virtually complete Kideris wheel had split along the grain at one point, and had been repaired by means of two battens or dowels (strictly speaking, wedges) held in dove-tailed mortices and uniting the split portions. The Pilkmose wheel was 90 cm in diameter, with a radiocarbon date of *c.* 2230 bc; other dates are *c.* 2230 bc for Kideris and *c.* 2260 bc for Bjerregaard, and on these grounds Rostholm points out that they are contemporary either with the end of the Danish Funnel Beaker culture (Middle Neolithic V) or, more likely, with the Single Grave culture (the Danish version of the Corded Ware-Battle Axe complex). In both instances, he points out, the radiocarbon sample was from a point in the tree-ring sequence fifty to a hundred years older than the manufacture of the wheels, which would then place this within the short bracket of *c.* 2200–2150 bc recently assigned to the Single Grave culture in Jutland (Malmros & Tauber 1975).

Three finds of disc wheels in imprecisely dated contexts, but possibly contemporary with the Dutch and Danish wheels are conveniently dealt with here. That from Schönsee, Braniewo (Poland), with integral nave and 70 cm diameter, Van der Waals thought might be attributed to the Haffküsten culture (Schneider 1952; Van der Waals 1964, 67, 75); that from Beckdorf, Stade (Lower Saxony) had an inserted nave, as did that from Aulendorf, on the Bodensee (south Württemburg), and such naves are not on

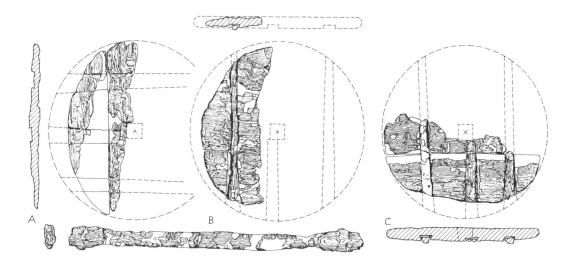

19 Tripartite wooden wheels and axle, Corded Ware culture, early 3rd millennium BC, from Zürich, Switzerland

20 Disc wheels and axle as excavated, 'Pressehaus', Zürich, Switzerland

present showing earlier than the second millennium, as at Glum, Oldenburg, described in a later chapter (Van der Waals 1964, 75).

One of the most important and surprising recent finds is that made in 1976 during excavations on the site of the 'Pressehaus' in Zürich. Here, in a waterlogged lakeside settlement of the Swiss Corded Ware culture, were found the remains of three wheels, two united by their axle and a third 3 m away, all just within the palisaded edge of the settlement. In contra-distinction to the Dutch and Danish wheels, those from Zürich are of tripartite construction and held by square mortices to a circular-section axle turning with the wheels. In considering the alternatives of their representing a wagon, or more than one cart, the excavator argues reasonably for the former. A radiocarbon date of *c.* 2340 bc was obtained from one wheel, and a further fragment of a centre plank with a square axle-mortice still containing the squared stump of the axle was identified among old finds from the well-known lakeside settlement of Vinelz on the Bielersee (Ruoff 1978). The Zürich wheels were *c.* 65–68 cm in diameter and in a fragmentary condition, but all appeared to have been made of three beech planks *c.* 5 cm thick, held together by transverse battens or dowels in dove-tailed grooves on one face, exactly in the manner of the repairs on the

and others including two dates from Vinelz of *c.* 2510 and 2220 bc (Ruoff 1978, 282). The Zürich vehicle is therefore well established as contemporary with the Dutch and Danish finds of individual or paired wheels and furthermore it (and probably the Vinelz fragment) are in the same Corded Ware context.

Technologically the Swiss finds are remarkable in being not massive single-piece discs with a perforated nave turning on a fixed axle, but of tripartite construction with relatively thin planks, fixed to a rotating axle by a square mortice. As such they are unique in prehistoric Europe. As we shall see, simple and tripartite disc wheels with massive integral naves are known on the south Russian steppe from the late third millennium bc, and in the second millenium in the Caucasus, while other forms of the tripartite disc wheel are present in the third-millennium Mesopotamia. It was these factors which led Van der Waals (1964, 77) to pose the question, in the context of the Dutch evidence, as to whether the simple single-piece types 'represent as it were the fossilized *Urform* of the disc wheel, that ought to have preceded the more sophisticated bi- and tripartite wheels with separate naves'. To his mind, taking into account the evidence from the Pontic steppe and the ancient Near East, this was not the case: the evidence 'makes it clear that these wheels also represent simplified imitations of Mesopotamian prototypes' and that composite disc wheels demand for their manufacture metal tools 'which certainly were undreamt of by Battle Axe people'. The Swiss wheels, and the technique of the Kideris repair, now show that composite disc wheels were within the technological capability of communities using stone tools, and the Mesopotamian derivation is part of a wider and crucial issue discussed at the end of this chapter.

To round off the Western European evidence, we must finally consider a small

21 Rock carving of ox-drawn wagon, 3rd millennium bc, *Rock 2, Cemmo, Val Camonica, Italy*

Kideris wheel just described, the dowels being of ash. A pair of these ran across the total width of the wheels and a single one ran radially to the axle-mortice and the Vinelz fragment suggests it was not duplicated on the other half of the wheel. There was no nave, as vertical stability was ensured by the rigid attachment to the axle. This was rounded in cross-section, with a worked collar and square tenon at each end, and appears to show signs of wear by rotation within the collars. The wheel-track (gauge) was *c.* 1.30 m. The Vinelz fragment, with axle-stump, square mortice, and the end of a morticed transverse batten on one side of this, comes from a site mainly of the same Corded Ware phase (that of Utoquai) as the 'Pressehaus' settlement, and as Ruoff notes, the same site produced a well-known ox-yoke (above p. 51). A number of radiocarbon dates have been obtained for Swiss Corded Ware sites: Utoquai itself *c.* 2235 bc,

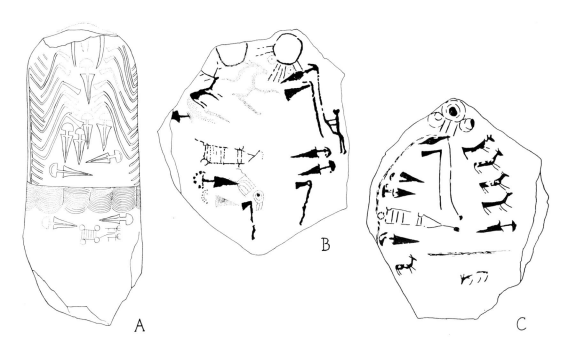

A

B

C

22 Wagon representations, 3rd millennium BC, *from* A, *Lagundo*; B *and* C, *Valtellina, Alto Adige, Italy*

group of representations of vehicles in north Italy, one on a rock face and three on decorated stelae, which have a claim to be considered as late third millenium in date. All have recently been discussed by Martine van Berg-Osterrieth (1972). Among the famous carvings in the Val Camonica, one rock face, Cemmo no. 2, of Anati's Phase III A (Anati 1975), has a large series of representations of bovids, caprids, men, daggers and a hafted axe, and, placed one above another on the left-hand edge of the composition, a pair of oxen drawing a four-wheeled wagon and another pair drawing a plough to which reference has already been made. The vehicle is shown with a roughly square body and four large disc wheels; the draught-pole bifurcates in a Y-form to join the front of the wagon and joins a yoke at the far end. The oxen are shown in profile, the wagon in plan with the wheels flat, the common convention of vehicle representations throughout European prehistory. This is the only vehicle representation of this phase at Val Camonica, though later examples occur in subsequent periods of rock carv-

ings: Anati dates his Phase III A to the end of the third millennium, the remainder to the second and first.

To the north of Val Camonica, in the Alto Adige, a remarkable series of decorated stelae, some anthropomorphic, have been recorded, and three bear representations of vehicles. Two stelae from Caven, Valtellina, are decorated in the Phase III A style, with caprids, a cervid, daggers, hafted axes and halberds, one held by a man, and on each a wagon without draught animals. Both vehicles are oblong or slightly trapezoidal, with relatively small wheels and cross-members indicated on the body, to which the draught-pole makes a Y-junction. Van Berg-Osterrieth sees such a type of vehicle as recalling those of Iron Age date, and would therefore reject an early date (1972, 107), but there seems no reason not to accept them as contemporary with Cemmo 2, since we have no independent evidence of vehicle types at this time. The last example is on a highly schematized but ultimately anthropomorphic stele from Lagundo, with complex patterning of hafted axes and dag-

21

22

gers and, incised towards the base, a rectangular, almost square wagon with disc wheels, cross-slatted body, Y-junction draught-pole, and a pair of yoked oxen. Schematic though they are, these north Italian wagon depictions are closely comparable with those from Bronocice, particularly in the Y-junction of the draught-pole to the body. This realistically represents a practical necessity of wagon construction, where the need to keep the body strictly horizontal on its four wheels calls for vertical mobility of the draught-pole to accommodate draught animals of varying sizes, and this demands a pivoted junction best achieved by bifurcation and flexible attachment at two points.

Confirmation of a third-millennium date is, as Barfield has pointed out (1971, 67), afforded by the stelae themselves which, stylistically and in their details of daggers and other features, are closely linked to those from the Petit-Chasseur site at Sion in Switzerland (Bocksberger 1971; A. Gallay 1972, a, b; G. Gallay & Spindler 1972). At this complicated site the elaborately decorated anthropomorphic stelae appear to have belonged originally to a trapezoidal chambered cairn designated Monument VI, and were later partly broken up and reused in burial cists of the Bell Beaker culture (A. Gallay 1972 b, 87). Monument VI overlies a Late Neolithic layer with a radiocarbon date of *c.* 2310 bc, and its burials were accompanied by blades of Grand Pressigny flint, the importation and use of which in Switzerland is characteristic of the Auvernier and Corded Ware cultures: Auvernier 1, with Corded Ware sherds, has dates of *c.* 2230–2012 bc, and other Corded Ware dates have been cited above (Strahm 1969; Suess & Strahm 1970; Sauter 1976, 61). The imported Grand Pressigny flint blades give us a correlation with the Dutch sequence, where they appear at the end of the Corded Ware horizon, the point which as we saw above is marked by the emergence of the All Over Ornamented Beakers in the last cen-

tury or so of the third millennium bc. The north Italian wagon representations just considered should then be broadly contemporary with the Dutch, Danish and Swiss wheels at the close of the third millennium, and in north Italy, with the metal-using culture of Remedello (Barfield 1971, 55).

The evidence from East Europe and Transcaucasia

Over a great tract of the south Russian steppe, north of the Black Sea and the Caspian, there extends a series of inter-related groups of third-millennium communities collectively known from their characteristic tombs as the Pit Grave cultures. Between latitudes 45° and 50°, the area of their distribution extends for some 2000 km from the River Dniester on the west to the River Ural on the east, its southern fringe bounded by the northern shores of the Black Sea and the Caucasian mountain range (Piggott 1968, 282, with refs; Merpert 1974; Coles & Harding 1979, 124–29). The graves (normally under a barrow or *kurgan*) 'are characteristically deep shafts in the ground, usually square or rectangular, lined with timber and covered with planks. Burial was by inhumation in a contracted position, and red ochre was used in quantity to cover the body; grave goods were poor'. Another (and as we shall see partly contemporary) tomb type is the Catacomb Graves between the Dnieper and the Volga, where 'burial is in a pit with a side chamber' and the grave goods may include a 'considerable quantity of metalwork, mostly of arsenical bronze, imported from the Caucasus' (Coles & Harding 1979, 125, 127). Alexander Häusler (1974; 1976) prefers to group Pit and Catacomb Graves together as part of an Ochre-Grave Culture (from the red colouring found commonly with the burials), with broadly an Earlier and a Later Phase, the latter with vehicle burials. He has recently published a comprehensive survey of the evidence for such vehicle remains, superseding that of the writer (Piggott 1968) and incorporating over thirty instances. In the following paragraphs documentation will be found in this source unless specific references are made (Häusler 1981). In burials of both Pit Grave and Catacomb Grave type vehicles, either complete or in a *pars pro toto* representation of individual wheels, are found, and constitute a valuable corpus of evidence. Unfortunately standards both of excavation and of publication have been deplorably low, and a large number of finds reported to have been made remain unpublished. In no published instance is there an adequately full technical description of the vehicle finds themselves.

The Pit Graves were divided into four phases by Merpert (1961) which are related to geographical areas within the general wide distribution-pattern already indicated, those of Dnieper-Azov in the west, the region between the middle Dnieper and the southern Donetz, the Lower Volga, and the Middle Volga respectively. In a more recent study of the Pit Grave culture between the Volga and Ural rivers these regions have been subdivided to make a total of nine (Merpert 1974 with map, fig. 1). He regards the original Lower Volga group as containing the earliest graves of Phase I; in Phases II and III the Dnieper-Azov group emerges, and by Phase IV all regions, including now the Middle Volga group, are represented. The Catacomb Graves (Piggott 1968, 300 with refs.; Coles & Harding 1979, 125) with a distribution largely coincident with the Pit Graves from the Dnieper to the Volga, were considered as subsequent to the latter, but, as we shall see, radiocarbon dates now show a considerable chronological overlap and they appear to be local variants of the Pit Grave theme.

Particularly owing to the work of Telegin (1977) in the western areas of the culture, a remarkably large series of radiocarbon dates has been obtained for the Pit Grave culture, showing it to range from *c.* 2500 to *c.* 1700

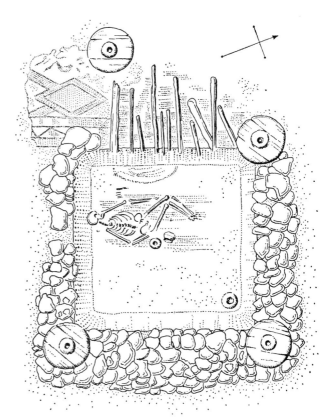

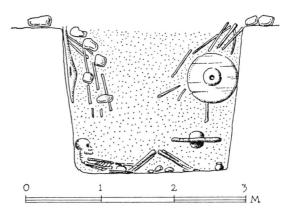

23 Burial with tripartite disc wheels, 3rd millennium BC, *Pit Grave culture, Tri Brata, Elista, Kalmyk ASSR*

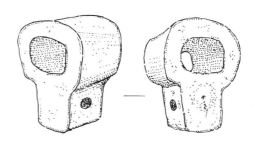

24 Pottery model of cart with tilt, Tri Brata

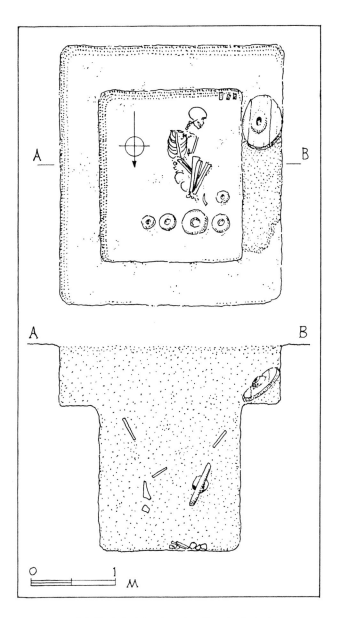

25 Burial with tripartite disc wheels, 3rd millennium BC, *Arkhara, Elista, Kalmyk ASSR*

bc, while a smaller group of Catacomb Graves runs from *c.* 2265 to 1910 bc (Coles & Harding 1979, 130, 153) showing them to overlap the Pit Graves in Merpert's Phase IV, a position confirmed archaeologically by the occasional presence in them of objects characteristic of the Catacomb Grave culture such as hammer-headed pins and cross-foot bowls. A direct radiocarbon date from

a vehicle grave, that from Balki Grave 57, is 2420 ± 120 bc (Kl-606), but the grave is stratigraphically later than Grave 40 in the same Pit Grave barrow, with a date of 2040 ± 110 bc (Kl-601). If the Grave 57 sample was from a roofing log, old wood may be involved. Another date, 1910 ± 80 bc (UCLA-1273), comes from one of the wheels of the Kudinov wagon, from a Catacomb Grave. Apart from these instances, the cumulative evidence suggests that the Pit Grave vehicle burials belong to a late phase of the culture, contemporary with those of the Catacomb Graves at the end of the third millennium bc.

The distribution of Pit or Catacomb Graves containing vehicle remains stretches across the North Pontic steppe from the River Dniester to the Kuban at the western end of the Caucasus range and the Kalmyk steppe on the River Manych between the Rivers Don and Volga. Assuming steppe conditions similar to those of recent times in the third millennium BC, the timber necessary for the simple and tripartite disc wheels, as well as the bodies of carts and wagons, would have to be sought in either the Woodland Steppe zone to the north or the Caucasus to the east. More than one identification of oak and at least one of pine has been made in surviving wheels.

The burial rite varied, from the deposition of a single disc wheel (Gerasimovka, Rostov, Elista, Perkonstantinovka) or a pair suggestive of a two-wheeled cart (Akkermen, Storozhevaya Mogila, Starosel'e Barrow 4 Grave 13, Pervokonstantinovka Barrow 1 Grave 6), to groups of wheels—three at Sovievka, six at Tri Brata *23* and seven at Starosel'e Barrow 1 Grave 8. No cart bodies have survived, though pottery models from Tri Brata, Elista Barrow 5 Grave 8 and Alitube, Rostov, represent square-plan bodies with arched tilts. The *24* remains of four-wheeled wagons have been recorded from at least sixteen sites, good examples being Elista Barrow 5 Grave 9,

56

Kudinov, Lola Barrow 4 Grave 7 and Novo-Titarovskaya.

Carts, therefore, apart from the models, can only be inferred from the burial of paired wheels, and with larger numbers ambiguity exists, as with the six wheels at Tri Brata which could represent three carts, a cart and a wagon, or symbolic wheels unrelated to any specific vehicles. More can be said of the wagons. As with all vehicle burials they raise the question of their original status as everyday structures used as a grave offering, or as funerary wagons or hearses: at Novo-Titarovskaya a quite 'impractical' vehicle made for burial purposes certainly seems present. In all instances the exigencies of space within the Pit Grave or the access shaft of the Catacomb Grave has led to the removal of the draught-pole, as in the later instance of Trialeti Barrow 5 or many Hallstatt burials described below. Remains of a draught-pole were thought to be present with the pair of wheels at Storozhevaya, and with the wagon in Lola Barrow 4 Grave 7 the dismantled pole was of forked composite construction, evidently similar to the second millennium BC Caucasian examples described later. Yokes are recorded from this grave and Starosel'e Barrow 1 Grave 8. The body of the Lola wagon measured 2.5 by 1.5 m; that from Kudinov was *c.* 2.20 m long: in the photograph published in Gimbutas 1970, 183, fig. 22 only the wheels are original, the rest reconstructed. The wagon framework in Elista Barrow 5 Grave 9 was 2.0 m long, the axles 1.7 m long with a square sectioned axle bed, and the wheel gauge 1.30 m. The wagons in Elista Barrow 8 Grave 7, and in Lola Barrow 4 Grave 7 had wickerwork arched tilts, the latter with back panels decorated with incised spirals, rectangles and triangles, both features again foreshadowing the Caucasian vehicles at Lchashen. At Lola it was also claimed that the front axle was pivoted by a wooden coupling bolt through a perforated wood disc, but the poor documentation in the original

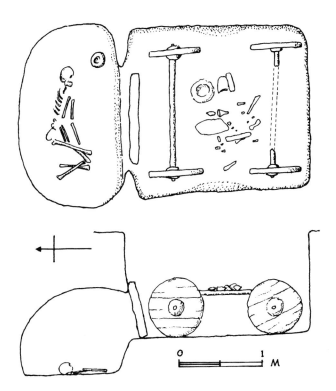

26 Wagon burial in Catacomb Grave, Barrow 5, Grave 9, Elista, Kalmyk ASSR

excavation report leaves such a remarkable technological anomaly insecurely substantiated.

The disc wheels are either single-piece (about a dozen records, including Akkermen and Storozhevaya in pairs) or tripartite with internal dowels (about an equal number of records, as at Tri Brata and in the Kudinov, Elista and Lola wagons). One two-piece wheel similarly dowelled comes from Starosel'e Barrow 4 Grave 13. All have massive integral naves rotating on a fixed axle, and those from Novo-Titarovskaya Barrow 1 Grave 9 were coloured white with paint or chalk. Diameters may be as small as 25 cm (Storozhevaya) but in general the single-piece wheels range from 50 to 78 cm, the tripartite discs from 47 to 92 cm, with most around 70–80 cm.

The tripartite wheels just described have important technological implications. Despite the absence of technical details and drawings, it is apparent that their construction, with the component members held together by rod-shaped wooden dowels in circular mortice-holes cut in the thickness of the planks, is precisely that of the second-millennium tripartite disc wheels from the Caucasus studied at first hand by the writer, who pointed out that such a technique was dependent on metal chisels, and more particularly, gouges (Piggott 1968, 288–89; 305–6). Such gouges, in arsenic-copper, have recently been shown by Chernykh (1976a) to be characteristic of his Circumpontic area in the earlier Copper Age, occurring in the Kura-Araxes (or Early Transcaucasian) culture and in the Maikop metalworking tradition of the western Caucasus, both within the later third millennium, and it was from the latter source, as we saw, that the Pit Grave and Catacomb Grave cultures were obtaining their metal equipment, including gouges (Coles & Harding 1979, 126, fig. 43), constituting Chernykh's Kemi-Oba metalworking tradition in the southern Dnieper and Crimea (1976a, map fig. 5 and metal types fig. 6). Further woodworking equipment includes shaft-hole axes, the well-known axe-adze from Maikop itself, and a range of true adze-blades with asymmetrical cutting-edges (Chernykh 1976a, fig. 6; Popova 1963). Shaft-hole axes and chisels or gouges in Hungary have been noted as of Caucasian derivation (Kalicz 1968, 46), and the same must apply to the hoards from Ezero in Bulgaria (Georgiev & Merpert 1966) and that of Staré Zamky in Moravia (Benešová 1956), both with dates of *c.* 2300 bc. Another copper type within the same western Caucasian metalworking tradition which may have a bearing on ox-draught, is the socketed hooked object, one of which was found with the male skeleton of the pair of burials of a man and a woman in the wagon from Grave 7, Barrow 4 at Lola and another in that from Grave 8, which the writer has suggested might be ox-goads. They appear, with double-hooked

variants, in Chernykh's Maikop and Kemi-Oba metalwork groups, from Catacomb Graves of the Donetz region and (with a shaft-hole axe and a gouge) in the Emen cave in Bulgaria (Popova 1963, pl. XVIII; Piggott 1968, 299–300; Chernykh 1976a, fig. 6; Coles & Harding 1979, fig. 43).

The last area to be considered is that of Transcaucasia in the context of the Kura-Araxes or Early Transcaucasian culture already referred to in connection with its copper metallurgy. Here unfortunately the evidence is really confined to pottery models of wheels, the potential ambiguity of which, when unsupported by collateral evidence, has already been commented upon. But the models in question have every appearance of representing wheels with well-defined naves, and at some sites at least (e.g. Kvatskhlebi and Geoy Tepe) it is possible to differentiate them sharply from contemporary spindle-whorl types (Javakishvili & Glonti 1962; Burton-Brown 1951). Burney, in a recent discussion of the problem, is unhappy about the acceptance of all such objects as wheel models, while acknowledging their morphological similarity. He points out the frequency of their occurrence at 'sites in Georgia, Armenia, the Nakhichevan region, Azerbaijan and eastern Anatolia... the very number of the discs, however, militates against their invariable interpretation as parts of model carts or wagons ...Though some of these discs belong to model vehicles, most of them could perfectly well have served as one type, though not the only one, of spindle-whorl' (Burney & Lang 1971, 74). This seems a very reasonable view and, for our purposes, the recognition that actual disc wheels for vehicles within the Early Transcaucasian culture formed the eventual prototype of the pottery models is the important factor. Burney, using available radiocarbon dates, has divided the culture into three phases, spanning the third millennium bc, Phase I *c.* 3000–2600 bc, Phase II *c.* 2600–2300 bc and

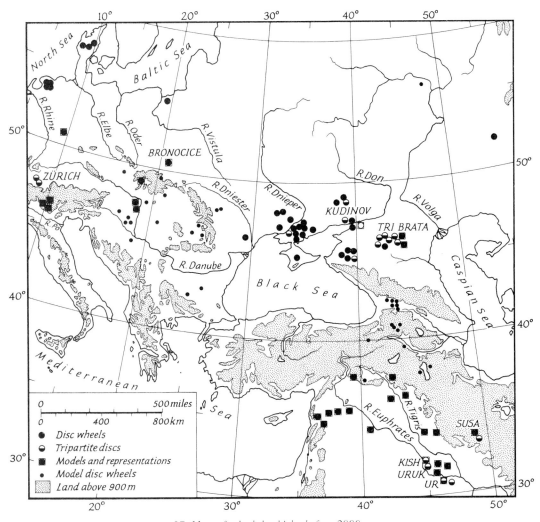

27 Map of wheeled vehicles before 2000 BC

Phase III *c.* 2300–2000 bc or perhaps later. The chronological range of the wheel models appears to run through all three phases, and the pottery tradition links Phase III with the culture represented by the Trialeti tombs of the second millennium, where actual vehicle burials occur. A knowledge of transport with disc-wheeled vehicles, presumptively ox-drawn, in third-millennium Transcaucasia (and perhaps from early in that millennium) seems a reasonable conclusion.

There are two possible claimants for actual vehicle burials of the third millen-nium bc in the Early Transcaucasian culture. The first is that of Zelenyy in the Tsalka province of Georgia, where, in a deep grave under a barrow, two grooves in the floor were interpreted as indication of the former existence of a four-wheeled vehi-cle with a gauge of about 1.75 m. The pot-tery was of Kura-Araxes type, some with the characteristic applied spiral motifs, and the grave is included in the Kura-Araxes group in a recent assessment of the Trialeti barrows (Kuftin 1948; Piggott 1968, 279; Zhorzhikashchvili & Gogadze 1974, 13). The second vehicle burial is that of Tetri-

Tskaro, with a gold pin and copper axe, adze and chisel (Burney & Lang 1971, 71). assigned to the Phase III of the Early Transcaucasian culture, but if this is the same grave as that with a more recently published radiocarbon date of 1380 ± 60 bc (TB-30) (Burchuladze *et al.* 1976, 356), it would be assigned to the second-millennium Trialeti series described in a later chapter.

Contact and transmission: Europe and the Near East

The evidence reviewed above shows the first appearance of wheeled vehicles in prehistoric Europe to have been a phenomenon with a remarkably restricted chronological range within the second half of the third millennium bc and an equally remarkably unrestricted range within a number of disparate archaeological cultures—TRB, Baden, Corded Ware, north Italian Copper Age, Pit and Catacomb Graves, Transcaucasia. Within these, only the Corded Ware contexts of the Dutch, Danish and Swiss finds form a continuum over an extended geographical area in north-west Europe. What we appear to have is the adoption over a few centuries of a technological innovation by a number of communities that were co-existent and contemporary, and that shared social, economic and agrarian patterns within which such an extension of the existing potential of ox-draught was acceptable, and with the requisite woodworking technology to make manufacture possible. In such circumstances, the question then arises as to whether these shared characteristics could have brought about the independent invention of wheeled vehicles as a result of modifying pre-existing sledge or slide-car transport in more than one place within a few centuries, or whether we are seeing a process of technological transmission and adoption from a single restricted area within which the primary invention was first devised. The current unpopularity

in some quarters of the long-standing model of diffusion as an agent of culture change, while a salutary warning against its uncritical use, surely does not entitle us to reject it absolutely, particularly in the instance of the adoption of an invention in all its detail. Early in this chapter we glanced at this problem and must now return to consider an explanatory model which has been advanced to cover precisely the period (the third millennium bc) and the subject (wheeled transport) with which we are concerned. This is the 'Kurgan Hypothesis' advanced by Marija Gimbutas that, while critically received by archaeologists, has proved attractive to many philologists concerned with the Indo-European language group, which forms an essential part of its content (Gimbutas 1970, 1974, 1978).

The thesis is briefly as follows. The name 'kurgan', the Russian for a barrow or tumulus, places its point of origin, which is that of the south Russian steppe. Here the Pit Grave and Catacomb Grave cultures are seen as the primary manifestation of a Kurgan culture or tradition, 'defined as collective socio-economic and ideological features observable over time and space. This tradition—characterized by a pastoral economy, an agnatically-linked, hierarchical social structure, seasonal settlements, small semi-subterranean dwellings and larger chieftains' houses, diagnostic burial rites including human and animal sacrifices and symbolic system with the sun as the dominant motif—can be traced through the millennia to each geographical region the Kurgan people settled or ' "Kurganized" the local population' (Gimbutas 1978, 278). It involves 'the hypothesis of the cultural change in Europe as well as in the Near East caused by the invasion of hordes of pastoralists from the North Caucasian—lower Volga steppe area' (1970, 155) as the result of 'several migratory waves of steppe pastoralists or "Kurgan" people that swept across prehistoric Europe. These repeated incursions and ensuing culture shocks and population shifts were concentrated in three major thrusts between 4400 and 2800 BC' (1978, 277). The invaders brought with them the domesticated horse and wheeled transport, and can be traced not only in continental Europe, but in Greece, Sicily, Syria and Palestine (1970, 182, 186–89), and were speakers of Indo-European languages.

This rather breathtaking concept contains many component elements—social structure, sun-worship, language—not susceptible of demonstration by direct archaeological evidence, but at best by second-order inferences or sheer assumptions. A recent archaeological critique (Coles & Harding 1979, 6–7) points out the fallacies inherent in accepting all grave-pits under barrows, and a few features such as the use of red ochre, as derivatives from the south Russian Pit Graves, and emphasizes that the large series of radiocarbon dates now available (many cited in this chapter) show not chronological priority for the Russian steppe cultures but contemporaneity with the supposedly derivative groups in Europe. That contacts existed between the late Pit Grave/Catacomb Grave phase and the Northern European Corded Ware cultures is shown by such objects as hammer-headed pins of Pontic type sporadically appearing in the north, the Bleckendorf (Saxony) find with a beaker of developed late third-millennium type being well known, and Van der Waals looked to the Pontic steppe for the nearest cognates to the Dutch single-piece disc wheels (1964, 64–67). It may be added that where direct archeological evidence is used to support the 'Kurgan' thesis, it is too often treated in an uncritical, if not tendentious manner. In its propagation, the fundamental question of alternative models of culture change is never hinted at, nor is there any awareness of the now long-standing debate on the 'Invasion Hypothesis' as a valid concept. It has a curiously old-fashioned air, reminiscent of the earlier writings

on the distribution of megalithic monuments and even at times of the hyper-diffusionism of Perry and Elliot Smith.

We are, however, left with a final, and very real problem, none the less important for being exceedingly difficult of resolution, that of the relationship between the European wheeled vehicles under review, and those of the ancient Near East. The long-standing view, going back to Gordon Childe (1951, 1954a), was that one of the best documented instances in European prehistory of diffusion was the primary derivation of wheeled transport from ancient Mesopotamia. The writer had occasion briefly to review this thesis recently (Piggott 1979), and simultaneously the Near Eastern evidence for wheeled vehicles was fully examined by Littauer and Crouwel (1979a). For our purposes it is sufficient to note the earliest evidence is the 'sledge-on-wheels' vehicle of the Uruk IVa pictographs of the late fourth millennium already referred to; in the third millennium and the Early Dynastic period models, representations and actual surviving remains of vehicles, two-wheeled and four-wheeled, become frequent. Some specialized forms of 'battle cars' and 'straddle cars' do not concern us, but the presence from early in the third millennium BC of tripartite disc wheels, their planks held together by external battens, is of prime importance. The problem can really be reduced to two aspects, technological and chronological: we are not involved in any general theory of wholesale cultural transmission of the old *ex oriente lux* or more recent 'Kurgan Hypothesis' type but with the widespread adoption of a rather specialized technological contraption within a limited period of time. From the outset, and on the limited evidence then available, Childe drew attention to the importance of the tripartite disc wheel in this connection, as a piece of complex technology, the wide distribution of which was difficult to explain 'unless the device had been diffused'. Present in Early Dynastic Sumer probably before the middle third millennium, he drew attention to its appearance on the one hand in Transcaucasia (Trialeti) and on the south Russian steppe (Tri Brata) and on the other hand in painted model form from the third millennium BC Harappa culture in the Indus Valley (Chanhu–daro). The recent Zürich finds have dramatically extended its distribution in prehistoric Europe at the same time.

When Childe wrote his studies of 1951 and 1954, the concept of isotopic dating had just been established in principle, but had not yet been applied systematically to archaeology. For Europe and the Near East a series of stratigraphic and typological sequences had been established, of varying degrees of reliability, giving relative but not absolute chronologies where historical dates were lacking. These local sequences could then be linked one with another over larger geographical areas, and attempts were made to relate these non-historic sequences to those of early historically documented civilizations in the Near East, and so produce chronologies in calendar years before the present. With what appeared to be a few fixed points, intervals were allotted by what can hardly be dignified as more than reasoned guesswork, and the same applied in the Near East before literacy permitted an historical record.

This summary of what to archaeologists is a familiar episode in the recent history of their subject is necessary to understand the position of absolute dating in the prehistory of Europe and the Near East today, crucial to our enquiry. For continental Europe before the last seven or eight centuries BC, when reasonable contacts with the historical Mediterranean can be established, some thousands of radiocarbon dates now provide an absolute time-scale that has superseded the conjectural chronology of a generation ago. Errors are quantifiable and can be subjected to appropriate statistical treatment, and dates can be checked against stratigraphical sequences and historical dates where they are relevant. In the Near East, and particularly in the Mesopotamian culture-sequence with which we are now concerned, the situation is different. A chronology derived from literate historical records can be constructed from approximately 2370 BC, the date of the accession of Sargon I as ruler of Akkad, but before that date the absolute chronology of the third and fourth millennia is really quite uncertain and lacks both the precision and the accuracy of that provided by radiocarbon for Europe at the same period. Mellaart's attempt to construct a coherent chronology from such radiocarbon dates as are available (1979) shows how few come directly from the Mesopotamian sites which concern us—only one from Uruk, believed to be of Level IVa, of 2815 ± 85 bc (*c.* 3580 BC) and four from the Royal Tombs of Ur, all from old samples of which three have margins of error of three hundred years. 'Until more radiocarbon dates are obtained for Mesopotamia this type of evidence can make only a limited contribution to the discussion of absolute chronology' (Munn-Rankin 1980), which for the prehistoric periods is therefore in the state of Europe thirty years ago, with absolute dates assigned to sequences by estimation which cannot, however, be quantified or checked by alternative means. The absolute chronology of Bronocice is more reliable than that of Uruk IV or the Early Dynastic periods of Mesopotamia in which the vehicle finds of Kish, Ur and Susa occurred, and is based on evidence of a different order. The firm dates in calendar years as given for instance in chronological tables (e.g. Porada 1965, 176–79; Lloyd 1978, Table III) or used in discussion of vehicle finds (Littauer & Crouwel 1979a) have, in fact, an appearance of accuracy which is spurious and cannot therefore be used in a like-to-like comparison with radiocarbon dates, whether calibrated or not. The calibrated range of date for Phase III at Bronocice, to which the cup with the wagon representation belongs, is *c.* 3530–3310 BC, but it would be improper to compare this date with that of 3200–3100 BC assigned to Uruk IVa, in which the sledge-on-wheels pictographs appear. The former is the expression in statistical terms of a radioactive process, the latter an archaeological computation of how far back in time beyond the beginnings of the Mesopotamian historical record, around 2400 BC, the relevant strata can be placed, on the grounds of changes in material culture and the succession structural features on many sites.

When we come to assess the relative chronological priority of the evidence of the earliest wheeled vehicles in the Near East and in Europe the problem today is more difficult than it was for Childe a generation ago. At that time the prehistoric chronologies of the two areas were both obtained by the same means, archaeological correlations extrapolated from fixed historical dates, but with the unequal application of isotopic dating methods this is no longer the case, and Europe has a time-scale independent of and better than that of the prehistoric Near East at the relevant time. For the period in question Mesopotamian dates can only be treated as theoretical approximations, and absolute priorities cannot be assigned. What can be said is that any time interval between the present evidence for the first wheeled transport in north-west Europe and Mesopotamia must be very short, and we can only fall back on inherent historical and technological probability in assuming a more likely initial invertion as a part of the complex innovations of this period in the Near East rather than in the simpler context of Neolithic Europe.

As the writer has said, such near-contemporaneity argues for some form of 'diffusion' rather than sporadic independent invention in more than one place, though 'whether "diffusion" is an apt term for the phenomenon we seem to be witnessing is another matter—"technological explosion" may be more appropriate'. If we ask whether (whatever name is given to the process) it is best accommodated in some form of 'diffusionist' model, 'if we mean the adoption of a technological novelty by a society or societies from outside sources, surely the answer must be "yes"; Switzerland and Sumer can hardly have invented, independently and simultaneously, the tripartite disc wheel around 3000 BC'. The area of invention may have been larger than Sumer and Elam, but they were central to it (Piggott 1979, 10, 15). The process of adoption in prehistoric Europe was as fast as that of the horse for riding by the American Indians: three centuries from the Caribbean to Calgary (Driver & Massey 1957, 284).

3 Wagons, carts and horse-drawn vehicles of the Earlier Bronze Age

Content and chronology

The evidence reviewed in the last chapter presents us with a picture of the likely origin of disc-wheeled vehicles as centred on Mesopotamia at the end of the fourth millennium (whether 'bc' or 'BC') and a subsequent remarkably swift adoption of this novel means of transport among a wide range of prehistoric European communities from Transcaucasia and Russia to Switzerland, and from north Italy to Denmark, within five centuries or so. It must be remembered that, owing to the limitations and imperfections inherent in all archaeological evidence, we are at present ignorant of what circumstances may have obtained in areas further to the west, including the British Isles. But within the very large area for which evidence of some kind or another is available, the pattern is consistent and points to the adoption, among disparate agricultural communities in varying ecological settings, of ox-drawn wagons and carts as an acceptable technological increment to their economies. Basically these communities were stone-using, with at best and in restricted contexts a limited copper technology, or the recipients of copper tools from outside sources. For all, antecedent traction-plough cultivation with an ard drawn by a pair of oxen may reasonably be assumed or on occasion demonstrated, and efficient woodcraft and carpentry is likewise attested, and the exploitation of herd animals for traction seen in terms of Sherratt's thesis of 'secondary products' of animal domestication beyond primary meat supply in the second half of the third millennium bc.

The common factor leading to the adoption of such a piece of complex technology as a wheeled vehicle, demanding a considerable investment of skilled craftsmanship, remains elusive, and any attempt to discover it would take us beyond the bounds of legitimate inference from archaeological evidence and into realms in which 'models' or 'paradigms' may be no more than euphemisms for unsubstantiated hypotheses or sheer guesswork. Among such guesses might be a need, for agrarian or economic purposes within some context of changing modes of agriculture or distribution of goods, for modes of bulk transport unfulfilled by existing sledge or pack-animal resources. In the subsequent history of wheeled transport, well documented in literate contexts and confidently to be inferred from non-literate antiquity, the element of prestige and status involved in the possession of an appropriate wheeled vehicle was clearly a social factor of importance among many peoples, later to be intensified by the adoption of the horse for traction as an animal of peculiar esteem. What we cannot know, from the negative archaeological evidence which alone could substantiate inference, is whether deliberate rejection as well as positive acceptance of wheeled transport technology operated among the varied communities of ancient Europe, either as a result of social or economic imponderables, or by reason of strikingly unsuitable terrain. The use of transport demanding more than minimal trails for humans or single pack animals certainly implies a degree of clearance of the original 'Wildwood' of temperate Europe in at least areas of permanent agricultural use, and this is itself arguably to be related to an increased development of ard cultivation. In a recent stimulating study, Bakker (1976) has argued for relatively extensive systems of recognized roadways in Northern Europe, in the area and as a part of the TRB culture, related to natural features in terms of the 'principle of least effort' which 'offered the easiest, quickest, most efficient route connecting two points', in which 'new means of transport, such as the ox-cart, may have necessitated certain alterations, but these resulted in only slight variations of the main theme'. Above all, an avoidance of wet soils where a vehicle could get bogged down would have become a paramount necessity, as with the inferred prehistoric roads in West Jutland, which 'tend to follow the borderline between loamy and sandy areas: hard when there is no rain, and dry during wet periods'. Fen causeways of wooden construction for foot passage were certainly in use in north-west Europe from the late fourth millennium BC, as the evidence from the Somerset Levels so dramatically shows, but the passage of wheeled vehicles would call for relatively heavy corduroy construction with transverse planks or logs. In the Netherlands, the Nieuw Dordrecht trackway, by which a single-piece disc wheel was found, is of this construction, with a radiocarbon date of *c.* 2140 bc, but 'one should realise that the trackway itself is suspected of being of ritual character' and 'almost certainly led nowhere, and simply ended in the bog' (Van der Waals 1964, 47). In the Oldenburg region of north Germany a functional trackway of this type has a radiocarbon date of 1465 ± 65 bc (GrN-3509) (Hayen 1972, 83), and other comparable dates are given in Coles and Harding (1979, 323).

In considering the development of wheeled transport in Europe from *c.* 2000 bc, the approximate date reached in Chapter 2, we encounter problems of terminology, classification and chronology. As technolo-

gically we move into full non-ferrous metallurgy, with tin-bronze standardized after a short prelude of copper-working, we come, in the conventional terminology inherited from the nineteenth-century Three Ages system, into the European Bronze Age. The term, with as we shall see certain chronological connotations, is still useful and difficult to replace by a satisfactory alternative; it has international currency among European prehistorians and has been used with success in the most recent survey of the whole field by Coles and Harding (1979). These authors have been followed in making a simplified and basic division of the phase into an Earlier and a Later Bronze Age, with conventional dates of *c.* 2000–1300 BC and *c.* 1300–700 BC, within which a multitude of regional and chronological systems of sub-divisions have been worked out, so that relative chronologies are fairly securely established. It is when turning to absolute chronology that we encounter something of a dilemma. Before the advent of radiocarbon dating, successive generations of prehistorians from Montelius onwards strove to construct an absolute chronology for the second and first millennia BC by establishing correlations in metal types, styles of ornament, and other features, between Europe north of the Alps and the Aegean, with its historical chronology based ultimately on that of Egypt. For what, following Coles and Harding, we will treat here as the Later Bronze Age, and beyond this in the iron-using contexts of Hallstatt C and D, and La Tène, Mediterranean correlations do, in fact, afford the best means of framing an absolute chronology up to the end of prehistory and the establishment of the Roman Empire. But for our Earlier Bronze Age of *c.* 2000–1300 BC, the main contention of 'Mycenaean' contacts between Central, Eastern, and north-western Europe around 1500 BC affording a fixed chronological horizon at this date has been seriously eroded by recent research and the writer, who supported this position in the past, would now align himself with the critics. The application of radiocarbon dates (very few from Eastern and Central Europe; more numerous in the north-west) does not unfortunately help to resolve the problem, for while they provide an internally consistent series 'in general the uncalibrated dates fit better with traditional notions of chronology' though these, of course, are not unequivocally substantiated. In terms of calibrated radiocarbon dates, an 'Earlier Bronze Age' would date from *c.* 2500–*c.* 1500 BC, and would render the 'Mycenaean' connections invalid in terms of absolute chronology alone (Coles & Harding 1979, 66–67). Harding (1980) has recently put forward arguments for a possible resolution of the problem. This chronological excursus is of direct relevance to our enquiry into the technology and development of wheeled transport in second-millennium Europe as it affects an issue of cardinal significance, that of the first appearance of light, horse-drawn, spoked-wheeled vehicles of generic 'chariot' type in areas beyond the oriental and Aegean world. In the past these have been seen in the context of a 'packaged transmission' from the world of Mycenaean Greece into that of Central Europe (cf. Bouzek 1966), but if these contacts as a whole are seriously called into question on grounds of archaeology and absolute chronology alternative explanations must be sought, and the problem is discussed at a later stage in this chapter.

Whatever absolute dates we use, Bronze Age Europe presents us, in technological terms, with a continuance of disc-wheeled heavy transport with ox-traction demonstrated or reasonably inferred throughout, and indeed this persists to the end of prehistory and (in some peripheral areas) up to modern times. Before the end of the Earlier Bronze Age a complete technological revolution is introduced by the development of the domesticated horse as a traction animal

in association with a light, fast, vehicle with a pair of spoked wheels, best known in its Near Eastern specialized form of a chariot for warfare, hunting, prestige and display from the beginning of the second millennium BC. In Europe such vehicles are known from the middle of the millennium in Mycenaean Greece, and about the same time in the Urals, on the Volga, and in Slovakia: it is in the interpretation of the relationships between these occurrences that the chronological problems just outlined become critical. In later prehistory the chariots of the continental and insular Celtic-speaking peoples constitute the final episode of this long-lived tradition. The use of the spoked wheel in Europe was, however, widely adopted for wagons and carts from the beginning of the Later Bronze Age onwards, certainly for horse traction and probably ox-draught as well, and was to become the norm, side-by-side with the archaic disc-wheeled vehicles. In view of this technological break it will be convenient to deal first with the continued development of disc-wheeled vehicles in the second half of the second millennium BC, and then to consider, as a separate problem, the horse-drawn, spoked-wheel series. Because of the abundance of actual surviving vehicles it is appropriate to deal first with the Transcaucasian evidence, taking into account here the remarkable series of rock carvings in Russian Armenia; moving then to the vehicle models of the Carpathian region; and finally to the scanty surviving evidence for the Earlier Bronze Age in north-western Europe.

PART I DISC-WHEELED OX-DRAUGHT

Transcaucasia and the Trialeti wagons

As we saw in the previous chapter in considering the evidence for wheeled transport in the Early Transcaucasian Culture, a burial at

Zelenyy in the Tsalka region of Georgia seems to have contained a wagon burial in a grave, represented only by the parallel grooves in which the wheels had been set. The burial rite, a pit grave under a barrow, not only immediately recalls the third-millennium south Russian Pit Graves with their vehicle burials described in Chapter 2, but leads us to consider the second-millennium vehicle graves in the Caucasus, the earlier in the Trialeti and Tsalka regions of Georgia, the rather later group on the shores of Lake Sevan in Armenia, all barrow or cairn burials and mostly with versions of pit graves, the log-roofed pit under the Trialeti 5 barrow being indistinguishable from a classic tomb of the Pit Grave culture. The writer, and more recently and decisively, Charles Burney, have stressed these connections (Piggott 1968, 282–84; Burney & Lang 1971, 94), citing evidence for at least one pit grave in Azerbaijan with radiocarbon dates before *c.* 2500 bc, and Burney seeing the Trialeti 'barrow-building' tradition as 'certainly imported from the north' as part of 'a larger and more permanent incursion' than any earlier contacts. The material was treated in some detail by the writer in 1968 and the Trialeti finds have now a published catalogue (Zhorzhikashchvili & Gogadze 1974).

Among the Trialeti group barrows in the steppe country around the River Khram in the Georgian SSR, surviving wagons were found in Barrow XXIX of the earlier (1936–40) excavations and Barrow 5 in the later (1957–63): it is convenient to retain the original roman and arabic numeration to distinguish the two campaigns. Traces of vehicles were said to have been observed in Barrows XVII and XLVI, and another vehicle burial, with no ascertainable details, is reported from a grave-pit at Avlevi. The dating problem presented by the Tetri-skaro (Bedeni) grave in Tsalka has already been mentioned in Chapter 2. The find was reported to the writer (V. M. Masson, pers.

28

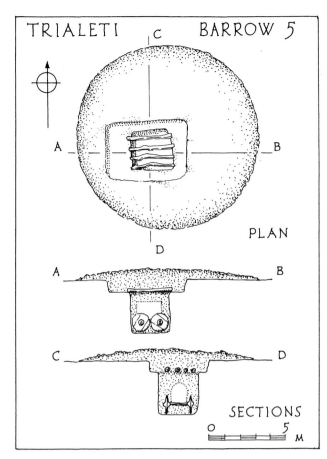

28 *Wagon burial, 2nd millennium* bc, *Barrow 5, Trialeti, Georgian SSR*

comm. Moscow 1971) as from Bedeni and as 'late Kura-Araxes' with gold and silver objects and a wagon comparable with that from Trialeti Barrow 5, and a second burial with a wagon alone. It has not proved possible to obtain further information: the radiocarbon date of 1380 ± 60 bc (TB-30) is said to be from 'wood, remains of tomb, 10 m deep, Bedeni upland, Tetri–Tskaro' (Burchuladze *et al.* 1976, 356).

The preservation of what remains of the two surviving Trialeti wagons was in both instances due to the fortuitous waterlogging of the lower parts of the pit graves in which they had been deposited. Photographs but no plan or section of the grave in Barrow XXIX have been published (Kuftin 1941,

30

67

pl. CVII; Piggott 1968, pl. XXI) but Barrow 5 was fully published with plans, sections and measured drawings of the wagon 31 (Japaridze 1960; Piggott 1968, figs 5, 11). The former seems to have had a large grave-pit under a cairn, the latter was very small, 32 m in diameter, covering a grave-pit 3 m deep with a log roof, within which the wagon was tightly accommodated, and was in all respects typical of the south Russian Pit Graves already described. Wood from the vehicle itself gave a radiocarbon date of 1420±60 bc (TB–26). In Barrow XXIX only the four wheels and their axles survived, but in Barrow 5 the lower framework and plank flooring of the wagon were also preserved as well as stumps of a deliberately broken-off A-frame draught-pole and the greater part of the yoke. In both vehicles the wheels were of massive tripartite disc construction, and their structures are virtually identical in size and proportions, as the dimensions given below, even when allowing the necessary imprecision of measuring wood which has been conserved after excavation in a wet state, clearly show. (The Barrow XXIX wagon was measured from the original in Tbilisi by the writer in 1966; the 30 figures for that from Barrow 5 are from Japaridze's scale drawings.)

	Wagon A (XXIX)	Wagon B (5)
Wheel diameter	1.15 m	1.15 m
Nave length	0.35 m	0.40 m
Gauge or wheel-track	1.40 m	1.45 m
Axle to axle length	1.50 m	1.20 m
Overall length over two wheels	2.65 m	2.35 m
Flooring of body	—	1.7 × 1.0 m

The coincidence of measurements, except for the intervals between the two pairs of wheels (*c.* 30 cm in Wagon A and reduced to *c.* 10 cm in Wagon B) argues for a mature and established tradition in technology and workshop practice, looking back to such Pit Grave vehicles as those of Kudinov and Lola and forward to the rather later second-millennium examples from Lchashen shortly to be described. Even the gauge or wheel-track spacing appears to be settling towards a norm which was to be perpetuated throughout European antiquity (Lola and Zelenyy 1.50 m, Trialeti 1.40 m and 1.45 m, Lchashen 1.60), and which appears to lie behind the recent English 'standard gauge' of 4 ft 8½ ins or 1.426 m. In the construction of the Trialeti wheels slight variations appear, but no more than might occur between individual craftsmen. 31 In Wagon A the shaping of the naves is slightly more sophisticated than the heavy truncated cones of B, and the arrangement of the internal dowels again differs in a minor degree. In both the component planks have internal tubular mortices *c.* 6.5 cm in diameter, gouged and reamed to shape (no trace of burning can now be detected), forming a pair, to either side of the nave, and while in A these pass completely through the centre plank, in B they each stop short of the centre. In both wagons the wheels turned freely on the axle-trees, which were of square section except at their bearing extremities, and in Wagon B fitting square mortices in the undercarriage. Rumyantsev, who carried out the conservation of this vehicle, was convinced, from the lack of wear on the wheels and the bearings of the axle, that it had never been used, but was completely new at the time of its deposition (Rumyantsev 1961). The nave perforations were *c.* 11.5 cm diameter in Wagon A, and a little larger in B. The ends of the axles had decayed in A but in B survived projecting and perforated, with ball-ended wooden pegs as linchpins.

Remains of the superstructure survived only in Wagon B. Here the undercarriage consisted of two lateral members, *c.* 10 × 15 cm and 2.0 m long, with simply carved scroll ends, morticed to the axle and carrying a floor of six transverse planks and at the front end a 'box seat' of planks *c.* 25 cm high. Along each side, morticed through the floor planks into the lateral members, were twelve slender rods of which the stumps alone survived and which Japaridze, on the analogy of certain of the Lchashen wagons, reconstructed as the framework of an arched tilt. The greater part of the composite draught-pole had been deliberately hacked off before the wagon was deposited in its closely-fitting pit but the stumps which remained, pivoted for vertical movement into the side-members of the undercarriage, show it, again on Lchashen analogies, to have been an A-frame draught-pole (as also suspected at Lola). The only other surviving fragment of Wagon B was the greater part of an ox-yoke, of Lchashen type, which morphologically is likely to represent a horn-yoke rather than a withers-yoke (cf. Fenton 1972). The wood of Wagon A has not been identified but the wheels of Wagon B are reported to be of pine (*Pinus* sp.). The Lchashen evidence

29 *Excavation of wagon in Barrow 5, Trialeti*

30 *Remains of wagon, 2nd millennium* BC, *Barrow XXIX, Trialeti, Georgian SSR*

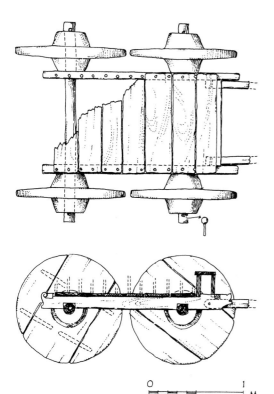

O   1
          M

31 *Scale drawings of wagon from Barrow 5, Trialeti*

would suggest that the bent-wood framework of the tilt is likely to have been of yew (*Taxus* sp.). The fixed axle-trees and the size of the wheels precludes any notion of a pivoted front axle and the vehicles, in common with all others in early antiquity, must have had a very large turning-circle, probably, to judge by modern experiments with a replica of a Sumerian battle-car, of the order of 20–30 m.

There remains the question of the absolute dating of the Trialeti burials. The writer (Piggott 1968, 281–85) originally proposed a scheme which, by including the Zelenyy grave of Kura-Araxes date, 'could span the last three or four centuries of the third millennium BC, and run into the beginning of the second', which was in accordance with the scheme of the Georgian archaeologists of three phases, *c.* 2300–2000 BC; 2000–1600 BC and 1600–1400 BC, with the wagon graves XXIX and 5 around 2000 BC, and was accepted by Burney (Burney & Lang 1971, 95) and, on the grounds of the jewellery, by Maxwell-Hyslop (1971, 74–76). The radiocarbon date for Wagon B was subsequently published (Burchuladze *et al.* 1976) as 1420 ± 60 bc, or when calibrated, *c.* 1730 BC, and this date would suggest that a position in the first half of the second millennium BC, rather than at its beginning, would be acceptable for the Trialeti vehicles. As a whole, the tombs would seem to represent the enrichment of local communities of Pit-Grave affiliations as a result of outside contacts, including areas of the ancient Near Eastern civilizations, in a manner well known and recurrent in European prehistory—the class of burial characterized as 'Royal Tombs' by Childe (Piggott 1978b). Trialeti XVII, for instance, contained among other items a silver situla with relief-decoration of animals, a silver dagger, eight silver pins and two of silver with gold heads, a gold cup set with turquoise and carnelian, and elaborate gold beads. The wagons take their place as status symbols in a society of which one part at least maintained a prestige economy supported perhaps by the copper, silver and gold resources of the Caucasus.

The Lake Sevan wagons and carts

On the southern shores of Lake Sevan in the Armenian SSR at a height of 2000 m, a very large number of tombs and cemeteries of the second millennium BC has been excavated, though the published record is sadly defective: Lalayan is said to have excavated about 1000 tombs in 1904–8 which remain almost unpublished. For our purposes, interest is centred on the Lchashen group of burials in pit graves under cairns, dramatically revealed in the 1950s when by accident the fall in the water-level occasioned by a hydro-electric scheme on the outfall river restored the lake to its prehistoric shoreline and exposed on its southern shores the hitherto submerged cemetery, the waterlogged graves preserving the wooden vehicles they contained (in the manner of Trialeti, some 200 km to the north-west). In addition, another vehicle burial of the same type from nearby Ner-Getashen (Adiaman) had been published by Lalayan in 1931. The Lchashen finds are still unpublished in full but from the short accounts available, and first-hand observations by the writer in 1966, we can assemble information on sixteen vehicles: seven wagons and six carts, all with tripartite disc wheels and presumptively ox-drawn; and two spoked-wheel chariots and a single wheel of a third. Here we will deal with the disc-wheeled wagons and carts, reserving the spoked-wheel vehicles (and the bronze chariot models, also found in the tombs) for a later section. The basic literature is contained in Mnatsakanyan 1957, 1960 a, b; Rumyantsev 1961; Esayan 1966; Piggott 1968; Burney & Lang 1971, 104–7; Hančar 1956, 173, pl. VI (Adiaman). Internal inconsistencies occur in the Lchashen reports at times, but fairly secure data can be

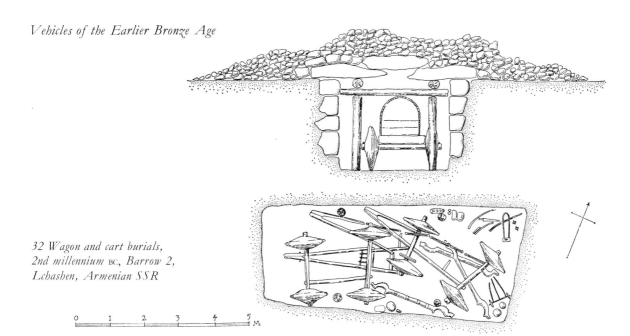

*32 Wagon and cart burials,
2nd millennium BC, Barrow 2,
Lchashen, Armenian SSR*

obtained. The burial pits were large and lined with massive walling of stone blocks, leaving, in the one published plan and section, that of Barrow 2, an area 7.5 × 3.0 m, within which a timber framing of logs was set, carrying a corbelled roof of massive stone slabs under a cairn surviving, in Barrow 2, to a height of *c.* 1.5 m. Some cairns were contained within a kerb of large stones, and most seem to have been provided with a dromos approach at the narrower end of the chamber, subsequently blocked by stone walling rather lighter than that lining the walls of the grave-pit. More than one vehicle was found in a single grave on occasion (a wagon, two carts and a toy cart for instance in Barrow 2; a wagon and a chariot in Barrows 9 and 11), and to summarize, wagons were found in Barrows 1, 2, 3, 9, 10 and 11; carts in Barrows 2, 6, 8 and 10. There were usually multiple burials of humans, cattle and sometimes horses, and abundant pottery in addition to bronzes and jewellery.

The wagons

The individual finds have been designated A–F for convenience of reference and are listed below with their significant associa-tions. Technical details of the constructional elements (e.g. wheels, tilts, draught-poles, yokes, yoke-ornaments) are then discussed. All have tripartite disc wheels and A-form draught-poles.

Wagon A From Barrow 1. With remains of tilt of unspecified shape, and buried upside-down, with the draught-pole and yoke removed and placed separately. Above the wagon were two ox-skulls and a human skull. Bronze yoke ornament in the form of a chariot group, bronze sword, axe, trident, bowl, etc. and gold beads.

Wagon B From Barrow 2. Wagon with tilt of 'C' rather than 'U' section at west end of chamber, with two ox-skulls under; two carts and a toy or model (carts A, B, C); yoke; two bronze yoke-ornaments, one a bird and one a bull; twelve human skeletons and rich grave-goods of bronze daggers and axes; gold, agate and carnelian beads; six bronze horse-bits with wheel-shaped cheek-pieces.

Wagon C From Barrow 3. Remains of wagon: 'part of the covered top of the body' and A-form draught-pole.

Wagon D From Barrow 9. Wagon with 38 'U'-shaped tilt and elaborately carved back and front boards. Human skeleton in

71

wagon; yoke with bronze bindings; bronze yoke-ornament with chariot group; also in grave spoked-wheel chariot (Chariot 2 below).

Wagon E From Barrow 10. Wagon with arched tilt, with two ox-skulls placed in front and draught-pole and bronze-bound yoke dismembered which 'had been lashed to the pole with cord'. Bronze yoke-ornament with chariot-group. In the same grave Cart F and a single spoked wheel (Chariot 3).

Wagon F From Barrow 11. Open wagon with carved panel back and wickerwork side-screens; box seat with front panel carved with cervids. Bronze yoke ornament in the form of a bull. In the same grave a spoked-wheel chariot (Chariot 1).

The grave inventories are given in varying degrees of detail in the published accounts, but all contained pottery, and most bronzes and ornaments.

The wagon inventory from Lake Sevan is completed by the Ner-Getashen grave published by Lalayan in an inaccessible Armenian monograph in 1931, and summarized in Hančar 1956, 173 and pl. VI.

Barrow 17 of the Nijni Adiaman group covered a burial chamber of Lcashen type, with a wagon on which lay a human skeleton with a socketed bronze bident by the right side, and by the yoke was a bronze yoke ornament in the form of a caprid. There were an additional thirty human skeletons (interpreted as sacrifices), four ox skeletons, pottery and bronzes, including two daggers, three arrow-heads, a knobbed mace, and armlets. Three wheels only survived and these, with the A-frame draught-pole, were reconstructed by the excavator as a three-wheeled vehicle (the third wheel within the apex of the 'A'). The bodywork, highly decorated with relief carving, was extremely fragmentary, and the reconstruction as an open vehicle with lateral rails with scroll-carved ends, each supported on four uprights, is open to question even in the revised form, as a four-wheeled wagon with A-form draught-pole, now exhibited in the Historical Museum in Erevan (personal communication in 1966).

Constructional details

The wheels of all the wagons were of massive tripartite disc construction, with the naves hardly differentiated but formed by a smooth progressive thickening (to *c.* 45 cm) towards the centre: at Ner-Getashen the naves were more pronounced. Diameters ranged from over a metre (1.1 to 1.7 m in Wagons A, B and F) to the 0.75 m for the pair of wheels at Ner-Getashen, the remaining single wheel measuring only 0.60 m. All had internal dowels in tubular mortices, with minor variations of arrangement (Piggott 1968, fig. 7). The axle-tree was square-sectioned and morticed into the undercarriage, with cylindrical ends on which the wheels turned, secured by simple wooden pegs as linchpins. The wheel gauge in Wagon F was *c.* 1.60 m. The excavator specifically noted that in all cases the wheels showed signs of wear and had been repaired, and in Wagon D a new unworn

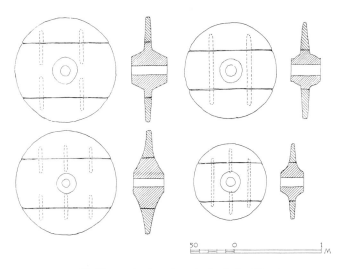

33 Types of tripartite wheels, 2nd millennium BC, *from Trialeti and Lchashen, Georgian and Armenian SSR*

72

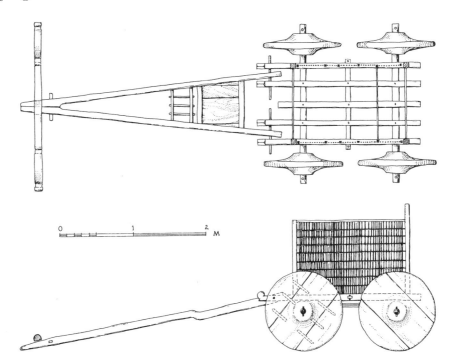

34 Scale drawings of wagon, 2nd millennium BC, from Barrow 11, Lchashen, Armenian SSR

left front wheel had been fitted. The wheels of Wagon E were said to have been 'charred or scorched, apparently for greater hardness'. The wood used was normally oak *(Quercus)*, with the axles of elm *(Ulmus)*.

Concerning the undercarriage and body we have full details of Wagon F, of which measured drawings have been published, but the attachment of the A-form draught-poles necessitated the same arrangement in all the wagons, with two pairs of lateral members, between the forward ends of which the vertical pivoting of the ends of the draught-pole was made on a transverse rod. In Wagon F the ends of these beams had a simple scroll moulding similar to those in Trialeti Wagon B. Flooring seems to have been of transverse planks, and the overall measurement of the body was 2.25 × 1.10 m. At the rear was a box-shaped seat, the front panel of which was carved with two cervids outlined by a narrow gouged groove, in a style quite dissimilar from the abundant cervids in the Syunik rock carvings in the Armenian mountains shortly to be discussed, but more resem-

bling those on the bronze belt-plates referred to later in Chapter 4. The exterior of the panelled back of Wagon F was carved with incised geometric patterns including running spirals and an openwork rail at the top, but the very elaborate carving on Wagon D, and at Ner-Getashen, was executed in relief with great technical skill and 38 elaboration, again employing a preponderance of spiral motifs, singly or in combination. Wagon F was unique among the Lchashen series in being open, with the panelled back already referred to and a foot-board. The sides consisted of a framework 1.05 m high, made of eight horizontal and 34 six vertical laths, 2 cm thick, through which pass lengthwise forty-three slender withies 1.0 to 1.5 cm wide. Some additional details 35 are available for other wagons. The body length of Wagon A was 1.45 m, with five longitudinal members morticed on to the axle, and the distance between the wheel rims only 70 cm. Wagon B had the body 'held on blocks or pads 1.08 m in length by 5 cm in width and 12 cm in thickness, which are fixed by dowel pegs to the axle, through

73

35 *Reconstructed model of wagon, 2nd millennium* BC, *from Barrow 11, Lchashen, Armenian SSR*

36 *Wagon with tilt, 2nd millennium* BC, *from Barrow 2, Lchashen, Armenian SSR*

which pass horizontal members which form the skeleton of the framework', and the floor measured 2.0 m by 0.78 m only.

The remainder of the Lchashen wagons had arched tilts of a framework of bent half-hoops, either 'C'-shaped (Wagon B) or 'U'-shaped (Wagon D) in cross-section. As we saw, such a tilt was reasonably assumed for Trialeti B, and the whole series falls into

line with the evidence from models, as the cart from Tri Brata described earlier or those of the ancient Near East (Littauer & Crouwel 1974; 1979a, 38). To these may be added another model, probably from Turkish Thrace and of second-millennium BC date, described later in this chapter. The tilt frames were made of bent yew rods, square or rectangular section, and elaborately slotted or morticed to take the transverse members of the construction.

The carpentry and craftsmanship throughout was of a very high order. It was reckoned that Wagon E, with an arched tilt, consisted of seventy separate pieces of wood held by 12,000 mortices including those in the tilt. In Wagon B the tilt frame was made of eleven pieces of bent yew, in each of which forty longitudinal mortices 5 × 2 cm had been cut with a chisel: in D the sixteen or so slighter rods had smaller mortices 2.5 × 1.5 cm. In the drawing of the panelled back of F at least fourteen mortice and tenon joints are shown.

The A-shaped draught-poles, of types hinted at in the Lola and Trialeti B vehicles, are extraordinary, and as we shall see shortly, explicable only as being, in fact, the copies of A-framed carts such as were found at Lchashen attached to the forward end of the wagon. That from Wagon F, published as a measured drawing, is typical and is made of twelve pieces of wood, the main members of which are two straight poles 3.6 m long, dowelled together at their apex by a massive peg and diverging to form a triangle with a base of 0.85 m, where their ends are transversely perforated to take rods running through the similarly perforated pairs of lateral members of the undercarriage, affording free vertical pivoting. The functional plank floor of the A-frame cart is retained, with a light openwork grid forward. In Wagon F, the distance between the yoke and the forward wheels was 3.10 m. Like the A-frame carts themselves, these A-form draught-poles, perhaps related, are

74

known from the ancient Near East and of wide occurrence in later European antiquity (Littauer & Crouwel 1977b).

Traction may be accepted as by draught oxen, and pairs of skulls (or skulls with lower limb-bones indicating hides) as a *pars pro toto* representation of this. The yokes, sometimes decorated with bronze bands and typologically suggesting as in Trialeti B horn-yokes rather than massive withers-yokes, would be lashed by cords (attested for Wagon E) or leather straps to the front end of the A-shaped draught-pole, where the projecting ends of the transverse dowel joining the side-poles would help in obtaining the secure but flexible connection always necessary between yoke and draught-pole in antiquity. It is in this context that we must take the bronze sculptures in the round of chariot scenes, bulls and a bird, all set on an upright moulded pillar springing from a double-hooked or anchor-shaped base, found with Wagons A, B, D and F, and in D recorded as lying at the junction of the metal-bound yoke with the draught-pole tip. These may be interpreted as yoke ornaments, standing upright and held in position by the criss-crossing of the lashings over the anchor-shaped bases, central to the yoke, and in the manner of the famous silver and electrum terret with an equid figure from Pu-Abi's ox-drawn sledge in the 'Royal Tombs' of Ur (Littauer & Crouwel 1979a, 26, fig. 10). This again has an alternative form of hooked base to secure attachment by the yoke-lashings in the same manner. The three Lchashen chariot scenes (with Wagons A, D and E) will be discussed in their own right when dealing with second-millennium horse-drawn chariots at a later stage; with Wagon B were two yoke ornaments, one with a bird and another with a bull flanked by a pair of goats, and with Wagon F a bull alone. At Ner-Getashen was a bronze figure of a caprid on a tapering pillar base, but no hooks for attachment. Other finds have been made in Armenian Bronze

Age contexts, notably in the richly furnished tombs at Artik, where one was in the form of a stag (Khachatryan 1963, fig. 30; Martirosyan 1964, pl. VIII). Other details of harnessing are unknown, but control by a cord to a nose-ring seems plausible and is suggested by some of the probably broadly contemporary Syunik rock carvings. With Wagon A was a bronze socketed trident, and at Ner-Getashen a bident by the skeleton in the vehicle. Such forked objects are not uncommon in the Armenian Bronze Age and they seem best interpreted as ox-goads, of similar function to the simple hooked forms noted earlier in Pit Graves and Catacomb Graves, and incidentally occurring in the rich grave, with gold and silver vessels ultimately recalling Trialeti, at Kirovakan in the Armenian SSR (Martirosyan 1964, fig. 31). The bronze types of the later Caucasian Bronze Age have been set out by Chernykh (1976a, fig. 4, I—III).

The carts

In addition to the wagons, five two-wheeled carts and one miniature or toy version were found in Barrows 2, 6, 8 and 10 at Lchashen. Such details as exist are as follows. All save one are of the A-frame construction described below.

Cart A From Barrow 2. This was found with Wagon B and Cart C, and the toy model B. A-frame.

Cart B From Barrow 2, as above. This is a miniature toy, *c.* 75 cm overall, with a pair of disc wheels on an axle and three sticks representing a rough A-frame. It lay under the bronze bowl, shaft-hole axes and trident or ox-goad.

Cart C From Barrow 2, with Carts A and B, and Wagon B. A-frame.

Cart D From Barrow 6. This is not illustrated but appears to be the only cart from Lchashen not of A-frame construction. It was poorly preserved, but 'the floor of the cart consisted of four parallel planks, the

37

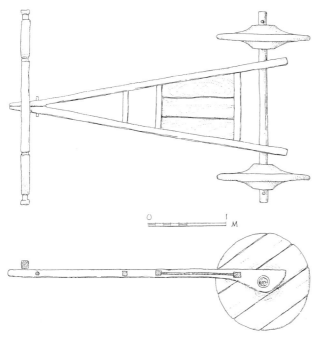

37 Sketch details of cart, 2nd millennium BC, from Barrow 2, Lchashen, Armenian SSR

sides were wicker-work and the rear solidly built of planks'. A crouched human skeleton lay on the cart.

Cart E From Barrow 8. It was fragmentary, the axle and wheels being missing. Two ox-skulls, two horse-skulls and bronze horse-bits with disc cheek-pieces were also found. A-frame.

Cart F From Barrow 10, with Wagon E and other associated finds. It lay near the entrance to the dromos of the tomb and was badly damaged. A human skeleton lay on it. A-frame.

Constructional details

Our knowledge of the detailed construction of the A-frame carts depends on Esayan's description of either Cart A or C from Barrow 2, with a perspective sketch with some measurements by Kochayan and others in the text from which the writer constructed a measured sketch plan and profile (Esayan 1966, 137, fig. 7; Piggott 1968, fig. 8). The

wheels of all carts were tripartite discs identical with those of the wagons, and 1.17 m in diameter. In the illustrated cart the wheels are now pentepartite, but the excavators regarded the additional planks as repairs to a wheel which would inevitably tend to wear more on the longitudinal than on the cross-grain parts. An English traveller in Inverness in 1734, commenting on the small tripartite disc cart-wheels then in use, said of such a wheel, 'having some part of the circumference with the grain, and other parts not, it wears unequally, and in a little time is rather angular than round' (quoted in Fenton 1973, 162). The Lchashen axle was 2.35 m long, circular throughout its length and the wheels, with a gauge of *c*. 1.70 m, secured by linchpins. The two beams forming the main frame of the cart were 3.50 m long, the far ends thickened to accommodate the axle through a perforation 12 cm in diameter. The planked platform is 1.0 × 0.80 × 0.60 m and the front end of the cart frame pegged as in the draught-poles of the wagons, and provided with a similar yoke for paired draught, with an allowance of 2.5 m between the yoke and the wheels as against 3.10 m in Wagon F, giving an indication of the size of the oxen employed.

These A-frame carts are known only from Lchashen and from the depiction of one, together with a spoked-wheel chariot, on a bronze belt-plate from Stepanavan, and a couple among the many vehicle representations among the Syunik rock carvings (Martirosyan 1964, 160; Piggott 1968, fig. 10; Karakhanian & Safian 1970, pls 45, 96). The rock carvings show the wheels (in one instance spoked) placed more or less centrally to the body, as they are in the two spoked-wheeled vehicles of similar type among the Peña de los Buitres rock paintings in Spain (Breuil 1933, 64–65; Almagro 1966, fig. 79). Typologically they are derived from the travois or slide-car type of vehicle and their dominance in second-millennium BC Transcaucasia is demonstrated

37

76

by their adaptation to form the draught-poles of wagons instead of the simpler and equally effective single-piece, or at most Y-shaped form. A-frame carts with tripartite disc wheels were in use in recent times in mountains of the eastern Caucasus, with well-known counterparts in Sardinia and Anatolia (cf. Piggott 1968, fig. 9) but prehistoric precursors are otherwise unknown. Outside Europe, models of two-wheeled vehicles of early historic date from India (early centuries AD) appear to be based on A-framed carts (Margabandhu 1973). It has been suggested that two carts represented in an eleventh-century English MS are of A-frame type, showing its presence in the Western European Middle Ages (Lane 1935, pl. III).

The dating of the Lchashen vehicle graves

Mnatsakanyan divided the totality of the Lchashen cemetery into four chronological phases, of which the fourth and probably the third phase contained the 'Chieftains' Graves' with rich grave-goods and vehicles, and dated Phase III *c.* 1400—1300 BC and Phase IV *c.* 1300—1100 BC, and this has been found broadly acceptable by other archaeologists (Piggott 1968, 285; Burney & Lang

1971, 104—7). Some of the gold beads from Barrow 2 might be equated with Maxwell–Hyslop's first phase of the Marlik cemetery, *c.* 1350—1150 (1971, 190—92), while the style of the incised cervids on the seat of Wagon F, and the running spiral carvings on the back panels here and in Wagon D, recalls that of the well-known bronze belt-plates of Armenia and Georgia of the later Bronze Age but pre-Urartian date. A single radiocarbon date of 1200 ± 100 bc (GIN-2) was obtained from 'wood remains of ritual

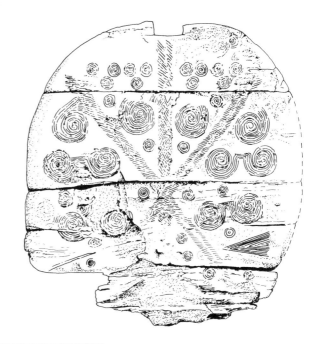

38 Carved decoration on panels of wagon from Barrow 9, Lchashen, Armenian SSR

chariot from burial' at Lchashen: if 'chariot' is taken literally in a translation of the original Russian, this would be from Barrow 9 or 11, but it could represent a less specific 'vehicle' (Cherdyntsev 1968, 423). In calibrated form it would be *c.* 1500 BC, and in whatever form, a date a couple of centuries later than Trialeti Wagon B would be reasonable enough.

The Lchashen 'Chieftains' Graves' could be seen as in the same tradition as that of Trialeti, of tombs richly furnished with prestige offerings, and belonging to the contemporary settlements, with stone-walled citadels and hillforts of pre-Urtarian date, in the mountains south of the lake and those to the north extending to the River Kura (Mikaelian 1968; Esayan 1976). Something comparable to a Pit Grave tradition could be said still to be present, but comparison can also be made with the cairns with boulder kerbs in the Persian Talysh.

The Syunik rock carvings

In the Syunik region of western Armenia, enormous numbers of rock carvings exist, at a height of some 3,300 m in the mountains, and have plausibly been seen as the products of an economy including transhumance, and a use of the rich upland pastures during the three months or so during the summer when they are free of snow: the carvings in the Maritime Alps above 2,000 m already referred to offer a comparable set of circumstances. The Armenian carvings, in common with all such, can only be dated in terms of the objects they may represent and have tentatively been assigned to a long span covering the fifth to second millennium BC. Of the fifty or so vehicle representations among the many thousand carvings published from the Oughtasar area, the seven chariots must belong to the last phase, and this may reasonably be held to apply to the vehicles as a whole, so that they are here treated as a broadly second-millennium BC

group. The corpus of Oughtasar carvings is admirably published by Karakhanian and Safian (1970), and reference is made to their plate numbers (e.g. KS 45, 332, etc.). The chariots have been made the subject of a special study (Littauer 1977) and are discussed at a later stage in this chapter, and the sledges present among the carvings have already been described. This leaves us with forty or so vehicle representations to consider here.

Conventions in prehistoric European two-dimensional art

As a brief prelude to this remarkable corpus of information, and to further linear representations of vehicles on rock-surfaces, pottery or other objects in later European prehistory, a digression on the artistic conventions they follow is necessary to their understanding. Twenty years ago Sir Ernst Gombrich, in a magisterial study which combined deep erudition with elegance and wit, made us aware of the fallacies inherent in applying the usual ideas of 'representation' in the modern Western world to earlier works of art (Gombrich 1960). Artists of all periods, he demonstrated, work within psychological mental sets which vary in time and place, but consistently and not capriciously, and indeed constitute an important element of style, whereby, for instance, we easily distinguish a second-millennium representation of a chariot in an Egyptian tomb-painting from one depicted on contemporary rock carvings in Armenia or Scandinavia. What is important is to recognize the artistic schema employed and to interpret the representation in its terms, and not in what may well be not only an anachronistic but a misleading importation of a modern and alien mental set into the past. In the instance of 'primitive' art, outside the mainstream of the Western European tradition, this theme has been briefly pursued by Jan Deregowski (1973).

To make a sweeping generalization, the representational linear art of the ancient Near East, Egypt, Crete and Mycenaean and later Greek art is normally lateral, or in profile: from these sources derives much of the later Western European tradition. Elements of the structure of a vehicle or of its draught-animals are to us, heirs of that tradition, depicted in a relationship and a convention which we find immediately comprehensible and, we feel, approximate to being 'true to nature'. The art of 'barbarian' Europe, which concerns us here, is conceived within a totally different schema. It is symbolic and enumerative, presenting a series of representational clues which can be mentally assembled—the vehicle, its wheels, its draught-animals and their harnessing. A plan view of the body of wagon or cart is preferred, but in such a view the wheels would be unrecognizable, and so for clarity they are shown full-face. The draught-animals are shown as plan views in the early Maritime Alps group, the heads of oxen with the horns invariably being so depicted, for purposes of identification, while horses' heads are shown in profile, as the bodies of oxen normally are. Here two conventions can be seen in operation, at Syunik and elsewhere. In the first, the draught-oxen are (apart from the horned heads in plan) shown as two profile views, one above the other (e.g. KS 5, 11, 144, 189 etc), and in the second a more symmetrical effect is obtained by setting the profile views on each side of the draught-pole as if (to modern eyes) the beasts were lying nose-to-nose (KS 2, 78, 134, 177, etc.) or back-to-back, not known at Syunik but on some of the south-east Spanish grave stelae (Almagro 1966) or in some Swedish rock carvings such as Frånnarp, in a group showing both postures (Althin 1945, pls 69–73). This last pair of conventions deserves special comment, for it has given rise to two apparently independent suggestions, one in respect of the Spanish stelae of the late eighth–early seventh century BC (Powell 1976, 168) and the other of the Syunik and similar carvings in Central Asia (Littauer 1977a, 261), that they depicted actual vehicle burials with slaughtered animals and even dismounted wheels, and were 'suggested to the artist by looking down into a tomb'. But this is to apply a modern mental set to an ancient and alien schema, and recalls the contention of a modern game-hunter and artist that the Palaeolithic cave paintings in most instances depicted not live animals, but dead beasts lying on their sides. A salutary comment on this stressed 'how important it is for the modern commentator to recognize the fact that he may not only look at a representation with quite different eyes from the artist responsible for the representation, but also react in a totally different way from that intended by the original artist' and that it is 'foolhardy to accept the conclusions derived from a twentieth century visual impression' without considering alternative schemata (Ucko & Rosenfeld 1967, 184–85). If the pair of draught-animals is considered a unit, we might have something comparable with the 'split-view' of animals noted by Deregowski in Africa.

The Syunik vehicles: classification

Excluding the sledge-type vehicles, already commented on above (KS 52, 96, 148, 317) and the chariots already discussed by Littauer (1977a) and to be dealt with later in this chapter, the remaining Syunik vehicles may be classified as follows.

1 Wagons, square-frame, large disc wheels (KS 11, 62, 105, 107, 120, 204, 207)
2 Wagons, square-frame, large spoked wheels (KS 122)
3 Wagons, square-frame, very small disc wheels (trolleys) (KS 2, 5, 134, 156, 177, 183, 187, 189, 223, 225, 226, 230, 232, 269, 317)
4 Carts, square-frame, disc wheels (KS 26, 61, 93, 190, 211)

39

5 Wagon, A-frame, disc wheels (KS 48)

6 Cart, A-frame, disc wheels (KS 96)

7 Cart, A-frame, spoked wheels (KS 45)

8 Miscellaneous unclassified (KS 12, 13, 78, 190, 207, 332)

The technique of execution is, in common with most European rock carvings of similar type, that of outlines pecked into the glacially smoothed rock surface by repeated percussion with a pointed stone tool. Sizes vary but most average from 1.0 to 1.5 m in overall length, with exceptional scenes which include a driver of up to 2.5 m (KS 2, 134). Where draught-animals are shown, these are invariably oxen. Wheels are shown by a circular outline, with crossed lines at right angles to denote spokes: the question as to whether these are to be literally interpreted as four-spoked wheels is discussed later. The wheels of the large group 3 are shown as very small knobs or pits of cupmark type at the ends of the axles, which extend well beyond the body and seem consistently to represent a vehicle of a distinctive type, which may conveniently be called a trolley: all are four-wheeled. The two A-frame carts (KS 45, 96) do not resemble the Lchashen vehicles but are close to the Syunik sledge representations (e.g. KS 148), and A-frame draught-poles to the wagons are never shown, though several poles have a Y-junction with the body. On the whole the rock carvings give the impression of depicting a variety of wheeled vehicles which do not have counterparts in the tombs.

The wagons

39 The outline of the wagon bodies tends to be square rather than elongated and the wheels are shown on axles or simply at the four corners. Little detail is usually shown of the floor, but three longitudinal and two cross-members of the frame are sometimes shown (e.g. KS 96, 204). KS 105 is quite exceptional, with a slightly trapezoid but broad

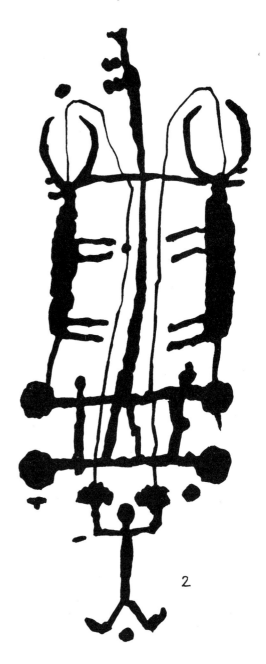

2

39 Rock carvings of ox-drawn wagons and carts, 2nd millennium BC, Syunik, Armenian SSR

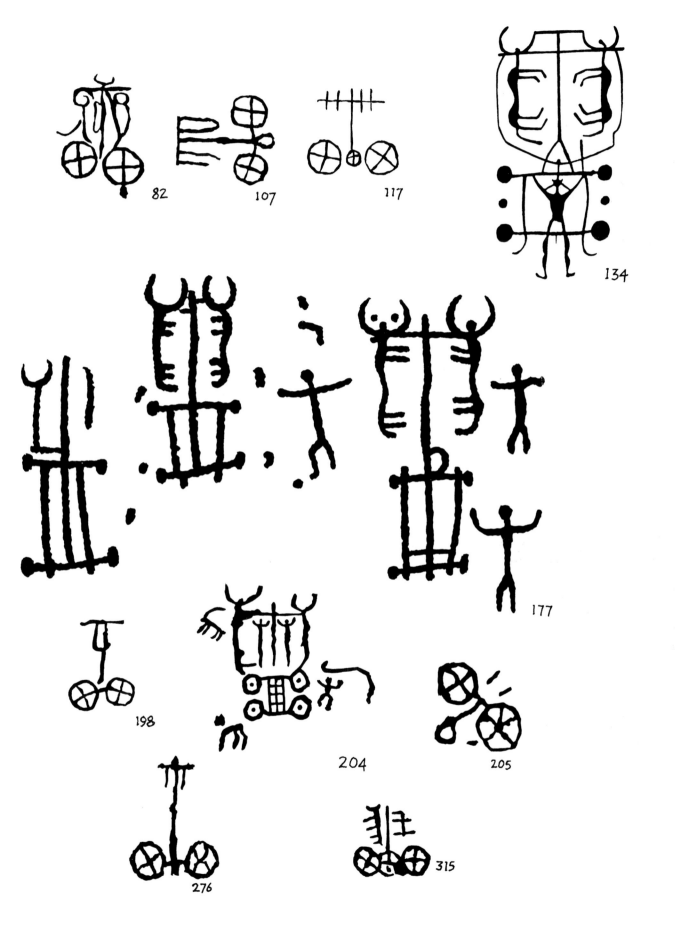

82 107 117 134

177

198 204 205

276 315

body, five cross-members in addition to the axles, and the ends of the side members curled inward, in a manner suggesting an exaggeration of the scroll-carved ends of the Ner-Getashen vehicle. Draught-poles are simple and either articulate direct to the front of the body or have Y-junctions in four instances (KS 96, 105, 107, 120). The yoke (as in all the Syunik vehicles) is shown as a straight bar at right angles to the pole joining the necks of both oxen.

The trolleys

39 Of these remarkable vehicles, known only from their characteristic and consistent representations in the Syunik rock carvings, no less than twenty-three individual representations singly or in groups, from fifteen locations, have been recorded. As noted above, their distinctive feature is the very small size of the wheels as depicted, normally as no more than 'knobs' at the ends of the long axles, which frequently protrude well beyond the sides of the vehicles. In two exceptional examples, KS 2 and 134, which are stylistically similar and show the driver standing behind the vehicle, the wheels are circular pecked 'cup-marks' up to nearly 20 cm across in large representations, but in all instances something distinct from disc wheels represented by a circular line seems to be indicated, and the word 'trolley' has therefore been used to underline this distinction. The vehicle bodies are shown as square or even wider than long, with three longitudinal members (KS 156, 162, 226). Some are reduced to a central bar, which becomes the draught-pole, and a pair of knobbed axles (KS 144, 187, 232, 317). The draught-pole has a Y-junction in some instances (KS 134, 144, 269) and in the exceptional scene with a driver, KS 2, it has a Y-junction not with the front, but the rear axle. This composition, and its companion KS 134, show the drivers holding guide-ropes which in KS 2 lead to the noses of the oxen, and in KS 134 more ambiguously to

their heads or the yoke. As with the other vehicles the yokes are featureless cross-bars.

The only parallel outside Syunik for these vehicles in rock carvings known to the writer is a single example, drawn by a camel, among the Karatau carvings in southern Kazakhstan (Kadyrbaev & Maryashev 1977, fig. 103). At the construction of the wheels in actuality one can only guess, and perhaps the trolleys of Pazyryk in the Altai, described in Chapter I, may be apposite, with their stubby wheels made of perforated sections of birch logs. The Syunik trolleys must have an important specialized function among the transhumant communities using the high mountain pastures, and transport of hay or other bulky fodder suggests itself.

The carts

The representations of vehicles with a pair 39 of disc wheels (rather than spoked-wheeled chariots) are few and many are ambiguous, KS 26 does, however, show a small square-bodied vehicle with a pair of disc wheels set at the back and a Y-junction draught-pole. Of the A-frame vehicles KS 48 is, in fact, a four-wheeler with very small wheels shown as circles but approaching a trolley's proportions. It, like the A-frame carts KS 45 and 96, has bowed sides like the sledges of Syunik (e.g. KS 52, 148). KS 45 has spoked wheels set about at the middle of the slatted body, and KS 96 disc wheels similarly placed in a more sketchy version. Neither resemble the long straight-sided, A-frame carts of Lchashen with their wheels at the rear.

An early second-millennium find in the Middle Volga area may conveniently be mentioned here, from a cemetery of the Fatyanovo culture at Balanovo. Grave 7 of the first group of burials here was that of a child, and contained among other grave-goods a pair of pottery models of disc wheels, presumably from a toy cart comparable with Cart B from Barrow 2 at Lchashen described above (Piggott 1968, 302 with refs; Häusler 1969a and Coles & Harding 1979, 129 on chronological position).

Vehicle models in the Carpathian Basin

Our next area providing information on wheeled transport in Early Bronze Age Europe is that of the Carpathian Basin, including the Hungarian Plain and Transylvania. The evidence is provided in the form of pottery models of four-wheeled wagons or, less satisfactorily, of wheel models alone, the potential ambiguity of which has already been commented upon. The presence of wagon models from a large number of widely distributed settlement sites of the earlier second millennium, however, gives us confidence in accepting presumptive wheel models in the same cultural and chronological contexts. Three general surveys of the material have been made which, if not always consistent in detail, give us a *41* sound picture of the general pattern (Boná 1960, 1975; Bichir 1964) and a few additional finds can be added (e.g. Kalicz 1968). Including fragments, and an exceptional find of eleven models found together in what has been interpreted as a shrine, at least a couple of dozen finds have been recorded. Apart from the Salacea 'shrine' (Coles & Harding 1979, 87) the models have been found among the normal domestic rubbish of potsherds and other artifacts in settlements, and in their interpretation poses the same problems as the well-known human figurines or 'Mother Goddesses' of Neolithic East Europe, alternatively secular toys or in some sense associated with cult practices.

While their general chronological position is clear among bronze-using communities of *c.* 2000–1400 bc (2520–1710 BC), the complications of detailed correlations in terms of local cultural groups have been emphasized by Coles and Harding (1979, 67–73), but need not concern us here. As indications of transport technology the models are important in demonstrating the continuance and maintenance of a tradition of disc-wheeled wagons with presumptive ox-draught attested from the third millennium bc in Baden-Pécel contexts, through the Vučedol-Zók phase at its end, and so into the period under review.

All the models are of four-wheeled wagons and no carts are represented. Most have been published without scales or dimensions, but overall measurements of *c.* 10 × 15 cm seem probable. The disc wheels are normally plain, occasionally decorated, but not in a manner to indicate structure as, for instance, the dotted ring on the wheels of the model from Pocsaj-Leányvár (Boná *40* 1975, 255). Tripartite construction is, however, indicated on two published wheel models (and if looked for, might well occur elsewhere): one, from Nyíreghyháza-Mongo, coming from a slightly earlier (Hatvan) context than the other, from Gyularvarsánd (Kalicz 1968, 158, pl. CXVI, 4; Boná 1975, pl. 144, 13). The wagon bodies are rectangular, pierced at the base or with four pierced lugs to take a wooden axle carrying the wheels. Some are plain, with

40 Pottery model of wagon, mid 2nd millennium BC, *from Pocsaj-Leányvár, Hungary. (Left wheel and axles restored.)*

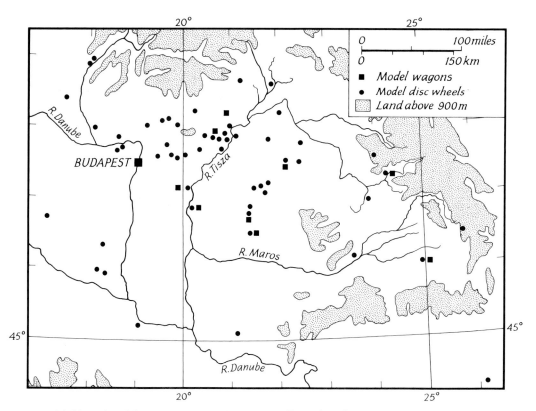

41 Map of model wagons and disc wheels in the Carpathian Basin, mid 2nd millennium BC

no high finish, e.g. Cuciulata and Otomani-Cetăţuie in Romania (Bichir 1964, fig. 1:2, fig. 2:1) but many are finely made, with elaborately ornamented sides and ends. While this may well be no more than a convention applied only to models, the possibility of decorated vehicles in actuality cannot be dismissed, as the carved panels on the Lchashen wagons show, and the use of colour is equally likely. The fragment from Alsovadasz-Vardomb, of the Hatvan culture in north-east Hungary, a phase which there follows on Vučedol-Zók, has on one long side two out of probably an original four circular holes surrounded by incised rings (Kalicz 1968, 117, pl. CXIII, 8). The four corners of the decorated models rise in peaks, recalling the wheeled cups of the Baden culture described in Chapter II, and imply that this must represent a structural feature in the original wagons. No draught-

pole is indicated, but in the model from Wietenberg in Transylvania there are at one end the broken stumps of what must have been two animal protomes in the manner of the Baden culture model from Radošina, and in another fragment from Lechinţa de Mures the one surviving peaked corner is modelled into a bovid head, presumably one of a pair (Boná 1960, pl. LXIV; Bichir 1964 figs 3, 4:2). In both, paired ox-draught seems indicated. At all events, the wagon models and those of disc wheels from at least forty sites (the lists in Boná 1960 and Bichir 1964 are inconsistent) show the *41* widespread adoption in the Carpathian Basin during the Earlier Bronze Age of ox-drawn wagons with disc, and sometimes at least tripartite disc wheels.

It is convenient here to draw attention to a model of a wagon with arched tilt of unknown provenance in the Archaeological

Museum at Istanbul, obtained from a dealer but which 'd'après la terre et sa patine, parâit provenir de la Thrace turque'. (Rollas 1961: the writer is indebted to Mr Sinclair Hood for this reference.) It is perforated for axles but the wheels are missing, and has three perforations at the base in front for attachment to some form of model draught, measures 12 × 7 cm, and is 12 cm high. The tilt is covered with a meaningless incised pattern, While undated, its counterparts are the Near Eastern models of the late third millennium BC recently studied by Littauer and Crouwel (1974), and if indeed its provenance were Turkish Thrace, it would be the first of its kind from Europe.

Disc wheels in North and West Europe

Beyond the Carpathian models just described the evidence of disc-wheeled transport in the Earlier Bronze Age is confined to a few finds of actual wheels which, by reason of their radiocarbon dates or archaeological context, can be placed before the end of the second millennium BC. They introduce us to a new technological development whereby the heavy integral nave of earlier simple or tripartite disc wheels is replaced by a separate tubular nave inserted in the plank.

In a peat bog at Glum, near Oldenburg, four single-piece disc wheels were found 42 close together by peat-diggers in 1880–83 and can reasonably be assumed to belong to a single wagon. They have been the subject of an exemplary technological study by Hayen (1972), and are among the best recorded finds of their type. Four radiocarbon samples were assayed, two from the edge and two from near the centre of the wheels, which had been cut from a single plank of alder *(Alnus)* and averaged *c.* 0.70 m in diameter, and from these it could be estimated that the wood came from a tree about two hundred years old when felled.

42 *Single-piece disc wheels with inserted naves, mid 2nd millennium* BC, *from Glum, Oldenburg, Germany*

The outer samples, approximating to the time when the wheels were made, gave readings of 1165 ± 140 bc (Hv-4058) and 1275 ± 90 bc (Hv-4057) or *c.* 1450–1570 BC. The wheels were almost flat discs, slightly thickening to *c.* 10 cm at the central hole which took the inserted nave, *c.* 14.5 cm in diameter. The adze-marks on the surfaces, *c.* 3.4 cm wide, must, as Hayen points out, have been made by a cross-hafted bronze 'axe', with a cutting edge of proportions paralleled in the flanged axes of the contemporary Earlier Bronze Age (for the type, cf. Coles & Harding 1979, fig. 111). The inserted naves were of birch *(Betula),* made from carefully chosen sections of stem still retaining their bark *c.* 26–28 cm long, and *c.* 15 cm diameter tapering to *c.* 12.5 cm, hollowed to a tube *c.* 7.5–8.5 cm in diameter, trued by thin wedges that must have been glued to the interior surface. They were presumably hammered into the central perforations of the wheels and held in place by friction. One wheel has had a crack along the grain repaired by a dowel in a dovetailed groove exactly in the manner of the repair on the late third-millennium disc wheel

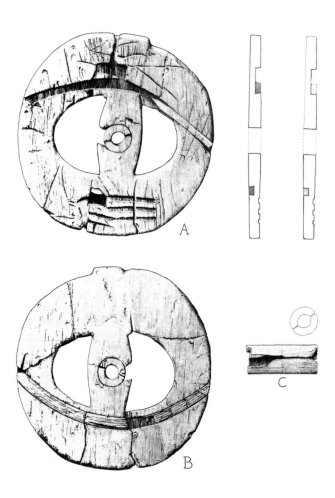

43 Tripartite disc wheel with inserted nave, external dowels and lunate openings, 2nd millennium BC, *from Mercurago, Italy*

which has a long time-span of *c*. 1450–1100 BC (Barfield 1971, 94) came fragments of a tripartite disc wheel of walnut planks *(Juglans regia)* *c*. 85 cm in diameter and with a central hole *c*. 10 cm in diameter, to take an inserted tubular nave. The component planks have been held together by two pairs of small dowels in mortices in the thickness of the planks, and with four transverse battens driven into dovetailed mortices on one face, in a manner comparable with the third-millennium Zürich wheels. From the same site comes a 15 cm diameter wooden disc with a square central perforation 1.5 cm across, which might be a wheel from a model or toy. The second wheel, again a nineteenth-century find, comes from Mercurago, where it may have belonged to the earlier (Polada) settlement of *c*. 1800–1450 BC, or to the later, contemporary with Castione (Barfield 1971, 74). This introduces us to two new modifications to the tripartite disc form which are to be found in later prehistoric Europe, the cutting of lunate openings in the side planks at their junction with the nave portion of the central member and the modification of the external dowelling arrangement, so that a single curved dovetail mortice is cut on each side of the wheel, symmetrically to each side of the nave, into which a flexible batten is sprung and held in lateral tension. There are two pairs of internal morticed dowels in the manner of the Castione wheel. The Mercurago wheel was *c*. 78 cm in diameter, with planks of walnut and dowels of larch *(Larix),* and the inserted nave was *c*. 25 cm long and *c*. 10 cm in diameter. A complete parallel, of Late Bronze Age Urnfield date (either Hallstatt A or B *c*. 1200–700 BC) comes from the Wasserburg at Buchau, south Germany (Piggott 1957, pl. XXII). The only other relevant Italian find comes again from Mercurago, where two planks with small lunate notches have been interpreted as the outer members of a tripartite disc wheel *c*. 1.32 m in diameter.

from Kideris in Denmark. Single-piece disc wheels with inserted naves of unknown date, are known from several sites in north Germany and Denmark, as are tripartite discs with similar naves (Hayen 1973, maps in figs 6, 8).

The prehistoric disc wheels from north Italy have been made the subject of two recent studies (Cornaggia Castiglioni & Calegari 1978; Woytowitsch 1978). From the nineteenth-century excavations of the terramara site of Castione de Marchese,

PART II SPOKED-WHEELED HORSE
DRAUGHT

Early horse domestication in Europe

The evidence for the domestication of the
horse in the fourth millennium bc in south
Russia has already been touched on, and it is
now necessary to examine the question of
wild and domesticated horses in Europe as a
whole, and in greater detail. The equids surviving into post-Pleistocene times comprise
a number of species, ranging from the true
horse, ancestral to all domestic breeds, to include the half-asses, asses and zebras, with a
geographical range from Mongolia to South
Africa. The Mongolian wild horse, probably still surviving in the wild and with some
45 two hundred representatives in zoos, stands
at the head of a genetic pattern reflected in
the geographical distribution of the equids.
This horse, *Equus przewalskii,* has 66 chromosomes, and the series continues with
46 *Equus caballus,* the domestic horse, with 64;
the wild ass *(E. asinus)* with 62, the half-
asses or hemiones *(E. hemionus)* with 56,
and so to end with the Cape zebra with 32.
The close relationship of the domestic to the
wild breed is thus clear. Another wild horse,
the Tarpan *(Equus gmelini)* survived on the
Ukrainian steppe until the middle of the last
century and a geneticist has recently suggested that the tarpan might be 'the missing
wild equid with 64 chromosomes, and derived from the Przewalski', and so a more
direct ancestor of the domestic horse. The
archaeological evidence of early domestication in the Sredny Stog culture of the
Ukraine might be seen as supporting this
hypothesis (Zeuner 1963; Bökönyi 1968,
1974; Benirschke 1969; Short 1975).

The east European evidence for early
horse domestication, with the emergence of
a characteristic 'steppe type' of animal, is
now well known (Kuzmina 1971; Telegin
1971; Bökönyi 1974). The area of distribution of wild horses in Western Asia from the

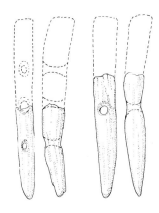

44 Antler cheek-pieces of bits, 4th millennium BC,
Sredny Stog culture, Dereivka, Ukraine, U.S.S.R

fourth millennium BC onwards has recently
been enlarged by the finds of bones from
Tal-i-Iblis in south-central Iran, and from
Norsuntepe and other Anatolian sites, so
that the Ukraine may now be seen as part of
a larger zoogeographical area (Littauer &
Crouwel 1979a, 24–25; J. Mellaart *in litt.*).
Horse-bones of presumed wild animals
hunted for food appear in several cultural
contexts of fourth-millennium bc dates from
the Ukraine westwards to Moldavia, but
domesticated animals are confidently identified in settlement sites of the Sredny Stog
culture, where horse averages 54 per cent of
all animal bones over all excavated sites, and
domestication for meat and probably milk
may be assumed. At one site, Dereivka on
the Dnieper in the south Ukraine, horse
bones approximate to 60 per cent of the
total, and the broken pieces of six antler
objects, morphologically close to what we
shall see can be confidently interpreted as *44*
the cheek-pieces of horse-bits in later contexts, were found, and further examples
were found in two other contemporary sites
of the same culture. Radiocarbon dates
range from *c.* 3565 to 2850 bc.

Sporadic finds of horse bones have been
made in several sites of the Hungarian Copper Age, and have been attributed to infil-
tration of domesticates from the east. But as
one moves into western and north-western

45 *Wild horses,* Equus przewalski, *mare and stallion, Whipsnade Zoo*

Europe a new factor must be reckoned with, that of the survival of post-Pleistocene populations of wild equids which would present opportunities for independent domestication in a manner comparable with the situation on the south Russian steppe. A palaeozoologist has recently remarked that the main work on early horse domestication has been carried out by those who tend to 'limit their researches to central and

46 *Representation of wild horse on silver bowl, mid 3rd millennium* BC, *Maikop, Kuban, USSR*

eastern Europe and do not go into detail as regards post-glacial western European wild horses. Small local populations of wild horses were certainly present in western Europe during Mesolithic and Neolithic times' (Van Wijngaarden-Bakker 1974, 345). Thirty years ago Lundholm demonstrated the continued presence of wild horses from Mesolithic times onwards in Sweden (cf. Zeuner 1963, 328), and the same can be seen in Britain (Grigson 1966, Van Wijngaarden-Bakker 1974), and wild horses survived in France, their bones being present at Roucadour in Lot, in Neolithic contexts of the late fourth–early third millennium bc. In Late Neolithic sites of the Altheim Group, in the Alpine Foreland of south Germany, however, the bones of domestic horses were found at, for instance, Altenerding and Pestenacker, and at the former settlement site amounted to over 11 per cent of the total bone remains (Driehaus 1960, 88–89). Altheim can be equated with various contemporary cultures, including the Hungarian Copper Age, Pfyn and TRB, the latter enabling radiocarbon dating to be used, and placing Altheim and its domesticated horses in the first half of the third millennium bc, a date in accordance with their occurrence in a Late Lengyel context in Slovakia and, even more significantly, at two TRB settlements in Poland, Ustowo and Ćmielów, the latter with a set of radiocarbon dates of *c.* 2825–2665 bc (Bakker *et al.* 1969, 12). The possibility of local domestication from wild horse populations beyond the Carpathian Ring by soon after 3000 bc has clearly to be recognized.

There is no evidence that these early domesticated horses served purposes other than providing alternative supplies of meat, but their presence in the third millennium side-by-side with the secondary exploitation (in Sherratt's terms) of cattle for draught with ploughs and wheeled vehicles made them potential candidates for similar use in traction or riding, even if this is not demonstrated by recognizable surviving evidence of artificial control. A recent study (Lichardus 1980) has interpreted certain perforated antler objects of Late TRB date, such as those in the graves of the Ostorf and Tangermünde cemeteries in north Germany, and in Seine–Oise–Marne contexts in north France, as the cheek–pieces of 'soft–mouthed' bits, implying horse control (though not necessarily, of course, for traction) in the later third millennium BC. Headstalls, bridles or nose-ropes of wholly perishable substances cannot, of their nature, be confirmed or confidently dismissed, and a recent study urging new study of Upper Palaeolithic evidence that might suggest some such horse management has been met with criticism (Bahn 1978; Littauer & Bahn 1980). At all events, by the closing centuries of the third millennium bc domestic horses, for whatever economic or social requirement, were becoming common among many communities widely spread over Europe, and in the west particularly associated with the makers of Bell Beaker pottery. The Csepel Island settlement north of Budapest had remarkably high proportions of horse bones (with a date of *c.* 2220 bc: Coles & Harding 1979, 69) and they occur with Bell Beakers in Moravia (Neustupný 1963; Ondráček 1961), and their first appearance in the late third millennium bc in Holland and in Ireland can similarly be related to Bell Beakers. In Spain also where the wild ass *Equus hydruntinus* survived into post-Pleistocene times, the appearance of the true horse in domestic form is attributed to the Bell Beaker culture, as in its wild form it would be unlikely to co-exist in the same zoogeographical territory as *E. hydruntinus*. In the Cerro de la Virgen stratigraphy it first appears with Bell Beakers, and horse phalanges decorated as anthropomorphic 'idols' are a well-known feature of the Los Millares and related cultures (Van Wijngaarden-Bakker 1974, 1975; Schüle 1969b; Maier 1961; Harrison 1980, 153). In all the forego-

ing contexts, from the Ukraine to the Iberian peninsula, the potentialities of the exploitation of the 'Secondary Products' of the horse in riding and traction were in existence by the beginning of the second millennium bc.

As the writer has put in a recent short survey, 'the new factor involved was speed provided by a new motive force, which in the instance of the small horses of antiquity could only be exploited by a combination of lightness and resilience of a new kind. To adopt a concept from structural engineering, the disc-wheeled ox-wagon might be seen as a slow, heavy, timber-built compression structure, and the chariot as a fast, light wood structure, largely in tension with its bent-wood felloes and frame' (cf. Gordon 1978, 145). Speed for human transport on land was suddenly multiplied by something like 10–from the 3.7 km (2 miles) an hour for ox-transport to the 38 km (20 miles) an hour reached with ease with a modern replica of an ancient Egyptian chariot with a pair of ponies, the chariot itself with its harness weighing only 34 kg (75 lbs) (Spruytte 1977). Any theoretical place or places or origin should then combine the availability of domesticated horses with flexible wood supplies, and perhaps a tradition of bent-wood construction in other structures such as dwellings, and an already established familiarity with heavy ox-drawn vehicles' (Piggott 1979, 11).

Figures have recently been published which stress the advantage of horse over ox-draught, using data from sources which include those of the colonization of the American West in the last century. The famous 'covered wagon' drawn by oxen moved about 24 km a day, and the absolute maximum was about 28 km; in South Africa in the Zulu War of 1879 no more than 16 km a day was achieved (Sandars 1978, 121). A horse team could cover 50 to 60 km a day 'alternately using a walking pace and a trot with every 10 km taking 45 minutes', but the ox has only a walking speed, which under a load is about 1.8 to 2.5 km an hour, while the walking pace of a horse is about 3.2 to 4.3 km an hour (Bökönyi 1980).

The basic vehicle form which we have to consider in earlier prehistoric Europe we may conveniently term, like its Near Eastern counterpart, a 'chariot' without in so doing assuming that it inevitably fulfilled the same military and ceremonial needs in barbarian Europe as it did in the contemporary literate civilizations of the ancient Orient. The chariot in these terms is an essentially light but strong and resilient vehicle built for the rapid transport or display of one or two persons, rather than a cart for heavy goods, with paired horse traction harnessed to a central draught-pole and yoke, in a tradition derived from the earlier harnessing of oxen. The publication of Lefebvre des Noettes's well-known study (1931) persuaded subsequent writers (e.g. Piggott 1950, 278; Vigneron 1968, 112) that such yoke-harnessing was necessarily inefficient and restrictive of the animal's capacity, but the practical demonstrations by Spruytte (1977, 1980) have shown that the disadvantage has been greatly exaggerated.

The construction of such a vehicle ideally consists of a relatively small platform to accomodate one or two persons, usually of D-shaped plan, upon an axle and pair of spoked wheels set mid-way or at the rear, with a minimal superstructure open at the back, with the draught-pole running under it from the axle, curved in such a way as to bring its far end, with the platform horizontal, to a height above ground approximating to the withers of the horses under the yoke, which was lashed with a flexible binding against a stout pin through the pole near its tip. In the light construction use was made of thin bent wood, metal fittings and leather, including interlaced straps as a flexible flooring, as can be gathered from surviving vehicles, linear representations and models from New Kingdom Egypt to

Etruscan and Roman times (cf. Littauer & Crouwel 1979, figs 42–47; Woytowitsch no. 175; Spruytte 1977, 1980). While all these features of the 'ideal' prototype are, of course, not expected to be found in every instance, they give an invaluable guide in the interpretation of, for instance, surviving metal fittings, or the fugitive traces in an excavation. Fine construction clearly demanded good metal tools, but Spruytte demonstrated by practical experiment that a full-size replica of the primitive type of chariot depicted in the Saharan rock paintings could not only be constructed with a very simple kit entirely of stone and wood tools, but effectively be driven with a pair of ponies (1977, 82, pls 20–22).

The areas with which we shall now be concerned will mainly be those of south Russia in the Lower Volga area, and the Carpathian Basin. In the former region, the cultures of the Pit and Catacomb Graves, ending (on available radiocarbon dates) *c.* 1700 bc, develop between roughly the Rivers Dnieper and Ural into that of the Timber Graves, still in the Pit Grave tradition but, as their name implies, a greater use of wood, in their lining and roofing. East of the Ural and the Caspian the counterpart of the Timber Graves is the Andronovo culture, extending to the River Irtysh. A few radiocarbon dates are available: Timber Graves are *c.* 1575 bc on the Lower Don and *c.* 1440 bc on the Lower Volga, while for Andronovo only dates from its later phase have been obtained, of *c.* 1410 and 1250 bc. Abundant evidence of domesticated horses and the bone and antler components of harness and 'soft' bits are common to both Timber Graves and Andronovo (Gimbutas 1965, 528–84; Piggott 1974, 1975).

The chronological problems of the Earlier Bronze Age in the Carpathian Basin have already been touched on at the beginning of this chapter, and the further complications arising from the use of several conflicting schemes of cultural terminology have been discussed by Coles and Harding (1979, 69–71). After an initial phase including in various localities the Bell Beaker, Vučedol, Zók and Makó cultural groups, a broadly tripartite sequence can be followed in the well stratified tell sites such as that of Tószeg in the Hungarian plain, giving rise to Moszolics's classification of Bronze I (Nagyrev), II (Hatvan) and III (Füzesabony); in the north-east the sequence starts with Hatvan (Moszolics 1969; Kalicz 1968). In Romania the Otomani culture is contemporary with Füzesabony. In terms of the few radiocarbon dates so far obtained this sequence would run from *c.* 2000–1500 bc or if calibrated, *c.* 2400–1700 BC, or with a subsequent final pre-Urnfield phase, to 1300 bc or 1500 BC. There is evidence of domesticated horses from, as we have seen, the Bell Beaker-Zók-Vučedol phase, if not before, though not apparently in settlements of the Zók culture itself. In immediately subsequent phases the horse becomes a common domestic animal, primarily for food, as Bökönyi stresses (1974, 248): at Tószeg itself it first appears in the A phase (Nágyrev) and reaches a peak of nearly 45 per cent of all animal bones at the end of this period, dropping to a minor peak late in Phase B (Hatvan) but rapidly mounting to 50 per cent in C (Füzesabony), (Bökönyi 1952). Pottery figurines of horses have on occasion been found (e.g. Kalicz 1968, pl. LIX, 7 in a Hatvan context), and a couple of entire stallions are represented on a gold disc of Earlier Bronze Age date from Grăniceri, Ottlaka (Transylvania), but not in harness (Moszolics 1965, pl. 24; Bökönyi 1974, fig. 90). Bone or antler elements of harness, especially a series of curved antler cheek-pieces for 'soft' bits, are abundant in Bronze III (Füzesabony) contexts but not certainly substantiated earlier, and continue into the subsequent Piliny (Bronze IV) phase (Moszolics 1953, 1960). The average withers height of the Hungarian Bronze Age horses was around 135 cm.

Spoked-wheeled vehicles

The evidence for spoked-wheeled vehicles in prehistoric Europe in the Earlier Bronze Age is scanty, nor can we in all instances demonstrate horse traction. We have at our disposal a cemetery with five vehicle burials reduced to soil stains, two pots with incised representations, some rock carvings, a number of pottery models of spoked wheels, and in Transcaucasia, on the edge the ancient Near Eastern world of second-millennium chariotry, a couple of actual vehicles and bronze models, the latter at least closely linked to oriental types.

In the southern Urals a cemetery of early Timber-Grave/Andronovo date was excavated on the Sintashta river near Chelyabinsk in 1972 and contained five timber-lined pit graves with vehicle burials but very imperfectly published (Piggott 1975; Gening 1977). The two wheels of the vehicles had been accommodated in a pair of pits or slots in the grave floor in a manner familiar from vehicle burials in Western Europe from the fifth century BC and traces of the decayed felloes and spokes remained as soil-replacement features, nowhere more than *c.* 4.0–4.5 cm in thickness and indicating wheels of very light scantling, 0.90–1.00 m in diameter, with ten spokes. In Grave 19, of which a plan and sections are published, the wheel-slots were *c.* 30 cm deep and the *47* gauge or wheel-track is *c.* 1.25–1.30 m, allowing for a body *c.* 0.90 m wide and of unknown length. There is no evidence for the draught-pole, but if present in Grave 19 it could have extended forward, passing a not quite central post supporting the roof, to the human burial which lay in the southern part of the grave, which was timber-lined and measured *c.* 3.0 by 1.7 m. Some graves are reported as having from two to seven horses buried above the roof in the covering barrow, and there were numerous 'head and hoofs' burials indicative of horsehides, some arranged in an elaborate 'cult

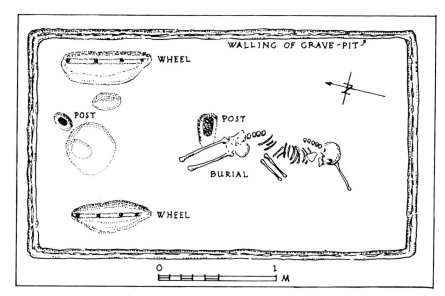

47 *Plan of chariot burial with wheels set in pits, mid 2nd millennium* BC, *Timber Grave culture, Sintashta River cemetery, Chelyabinsk, Urals, USSR*

setting'. Cheek-pieces of bits are of the spiked disc type discussed below. The light structure of the wheels allows of a reasonable inference of bent-wood techniques, and whatever the exact nature of the vehicles, their deposition in graves argues for their having had a prestige function when in use, comparable with the discwheeled carts and wagons in earlier Pit Grave and Catacomb Grave contexts.

In a Timber Grave barrow cemetery on the Lower Volga near Saratov was a burial with a pot incised with a representation of a vehicle with two four-spoked wheels, a *48* draught-pole and yoke and a single draught-animal, certainly not a horse and probably a bovid (Galkin 1977). The details of the body are not easy to interpret, but it has been suggested that we should see it as having a wide oblong floor plan and a high front rail, represented as lying over the pole, 'reminiscent of the traditional high front with depressed centre of the old "battle" and "platform" cars of the ancient Near East' (Littauer & Crouwel 1979a, 70). At all events we have a representation of a light vehicle having wheels with four or more spokes: the crossed circle may here, as in other representations such as rock carvings,

represent reality, or it may be an artistic short-hand convention for the idea 'spoked wheel' in general. The single yoked bovid is unexpected, and unharnessed horses are known from other decorated Timber Grave pots, as for instance one from Polyanki, Kazan (Gimbutas 1965, fig. 357). The question of a high front with Near Eastern counterparts raises wider issues discussed in a later section.

We now return to the area of the Carpathian Basin, Transylvania and Slovakia where, as we saw, numerous pottery models attested the use of disc-wheeled vehicles, presumptively ox-drawn, around the middle of the second millennium, and domestic horses were equally present in the same archaeological contexts as the models. Among the pottery wheel models is a significant group which represents spoked rather than disc types, the spokes being indicated by radial incisions or by modelling to obtain an openwork, invariably four-spoked, representation, in some examples the effect being achieved by making four circular per- *49* forations in the disc. As with the linear representations, while these may genuinely represent four-spoked originals, a certain reserve must be expressed: on one broken

92

model wheel from Gyularvarsánd an original six spokes appear to have been incised. When compared with the disc-wheel models and their accompanying wagon models, spoked-wheel models are rare, and with very few outliers, for instance south of Lake Balaton, are confined to a northerly area in Slovakia and Moravia between Bratislava and Brno. Their context in Hungary and Transylvania is (like most of the model disc-wheel wagons) that of the Füzesabony and Otomani phases of the Earlier Bronze Age, and in Czechoslovakia to the Mad'arovce and Věteřov cultures, broadly contemporary with the later half of the Hungarian phases in question: at Nitriansky Hrádok in Slovakia, which has produced spoked-wheel models, a radiocarbon date of 1400 ± 200 bc is available for Mad'arovce, but the large margin of error limits its chronological value (Coles & Harding 1979, 69–79). Three Věteřov dates range from *c.* 1700 to 1530 bc (Harding 1980, 182). The spoked-wheel models have been listed and discussed in Tihelka 1954, and Boná 1960 and 1975 (with map X); other examples in Točik 1964, 237 and Dušek 1960, 295. While presumably belonging to model vehicles, there is no evidence as to whether these were two or four-wheeled, though the well-known pottery cinerary urn of Later Bronze Age date from Kánya in Hungary is carried on four wheels (Csalog 1943). Nor is it known whether the original spoked-wheeled vehicles implied by the models were horse-drawn, though the models are within the geographical area and cultural contexts in which, as we shall see, surviving bone and antler elements of horse-bits and other harness are an outstanding feature.

Two finds of actual bronze-sheathed four-spoked wheels, in each instance an unassociated pair, have been made in the Carpathian area in former Hungarian territory, from Abos (now in Slovakia) and Arokalya (now in Romania). Their affinities, however, would seem to lie with a large series of

48 *Representation of chariot on pot, mid 2nd millennium* BC, *from Timber Grave culture burial at Saratov, Lower Volga, USSR*

Later Bronze Age date and they are therefore discussed in the next chapter. In themselves they imply light two-wheeled vehicles, but unambiguous evidence of such horse-drawn carts or chariots is provided, at the end of the period here regarded as the Earlier Bronze Age, by a find of the Piliny culture at Vel'ke Raškovce near Trebisov in eastern Slovakia (Vizdal 1972). The Piliny

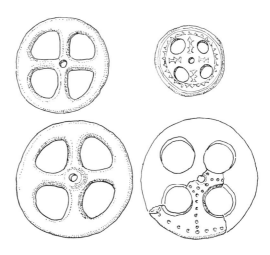

49 *Pottery models of spoked wheels, mid 2nd millennium* BC, *Mad'arovce Věteřov cultures from Slovakia*

phase, succeeding that of Füzesabony and in some part contemporary with Mad'arovce, is largely represented by cremation graves with distinctive pottery (Coles & Harding 1979, 71, 92) and in such a grave at Vel'ke Raškovce a two-handled amphora was decorated with an incised frieze of four vehicles with two four-spoked wheels, drawn by a pair of horses, and with the driver standing outside and behind the equipage. The body of the vehicle is shown as small and D-shaped, open at the back, and bisected by a line which continues as the draught-pole between the two profile views of the horses. No yoke is shown. The driver, a highly schematized rod-like figure, stands with arms outstretched as if holding reins, and the treatment of the head could conceivably be seen as indicating some form of helmet. The frieze clearly depicts scenes of prestige and possibly ritual significance, perhaps funerary display, as Powell suggested (1976), looking significantly towards Dipylon vases, and the repetition of the vehicle motif as an ornamental scheme ultimately recalls the disc-wheeled wagons on the TRB cup from Bronocice. At all events we have unequivocal evidence for spoked-wheeled vehicles of prestige that can only be called

chariots in Central Europe early in the second half of the second millennium BC.

In the Earlier Bronze Age Western Europe has practically no evidence to offer for a knowledge of spoked rather than disc wheels. Italy provides examples of the peculiar type of 'cross-bar' wheel of openwork construction, discussed later in this chapter, and a single carved wooden object, *c.* 17 cm in diameter and perhaps a spindle-whorl, in the form of an eight-spoked wheel from Barche di Solferino on Lake Garda, a settlement of the Polada culture dated by Barfield to *c.* 1800–1450 bc. Instead of a normal nave it has a rectangular slot 4×0.75 cm at the centre (Barfield 1971, 74; Cornaggia Castiglioni & Calegari 1978, pl. III; Woytowitsch no. 4), and it cannot be taken as certain evidence of the use of full-sized eight-spoked wheels in north Italy at this time. The rock paintings of Peñalsordo and Peña de los Buitres in the Badajoz region of Spain, undated but presumptively of second-millennium BC date, include rough representations of A-frame or rectangular carts, and one wagon, with spoked and in one instance 'cross-bar' wheels of a type discussed below (Almagro 1966, fig. 79). Rock carvings at Val Camonica and in Scandinavia appear to

50 Representations of chariots and horses on pot, later 2nd millennium BC, *Piliny culture, from grave at Vel'ke Raskovce, Slovakia*

94

be of later date and are dealt with in Chapters IV and V below.

The final area for spoked-wheel transport in the second millennium BC is Transcaucasia, in contexts already discussed in connection with the disc-wheeled vehicles from the same sites: representations among the *39* Syunik rock carvings and actual surviving vehicles from the waterlogged tombs at Lchashen on Lake Sevan. The former, originally published in Karakhanian & Safian 1970, have been discussed in detail in Littauer 1977; the latter in Piggott 1974 with references to the original sources. In her study, Mrs Littauer makes an important distinction, applicable to the whole question of two-wheeled vehicles, between carts and chariots: the former, 'with solid or spoked wheels, were confined to the same uses' (transport of men and goods) as four-wheeled wagons; 'they usually had a central axle, and the passengers normally sat in them. Chariots had spoked wheels, and with certain notable exceptions, a rear axle; the occupants stood in them, and they were used for military, sporting or ceremonial purposes'. On the Oughtasar rocks of Syunik, among the very large number of ox-drawn wagons described earlier in this chapter, there are only seven two-wheeled vehicles with spoked wheels that can be classed as chariots, while one wagon and one A-frame cart is shown with spoked wheels. In the two instances where draught animals are shown with the chariots they appear to be oxen, certainly not horses. The wheels are shown as four-spoked except in one instance where one of the pair has five spokes (as on the A-frame cart). On the spoked-wheel wagon (KS 122), the representation, if taken literally, would depict one disc, two four-spoked and one five-spoked wheel, emphasizing the limitations of accuracy in such schematic renderings. The chariot bodies, where shown at all, are crossed discs and the yokes have long loops or transverse bars suggestive of ox-harnessing.

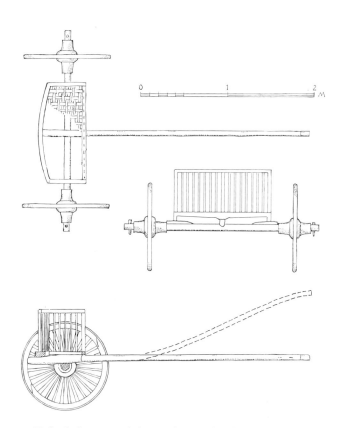

51 *Scale drawings of chariot, later 2nd millennium* BC, *from Barrow 9, Lchashen, Armenian SSR*

The Lchashen tombs produced two-wheeled vehicles with spoked wheels, Chariot 1 from Barrow 9, with Wagon D, Chariot 2 from Barrow 11, with Wagon F, and a single spoked wheel (Chariot 3) from Barrow 10, with Wagon E and the A-frame cart F. Chariot 1, published with measured drawings, is virtually identical with Chariot 2, with a small light body 1.10 m wide and 0.51 m deep, straight in front and slightly curved behind, with a rail on slender *51* uprights at back and sides. Slots in the framework suggested a floor of interwoven leather straps. The axle is central, 2.25 m long, the wheels rotating freely, with wooden linchpins. The wheels, 0.98 in diameter in Chariot 1 and 1.02 m in Chariot 2, had turned naves 43 cm long into which were morticed no less than twenty-eight spokes, again morticed at their outer ends into a felloe of two half-circles of bent wood

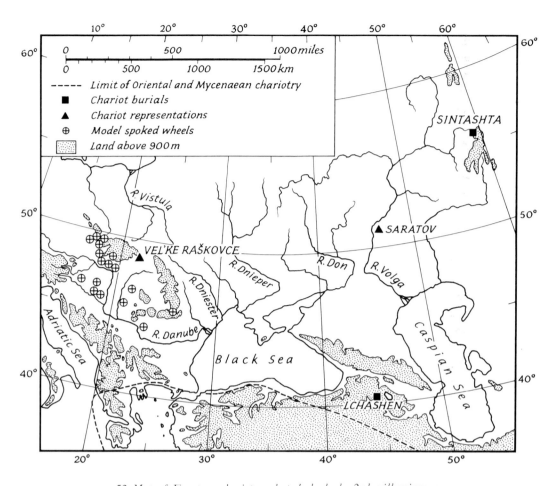

52 Map of European chariots and spoked wheels, 2nd millennium BC

with scarf joints, probably originally *c.* 4 to 5 cm thick, with a tyre of thin wood or leather. The original draught-poles were missing in both vehicles, and the straight pole of the reconstructions is subject to grave criticisms, as, with the body horizontal, this would only provide for draught-animals about 0.5 m high at the yoke, a clearly impossible figure, as horses would have had a withers height of around 1.30–1.40 m, a size obtainable only by a bent pole in the manner familiar from ancient Near Eastern chariots (M. A. Littauer *in litt.*). The open-fronted Lchashen vehicles can hardly, in the terms quoted above, be classed as chariots in the normal oriental sense, but as light carts in which the passengers sat rather than stood as charioteers, whatever prestige such

a vehicle may have had for its occupants. The writer has compared the peculiarities of wheel construction, with a very large number of spokes and two-piece bent felloes, with Chinese chariots from Shang and Zhou contexts and the four-wheeled carriage of the fourth-fifth century BC from Pazyryk in the Altai (Piggott 1974, 1978a) and the problem is further discussed in the final section of this chapter.

Reference has already been made to the bronze yoke ornaments from the Lchashen tombs 1, 9 and 10 (with Wagons A, D and E) in the form of charioteering groups. The poor published photographs have been supplemented by personal examination by the writer and Mrs Littauer (Mnatsakanyan 1957, fig. 8; Piggott 1974, pl. V; Mnatsakanyan 1960, figs 5 and 7; Littauer & Crouwel 1979a, 75, 78, 80). The model from Barrow 1 shows a body with front rail and open at the rear, and the axle mounted at the back, with eight-spoked wheels, a high arched draught-pole with four bird-figures at the yoke-junction and a pair of horses with arched necks. It carries the figures of two bearded warriors side by side and hand in hand, armed with daggers and wearing crested helmets. The body (as in the other two models) has a vertical loop as a mounting hold at the rear and a horizontal running from it to the vertical support of the front breastwork, dividing the body into two. Mrs Littauer thinks it possible that 'an attempt to indicate yoke saddles' was present in the models (*in litt.* 13. xi. 1978). These intermediary members between the yoke and the animals' necks occur as a harnessing technique in ancient Egypt, China and Pazyryk in the Altai, and are referred to again below (Littauer 1968a). The Lchashen chariot models from Barrows 9 and 10 repeat the features of that from Barrow 1, with six-spoked wheels, and are virtually identical, with the figure of a stag set forward of the horses, implying a hunting scene rather than the use of the chariot in war.

The cross-bar wheel

A peculiar form of wheel, of openwork construction but not with radial spokes, has long been recognized in antiquity as the 'cross-bar' wheel, but its significance in the technological history of the wheelwrights' craft has only come to be recognized in recent years. It has been characterized as having 'a single diametric bar, thick enough to accommodate the nave, and by two slender "cross-bars" on either side of the nave which traverse the central bar at right angles to it, and the ends of which are morticed into the felloes. There may alternatively be four (or more) shorter cross-bars in

similar relation to the central bar, but with their inner ends morticed into it' (Littauer & Crouwel 1977a). Our immediate concern here is with two north Italian finds of the earlier second millennium BC, but the type has been shown to have been in use as early as the late third millennium BC in the Near East (at Tepe Hissar in Iran), and continuing in use at least to the second, and with a notable frequency of use in Greece from the sixth century BC, as Miss Lorimer demonstrated in a pioneer study of eighty years ago (Lorimer 1903), as also in Etruscan contexts. A sixth-century example of an actual wheel comes from Gordion in Asia Minor (Kohler 1980, fig. 32). Its recent ethnographical distribution is remarkably wide, and includes the Iberian Peninsula (and thence to Mexico), England, China and outer Mongolia.

The Italian examples both come from the lakeside settlement of Mercurago in Piedmont already referred to in connection with the tripartite disc wheel with lunate openings from the same site and, like it, attributable to either the earlier (Polada) or later Middle Bronze Age phases of the settlement, with a chronological range of *c.* 1800–1100 BC (Barfield 1971; Cornaggia

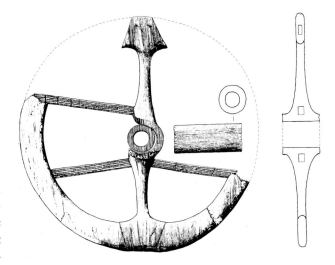

53 Cross-bar wheel, 2nd millennium BC, *from Mercurago, Italy*

Castiglioni & Calegari 1978; Woytowitsch no. 3). The first, known since its publication in the last century, is of walnut wood and of elegant construction, *c.* 88 cm in diameter, with the central cross-bar carrying an inserted tubular nave 25 cm long and with spade-shaped ends morticed into a bipartite or quadripartite felloe and, on each side of the nave, two slightly curved rods passing through the cross-bar and morticed into the felloe. The second consists of one quarter of a quadripartite felloe of oak *c.* 8 cm thick, with an original diameter of *c.* 90 cm, with a pair of parallel cross-rods morticed into the felloe-segment and originally into the now missing central cross-bar, the whole of more massive construction than the first wheel.

These Mercurago wheels are unique survivals in second-millennium Europe, but in the rock paintings at Peña de los Buitres in the Badajoz region of Spain (already referred to in connection with representations of sledges and A-frame carts) one cart is shown not with spoked, but with cross-bar wheels (Almagro 1966, fig. 79 m). While undated, as we saw, the second millennium BC seems a reasonable assumption, but, as we shall see, the cross-bar wheel is represented in model form from Portugal in the Early Iron Age. The same reservations as to absolute date apply to examples of what appear to be such wheels among the rock carvings in Scandinavia, further considered in the next chapter (Marstrander 1963, 282, fig. 70; Hayen 1973. fig. 13).

Technologically, the derivation of the cross-bar wheel from a tripartite disc with lunate perforations (as at Mercurago itself) was first put forward by Lorimer and accepted in all subsequent discussion as a convincing sequence, rather than seeing it as a clumsy or incompetent version of a normal spoked wheel (cf. Atkinson & Ward 1964 on recent Yorkshire examples; Lucas 1972; Fenton 1973; Littauer & Crouwel 1977a). Although our earliest evidence is from the ancient Near East, it could be

argued that the potential for the independent emergence of the form existed wherever fenestrated tripartite discs were developed, and so far as we are concerned this would include several areas of Western and Northern Europe. In view of the conservatism of wheelwrights' techniques, with classic tripartite discs still being made in Ireland in the 1950s and a wide variety of discs with lunate openings, and occasional cross-bar wheels, still in use in the Iberian Peninsula (Galhano 1973), it is worth while dwelling for a moment on the presence of cross-bar wheels surviving in earlier nineteenth-century England. One actual pair, morticed to a turning axle, survives from Sedbergh in Yorkshire, and a third (very near to the assumed tripartite disc prototype) from Yorkshire without a precise provenance (Atkinson & Ward 1964), and iconographic and literary sources demonstrate its wider distribution in north-western England. W. H. Pyne's illustration (reproduced in Littauer & Crouwel 1977, fig. 7) belongs to a group of drawings depicting slate quarrying, probably in the Lake District. Rather than seeing, with Fenton, these wheels as a local and recent development 'presumably to be interpreted as due to the influence of the spoked wheel', or with Littauer and Crouwel looking to the trade in 'Spanish jennets', or soldiers returning from the Peninsular Wars to link Iberia and England, it may be permissible to look back to prehistory. If so, the most likely context would be that of the well-known contacts between Iberia and the British Isles implied by the distribution of bronzes in the earlier first millennium BC (Coles & Harding 1979, with refs), though later, Early Iron Age contacts between the two regions are not excluded.

Horse harness

Over the whole area where we have evidence of Earlier Bronze Age wheeled transport in the form of carts or chariots in the

Earlier Bronze Age, from the Urals to the Carpathians, we also have evidence not only of domestic horses, but of varying forms of bits with 'soft' mouth-pieces and other elements of harness represented by their non-perishable parts (notably bit cheek-pieces) in bone or antler. Such harness does not, of course, provide unequivocable evidence of draught-animals for vehicles, as it could equally well be used on horses for riding, but there is a strong likelihood that a correlation between the harness and horses used for paired draught does, in fact, exist. More than one prehistorian has stressed the close relationship between prestige harness and prestige vehicles (e.g. Hüttel 1977). A series of studies of this harness gear has been made; Smirnov (1961) and Gimbutas (1965, 539–41) deal with Timber Grave Andronovo finds in the Volga-Ural steppe area; the more westerly Timber Grave and earlier (Catacomb Grave) cheek-pieces extending to the Pontic region have been discussed by Leskov (1964), Movsha (1965), Littauer and Crouwel (1973) Smirnov and Kuzmina (1977), and Kuzmina (1980). In the Carpathian Basin the basic studies are those of Moszolics (1953, 1960), Bandi (1963) and Boná (1975, with map V). These three geographical areas, while showing evidence of contact within a common continuum of cultural traditions in the second millennium, do, in fact, also show characteristic differences in harness typology, particularly in the cheek-pieces of bits with mouth-pieces of perishable materials. An important and recent general survey is that of Hüttel (1977).

Smirnov made a classification of the bit cheek-pieces of the eastern, Volga-Ural area into five types, running chronologically from Timber Grave into Scythian times, and only his Type I and II concern us here. The bone cheek-pieces of Type I are rectangular plaques with blunt serrations on the inner edges, as at Komarovka on the Volga near Kuybyshev (Gimbutas 1965, fig. 363), which he compared with bronze examples

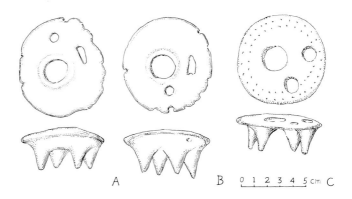

54 *Antler and bone spiked disc cheek-pieces;* A, B, *3rd millennium* BC, *Catacomb Grave at Trakhtemirov, Ukraine, USSR;* C, *mid 2nd millennium* BC, *Monteoru, Romania*

of similar form in the Near East, the Type I of Potratz (1966) but which do not seem to have counterparts in Europe west of the Volga. Smirnov's Type II, a stemmed disc, instanced by one from the Andronovo cemetery of Alakul in the Kurgan province of west Siberia, does, interestingly enough, have its counterparts in the Hungarian Bronze Age, with two examples from Tószeg (Moszolics 1953, fig. 20, pl. XIV, 14, 15; Kuzmina 1980), presumptively of Bronze III (Füzesabony) date.

The cheek-pieces of the next type under consideration consist of perforated discs of bone or antler (some at least made from the epiphysial ends of cattle metatarsals and others from the burr end of red deer antlers) with a flat face, sometimes decorated, and three or more usually four internal blunt spikes or pointed studs and range from *c.* 9.0 to 4.0 cm in diameter. They have a wide geographical distribution mapped by Kuzmina (1980) with a dozen finds from the Volga to the Dnieper and the Crimea, and range in date from the Catacomb Grave phase of the early second millennium (the two western finds, from Kamenka in the Kerch' peninsula and Trakhtemirov near Kiev) to the Abashevo phase of the Timber Grave culture. In two instances (Trakhtemirov and Staro-Yur'evo) they were found in pairs. In the

99

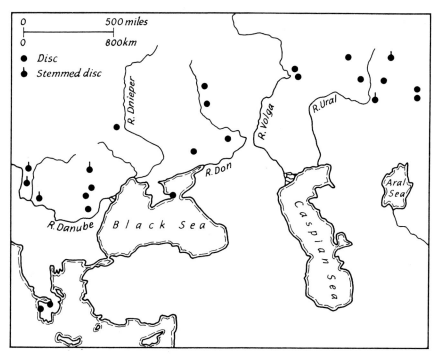

55 Map of bone and antler disc cheek-pieces, 3rd–2nd millennium BC

Carpathian-Transylvanian region there is a fine example with internal spikes, 7 cm in diameter, from the Monteoru settlement in Romania (unpublished) and analogous perforated discs without spikes from Tószeg and Füzesabony, all of comparable date within the Hungarian Bronze Age III phase (Bandi 1963, figs 1, 2; Hüttel 1977, fig. 8). Leskov in 1964 drew attention to the similarity of the Russian spiked discs to finds from Shaft Grave IV at Mycenae, and this led to a thorough discussion and a new publication of the Mycenae objects by Littauer and Crouwel in 1973. There are four discs from Shaft Grave IV, of bone and 4.8–4.9 cm in diameter, two with a slight projection on the edge, decorated with a band of running spirals or wave-pattern on the face, each with three blunt spikes, two of which are perforated. They have central perforations in which are tubular bronze and bone insets and near the rim two small circular perforations and a slot. After a detailed review of their features, the authors dismiss

the cheek-piece interpretation and suggest alternative explanations for their function. Despite this authoritative statement, the writer remains impressed by the close morphological resemblance of these discs to the Russian series under discussion. Their small size is within the range of diameters—4.5 cm only at Tavlykaeva—and the hole and slot is paralleled at Trakhtemirov, and the side projection at Staro-Yur'evo (Smirnov & Kuzmina 1977, figs 11, 3, 4, 8, 9; Hüttel 1977, fig. 10b; Kuzmina 1980). Whatever the function of the Mycenae discs, in the writer's view they must be considered in relationship with those from south Russia and Romania, for which Littauer and Crouwel allow 'there is certainly a distinct probability that they are cheek-pieces' or 'burrs' 'fitted inside the cheek-pieces over the mouthpiece to exert lateral pressure on the horses' lips'. They have in turn been compared with the second-millennium bronze bits with spiked wheel-shaped cheek-pieces of the Near East, and like the

100

antler time cheek-pieces of the Carpathians next to be considered, 'it seems very likely that these inspired the all-metal bits that first appeared in the Near East in the 15th century BC' (Littauer & Crouwel 1979a, 89; Moorey 1971, 104).

In the Earlier Bronze Age of the Carpathian Basin, contemporary with and often from the same contexts as the model wagons and model disc and spoked wheels already described, bit cheek-pieces of a distinctive type appear, crescentic and normally made from the brow-tine of a red deer's antler. These have a central hole or slot for the 'soft' mouth-piece of rope or leather, with perforations above and below for a bifurcate rein, and are on the one hand related to the type represented in the earlier Sredny Stog examples as at Dereivka, and on the other hand are the progenitors of a long series of such cheek-pieces in European prehistory, either in antler or as early in the Near East and later in Europe, in bronze. They are frequently decorated, and stray examples have been found as far afield as second-millennium sites in Asia Minor such as Bogazköy, Alaca Hüyük and Beycesultan (cf. Littauer & Crouwel 1979a, 88). There is no certain record of their appearance in the Carpathian Basin until the Bronze III (Füzesabony) phase, and they then continue into later contexts. With them may be taken a group of Y-shaped perforated objects of antler regarded as strap-dividers for the bifurcated reins demanded by the crescentic cheek-pieces (Moszolics 1953, 1960; Bándi 1963). Much play has been made of the few Anatolian antler cheek-pieces, and especially that from Beycesultan, as providing prototypes, and an historical date, for the Carpathian

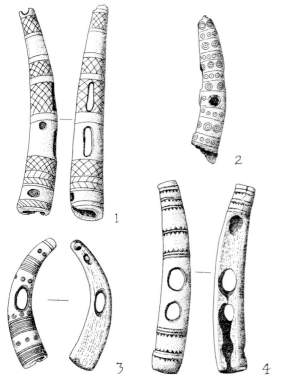

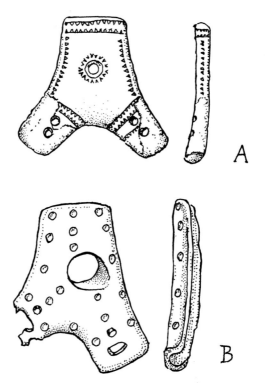

56 *Antler tine cheek-pieces, mid 2nd millennium* BC; *1, 2, Tószeg; 3, Pákozdvár; 4, Köröstarcsa, Hungary*

57 *Bone harness strap-dividers, mid 2nd millennium* BC; *A, Nyergesújfalu-Téglagyár; B, Pákozdvár, Hungary*

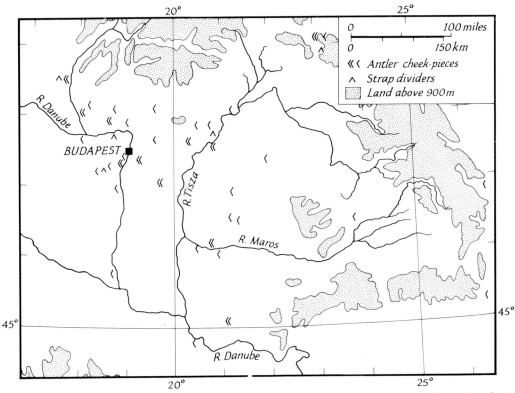

58 *Map of antler tine cheek-pieces and bone strap-dividers, mid 2nd millennium* BC, *in the Carpathian Basin*

series, but they should be regarded either as evidence of a widespread, if sporadic, use of 58 the type both in Europe and Asia Minor, or if not indigenous, as strays from Europe rather than the contrary.

Woytowitsch (1978, 117–19) has drawn attention to over thirty finds of antler and bone cheek-pieces of types which he has compared with the Carpathian series just mentioned, from settlements in the Po Valley of the Bronze Age. Typologically some at least of the cheek-pieces could be early, and one, from Montale, has a band of flattened spiral wave-ornament, and others have ring-and-dot motifs which resemble the Hungarian examples just described. The numerous finds from the same area of little bronze wheel-models and pendants might be taken with the cheek-pieces to indicate spoked-wheeled vehicles with horse traction in north Italy around the middle of the second millennium BC.

It should be remembered, however, that antler cheek-pieces of identical or closely related forms continued to be used well into the Late Bronze Age in Hungary, Czechoslovakia and (significantly for the Italian finds) Switzerland (Moszolics 1953; Jelinkova 1959; Balkwill 1973; Britnell 1976). The unstratified fragment of such a cheek-piece from an Early Bronze Age (Reinecke A2/B1) settlement at Waldi in Switzerland (Kubach 1977, fig. 6) is, in fact, closely paralleled by a Late Bronze Age Hungarian find (Moszolics 1953, fig. 36) and so must be a later stray on the earlier site. It might be pointed out, in connection with the bronze wheel models, that the 'wheel-headed' pins of the German Early Bronze Age, sometimes sometimes quoted as analogues, have no relevance to spoked wheels in actuality, though they are occasionally referred to in such a context. Crossed circles are common decorative motifs, possibly sometimes connected with solar symbolism, but in themselves are no more.

Carts, chariots and the Near East

There remains for discussion a problem akin to that which faced us at the end of the last chapter, that of the relationships between the adoption and development of the light vehicle with two spoked wheels, normally horse-drawn, in second-millennium communities of the Near East and of Europe. Whereas a technological progression from fenestrated tripartite disc wheels to crossbar types seems likely, Childe's view that 'though naturally an expression of the wheel idea, the spoked wheel was a new invention rather than a modification of the tripartite disc' (1954b, 214) remains a valid hypothesis. If we accept it, our next step must be to examine the evidence for a place of origin for the new invention, and the alternative, that it could have been independently developed in more than one area where the antecedent condition of domestic horses available for secondary use as traction animals, was present.

The latest statement of the problem is that of Littauer and Crouwel (1979a, 68–71). They consider that the Near Eastern evidence they have assembled in exemplary detail 'strongly suggest the possibility of a local evolution of the light, spoked-wheel horse-drawn chariot in the Near East itself, in contrast to the long-held theory that this was introduced from outside in an already evolved form by Indo-European steppe tribes'. They would see the technological genesis of the wheel with nave, felloe and radial spokes, in the cross-bar wheel, both forms being documented from the very beginning of the second millennium BC, with bent-wood techniques also of early origin, together with equid draught with onagers. Horses too were becoming familiar to Mesopotamian peoples from equally early in the second millennium BC, and the linguistic and lexical evidence adduced for Indo-European influences is regarded as too late to be relevant to origins. 'Whatever the role of particular peoples in the origin of the spoked-wheeled, horse-drawn chariot, there appear to be no cogent archaeological or linguistic arguments against its development in the Near East itself.' If this thesis were accepted, all the European instances of light spoked-wheel vehicles surveyed in this chapter, from the Urals to Slovakia, would be accounted for in terms of the model of diffusion from 'higher' to 'lower' cultural contexts, from the Orient to Europe, which Childe developed and maintained so persuasively in the past, and within which we have seen the development of the tripartite disc as best explicable. The subsequent spread and adoption of the chariot or light prestige vehicle with two spoked wheels and horse draught would be a repetition of the earlier circumstances, now drawing on a wider geographical area of literate civilizations which by the mid second millennium BC could include the chariot-using peoples of Hittite Asia Minor and Mycenaean Greece.

The writer recently restated the case for an initial development of the light, two-wheeled, horse-drawn vehicle beyond the ambit of the ancient Near Eastern civilizations, and saw its development into the familiar oriental war-chariot as 'the result of a ready social acceptance of the light, spoked-wheel, horse-drawn vehicle from alien, non-urban, non-literate communities to the north . . . within the natural territory of the wild horse'. This was in the course of an enquiry into the origins of chariotry in Shang China (and incidentally, Vedic India) in the late second millennium BC, and it was suggested that these might better be found in the Timber Grave and Andronovo continuum rather than in the metropolitan ancient Near East itself (Piggott 1978a). It is still felt that this area, with its westward extension into the Carpathians, cannot be excluded from a wider concept of a diffuse region within which experimentation with light, probably bent-wood, fast horse-drawn carts for speed and prestige took place; an area which included the ancient Near East, in which the development of the cart into the chariot, with its institutionalized use in warfare, took place. Hüttel (1977) has suggested something of the same concept when seeing the Carpathian Basin as a 'distant outpost' of an 'international world' of chariotry. War chariots, and chariot warfare are specialized products of social situations not demonstrably or necessarily obtaining in Europe of the second millennium BC, but the horse-drawn cart, fast and if needs be elegant in construction and decoration, can be a prestige vehicle in its own right: the status of the Lchashen vehicles as carts for seated passengers, as opposed to the chariots in the bronze models (here surely derived from Near Eastern chariotry) has already been pointed out.

The Carpathian area demands special attention, in view of the fact that many archaeologists over the past twenty years have looked to Mycenaean Greece for the origin of spoked-wheeled vehicles and presumptive chariots in that region, as a part of a wide range of objects, ornament, and cultural traits deriving from that region as a result of Mycenaean trade activities in Eastern and Central Europe. Bouzek, in a general review of the evidence, writes 'clay models of four-spoked wheels from South-Moravian, South-Slovakian and Hungarian localities prove the contemporary spreading of the two-wheeled war chariot to Central Europe' from the Mycenaean world (1966, 256). The thesis of Mycenaean contacts was of wider implication, and included Western Europe and the British Isles: it was favoured by many, including the writer, who reviewed the whole postulate and

endorsed its general soundness in the same year as Bouzek (Piggott 1966). Unfortunately, a critical examination of the evidence over recent years has successively undermined the main props of the structure of theory supporting the thesis—faience beads, bronzes, spiral and wave-pattern ornament, for instance (Harding 1971, 1975; Harding & Warren 1973; Coles & Harding 1979, 150–53). Today the writer would take his stand with the critics, and with them regard the concept of any significant Mycenaean or Aegean 'presence' in second-millennium continental Europe as illusory (apart from the question of the amber space-plate beads, which is a Western European phenomenon unrelated to our chariot problem). The complex of elaborately harnessed horses with concommitant light horse-drawn vehicles of chariot type in East and Central Europe must be regarded as of indigenous development over a 'technological *koine*', within which the Near East, the Levant, Egypt, Asia Minor, and Mycenaean Greece all came to play important and often dominant roles. It was in these contexts of literate, highly organized states that social conditions allowed not only of the development and maintenance of the vehicles and their horses, but the emergence of a professional chariot-warrior class as a military cadre in disciplined warfare. In the more rustic ambience of barbarian Europe of the second millennium BC the prestige of the horse-drawn vehicle which was not only a potential chariot in the Near Eastern sense, but the ultimate ancestor of the tilbury, cabriolet or curricle among Victorian pleasure-carriages, played its part in the demonstration of status in an hierarchical society by the possession of a socially admired fast vehicle, to prove a persistent psychological factor in human societies.

4 Vehicles of the Later Bronze Age

Chronology and culture

Following the division of the European Bronze Age, outside the Aegean and Mediterranean world north of the Alps, into an Earlier and Later phase as set out at the beginning of the last chapter, we now come to evidence for wheeled transport in the second of these divisions. The chronological framework within which we now have to set our evidence spans some six centuries and is based on an elaborate series of correlations of elements of material culture, particularly in the products of the bronze-worker's craft, which now reaches a high degree of sophistication. Such bronze-work in particular enables a network of connections between Europe north of the Alps and the Mediterranean world, to be established, and from this an historical chronology, based ultimately upon that of the literate civilizations of Egypt and the Near East, can be in turn applied to more northerly transalpine regions. This was demonstrated in a classic study by Müller-Karpe (1959) which, after twenty years of active investigation, criticism and new evidence, still remains viable and basically acceptable. Müller-Karpe took over the long-standing classificatory and chronological system of Reinecke, who, initially on the basis of south German finds, defined a Bronze Age divided into four periods, A–D, in the last two of which (C and D) iron-working was fully established as a dominant technological factor, but, as Reinecke thought on evidence we now know to have been misinterpreted, linked to the preceding A and B ones by tentative beginnings in the earlier phases.

In the traditional technological terms we would now end a 'Late Bronze Age' with Ha B, taking Ha C and D as the earlier phases of an 'Iron Age'. At the other end of the chronological scale, Reinecke himself perceived that his final, D, division of the Bronze Age (Bz D) had features that looked forward to his Ha A, and, unable to escape from the old terminology, we now would make our Late Bronze Age (or, from the dominant burial rite of cemeteries of cremations in pots, an Urnfield period) begin with and include Bronze D, and end with the final sub-phase of Hallstatt B. Familiar to the prehistorian and accepted with all its inconsistencies of nomenclature, the historical accidents which led to its persistence in European archaeology need to be explained to the non-specialist reader. We will then follow international practice and in our present chapter use the scheme with its subdivisions for the Later Bronze Age: Bz D, Ha A1, Ha A2, Ha B1, Ha B2, Ha B3. For North Europe and south Scandinavia we have, however, to correlate this scheme with that of the six divisions (I–VI) devised for the Bronze Age of that region by Montelius (Müller-Karpe 1959; Coles & Harding 1979, 491, 516; Thrane 1962a). Where precision is unobtainable or unnecessary, the convention of 'Early Urnfield' (B2 D, Ha A1, A2) and 'Late Urnfield' (Ha B1, B2, B3) can conveniently be used.

The Müller-Karpe scheme provided not only a relative sequential ordering of the archaeological material over a large area of continental Europe, but an absolute chronology in which the six periods defined were each boldly allotted a century apiece, beginning with B2 D from 1300 BC to Ha B3 ending at 700 BC. Even if this seemed at first sight 'startlingly, even scandalously, mnemonic' (Cowen 1961, 42) it has proved to be a useful and on the whole convincing scheme within which to operate and is broadly used here. Sandars (1971) has, however, argued, partly on the grounds of more recent revisions of Late Mycenaean chronology, for an absolute dating in which Bz D begins a century later, and with the beginnings of Ha A, occupies the twelfth century, Ha B starting in the late eleventh. This lowering of the dating has seemed 'too extreme' to Coles and Harding (1979, 383) and for our present purposes of the study of the use and typology of wheeled transport among 'Urnfield' communities a relative chronology is of greater moment than absolute dating within a century or so. Radiocarbon dates are few and their inherent margin of error diminishes their value in contexts demanding a close time-scale within a century or so: they are listed and discussed by Coles and Harding (1979, 379–80).

Whatever the precise dating, the Urnfield phase of later prehistory in continental Europe from the thirteenth (or twelfth) to the eighth century BC, not only forms a distinctive and cohesive whole, but marks a period of change, innovation and in some sense an inferred realignment and movement of peoples which differentiates it from the equally recognizable world of the earlier second millennium, even though elements of continuity are undoubtedly perceptible. From the point of view of the present enquiry into the development of wheeled transport, the evidence at our disposal largely derives from geographical areas westward of those which have provided material for earlier phases in this technology, and though we start within the Danube basin, our main concern will be with Central and Western Europe, as far north as southern Scandinavia, and by Late Urnfield times including Britain for the first time. By Early

Urnfield times, horse traction is well established side by side with the more ancient tradition of ox-draught, and develops an increasing importance, and with an extensive and sophisticated background of abundant bronze-working is reflected in the archaeological evidence of the metal elements of harness, especially a complex series of horse-bits. The funerary cremation rite may rob us of wooden vehicles accompanying the dead, but the practice continued, and the development in the period of vehicles of prestige increasingly decorated with bronze fitments and overlay enables us to recognize some features of the full-size or model versions of these from the metal that has survived the funeral pyre. Rock carvings, in Scandinavia and north Italy, offer a valuable commentary on current types of wagons and chariots, and it is an indication of the increasingly important status of the horse that of twenty-four Scandinavian representations of vehicles with their draught-animals likely in the main to be of the Later Bronze Age, only one and a probable second show oxen, the remainder horses. In the Val Camonica, where vehicle representations continue into the Iron Age, no oxen are represented, contrary to such early examples as Cemmo 2, described in Chapter 2. An important series of bronze models, many carrying sheet-bronze vessels on wheeled carriages (the *Kesselwagen* type), or birds on wheels *(Deichselwagen)* give us some idea of current wheel types. It will be convenient to discuss the material typologically, beginning with disc wheels, represented by only two finds, and then the abundant spoked-wheel evidence which involves a further important question, that of two-wheeled chariots co-existing with four-wheeled wagons or carriages.

Disc wheels

The continued use of the by now traditional disc wheel, solid or tripartite, in Later Bronze Age Europe is attested by two actual

59 *Tripartite disc wheel with inserted nave and external battens, early 1st millennium* BC, *from the 'Wasserburg', Buchau, Germany*

finds and the sometimes ambiguous evidence of the rock carvings. The first find of an entire tripartite disc wheel is from the waterlogged fortified island settlement of the 'Wasserburg' in the Federsee near Buchau in Württemberg, which is of two main phases of construction, Ha A and Ha B: it is not known to which the wheel should be attributed and so chronologically it lies between *c.* 1200 and 700 BC (Coles & Harding 1979, 354). It has not been published in detail (but see Piggott 1957, pl. XXII) and is typologically very close to the Mercurago tripartite disc described in the last chapter, *c.* 78 cm diameter, with lunate openings, small internal dowels and curved exterior battens in grooves, and an inserted tubular nave *c.* 35 cm long. From the same site but not in association with the wheel is an axle-tree, with central square-sectioned axle-bed to carry the body, and cylindrical arms to carry the wheels, with slots for linchpins. This also remains unpublished. The second wheel is again from a waterlogged island fort, that of Biskupin in Great Poland, which dates from the end of the Lausitz

phase of Urnfield culture, and may indeed be chronologically within the Iron Age Ha C and D periods, though it is culturally within the Urnfield tradition and has radiocarbon dates from *c.* 720 to 560 bc (Coles & Harding 1979, 356, 380). The wheel is apparently in one piece with large lunate openings. There is no nave but the large central perforation is roughly square, and taken as the recipient of an axle which turned with the wheel (Kostrewski 1936 a, b).

The Val Camonica rock carvings offer a series of vehicle representations of stylistically late prehistoric date, and of the four-wheeled wagons most are spoked with a distinctive long body and are better considered as Early Iron Age, and will be dealt with in a subsequent chapter. One representation, however, stands apart stylistically and typologically, that of the disc-wheeled wagon on Naquane Rock 47 (Berg-Osterrieth 1972, 49–53; Anati 1975, 100) assigned to the transition from Period III to IV in the local stylistic sequence, about 1000 BC. The wheels are shown as large outline discs with no indication of naves, at the four corners of

60 *Rock carving of disc-wheeled wagon, early 1st millennium* BC, *Rock 47, Naquane, Val Camonica, Italy*

a rectangular body and the pair of draught-animals are equids (Anati has suggested asses on the grounds of their long ears), demonstrably male, each with an exagerrated penis. A similar, smaller and less distinctive vehicle representation is that on Naquane 57 (Berg-Osterrieth 1972, 34–37), roughly executed and adding no details to the Rock 47 vehicle. The two-wheeled vehicles from Val Camonica, one of which, if interpreted literally, appears to have disc wheels (Campanine) are considered as a group, with their spoked-wheeled counterparts, in a later section of this chapter.

The Scandinavian rock carvings are exceedingly numerous and widely distributed from southern Norway (Rogaland and Østfold) to Sweden (Bohuslän, Uppland, Blekinge and Skåne) and Denmark. Some of them, notably some of the several thousand ship representations, can be shown to belong to the Earlier Bronze Age, but it is convenient to discuss the vehicle representations in a Later Bronze Age context, partly on the grounds of the predominance of horse traction. A general statement is given by Coles and Harding (1979, 317–21) and regional surveys include Baltzer 1881 (Sweden), Althin 1945 (Skåne), Marstrander 1963 (Østfold), Glob 1969 (Denmark) and Fredsjö 1971, 1975 (Bohuslän). The artistic conventions employed are consonant with those discussed in the last chapter in the instance of the Transcaucasian rock carvings, but an element of ambiguity is introduced in the wheel-types, where a crossed circle or an open circle seems to have been used indiscriminately in depicting four- or two-wheeled vehicles. Single detached crossed circles occur, sometimes (as in Denmark) where no other vehicle representations are present, but they cannot in themselves be used as evidence of local wheeled transport, when alternatives such as solar symbols could provide an

explanation. Of the rare examples of ox-draught already referred to, the quite unambiguous example, at Rished (Askum) in Bohuslän, has a spoked-wheel wagon (Berg 1935, pl. XXVII, 1), and the other, with the bodies of the draught pair damaged, but with what appear to be two pairs of horns intact, has open circles for wheels which might be taken at face value as disc wheels (Marstrander 1963, fig. 44, 6, from Lille Borge in Østfold). The two-wheeled vehicles with open circles for the wheels (e.g. Marstrander 1963, fig. 44; Hagen 1967, pl. 38; Littauer 1977, fig. 24) may depict disc-wheeled farm carts, but are here taken with the spoked-wheeled representations of 'chariots' discussed below, as are a couple of examples of what appear to be cross-bar wheels.

In general, the slight evidence for disc-wheeled vehicles in the Later Bronze Age outlined above must reflect the imperfections of the archaeological record rather than the agrarian realities of the period. The traditional use of such carts and wagons, well established in prehistoric Europe, as we have seen, from the third millennium BC, was to continue down to and beyond the Roman period, with wheels of both simple disc and tripartite construction, and radiocarbon dating of the many unassigned north European examples, for instance, might help to bridge the gap (cf. the maps in Hayen 1973, figs 4, 6, 8). The persistence of the axle-tree type from Buchau into the last few centuries BC, as in the finds from the bog trackways of the Lengener and Ipweger Moors near Oldenburg, points in the same direction (Hayen 1973, fig. 2). All such vehicles we should probably regard as utilitarian farm equipment, but the bulk of our Later Bronze Age evidence relates to spoked-wheeled two- and four-wheelers, many clearly vehicles of cult or prestige in some form or another.

108

Spoked-wheel vehicles: bronze fittings

The bulk of our evidence for vehicles in the period *c.* 1300/1200–700 BC is for spoked-wheel constructions, in the form of models of varying sizes, mainly of bronze and sometimes of pottery, and the bronze sheathing or mountings of elaborate vehicles which can be inferred to be vehicles of secular prestige or associated with cult practices. The high technology of bronze-working, and the large metal resources which lay behind it, enabled it to be used lavishly in the decoration of woodwork, and this ensured the archaeological survival of at least the decorative elements of vehicles which, deposited in burials, have been otherwise destroyed in the funeral pyre incident to a predominantly cremation rite. Other bronze-work, such as the sheathing of naves, spokes and felloes of wheels, has survived in hoards or as chance finds, and finally rock carvings form an important source of contemporary depiction.

The cult practices, inferred but not understood, form a part of a weird world of elaborate symbolism represented in the decorative motifs, largely surviving in bronze-work, in which birds (especially apparently water-birds), bulls or other horned animals, and wheels or miniature wheeled vehicles all play a part, throughout Urnfield times and into the Hallstatt Early Iron Age which follows. Fabulous creatures such as horned birds, or 'bull-birds' on four legs but with birds' heads, sometimes set on wheels, appear, and this element of magic and fantasy makes the interpretation of the vehicle elements involved, in anything approaching practical everyday terms, difficult. However, they cannot be ignored or dismissed, and with caution can be used to provide useful clues.

We may start with a well-known find of models from Yugoslavia, the two pottery figurines in wheeled vehicles from a ceme-

61 Pottery model of bird chariot, early 1st millennium BC, *from Dupljaya, Yugoslavia*

tery of the Dubovac-Žuto Brdo culture at Dupljaja in the south Banat (Alexander 1972, 77, pl. 29). The complete example is a good illustration of the mixture of fact and fantasy just referred to, as taken literally, it represents a bird-headed woman conveyed on a tricycle drawn by a pair of swans, a set of circumstances unlikely to occur in real life. But its basic elements are clear: a vehicle or chariot with a pair of four-spoked wheels and a bird protome in front, with paired draught (of magic birds) provided with an extra wheel to make the whole model mobile. The second model similarly shows a woman standing in a chariot with two four-spoked wheels, with a broken central draught-pole which it would be reasonable to assume originally terminated in a bird-head: as we shall see, there is evidence for such bird-protome fittings in bronze for model, and probably full-sized functional chariots, in Urnfield contexts. We may then accept the Dupljaja models, shorn of their mythical attributes, as evidence for chariots of a normal if ceremonial use. Their dating is imprecise: Alexander (1972) and Powell (1963) would put them as approximately contemporary with the Piliny culture of

61

Czechoslovakia and so with the chariot representations on the Vel'ke Raškovce pot described in the last chapter. Coles and Harding, however, incline to a date late in the Dubovac-Žuto Brdo culture, bringing them in closer alliance to Urnfields and their bird symbolism (1979, 408). A further point might be made in connection with another well-known figurine in the same regional art-style, that of Kličevac (Alexander 1972, fig. 44) with breast-stars which on the one hand recall those on the bronze corselets from Čaka and Ducové, of late Bz D or early Ha A1 (Snodgrass 1971), and on the other the treatment of the eyes of the horse on the famous Trundholm vehicle model, of comparable date and discussed below (Sandars 1968, 185).

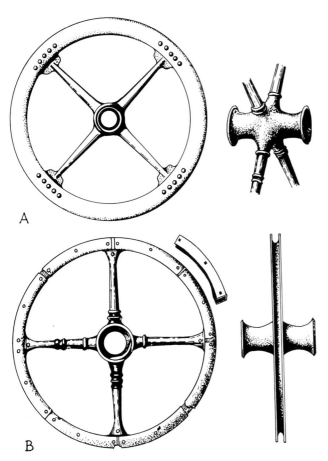

A

B

62 Bronze-sheathed wheels, early 1st millennium BC; A, *Abos, Slovakia;* B, *Arokalya, Romania*

In the Carpathian area, where as we saw in the last chapter model spoked pottery wheels occur in context of the Earlier Bronze Age, a decorated bowl, serving as the recipient of a male cremation, and carried on four such wheels, came from a tumulus-burial at Kánya, Somogy in Hungary and is dated to Ha B (Csalog 1943). This not only carries on the tradition of earlier model pottery vehicles but, in this instance at least, indicates the existence of four-wheeled wagons or carriages. The Kánya wheeled pot is linked to a well-known series of bronze bowls on wheels from Central Europe and south Scandinavia (the *Kesselwagen* type) which is discussed at a later stage. It is convenient, however, first to discuss other evidence from the Hungarian area for wheeled vehicles early in the Later Bronze Age.

The bronze-sheathed wheels from Abos and Arokalya have already been mentioned in passing, and, though undated by association, their typological resemblance to the miniature wheels of the *Kesselwagen* series (and to those of the Trundholm model) justifies their discussion at this stage. Both were found as isolated and unassociated finds in pairs, which suggests that they represent two-wheeled vehicles. The story of the Abos wheels is an archaeological tragedy. They were 'dug out of a cornfield' at Abos, then in Hungary but now in Slovakia, north of Kosice, sometime before 1853, when they formed part of the Fejerváry Museum bought by Joseph Mayer of Liverpool, and given by him, with the rest of his remarkable archaeological collections, to the City of Liverpool Museum in 1867. Thereafter, no record whatever was made by the museum of the wheels, which were destroyed by enemy action in 1941 (Nicholson, 1980, 60 and *in litt.*). When Josef Hampel was compiling his illustrated corpus of Hungarian Bronze Age antiquities in the 1880s, he used a drawing of one of the wheels preserved among the Fejerváry

62

papers in Budapest, and this constitutes our only record of the find (Hampel 1890, pl. LIX, 1; Kossack 1971, 147, fig. 30, 2). From this scale drawing the wheel appears to have been *c.* 98 cm in diameter, with four spokes, all of bronze or of bronze over wood, and on analogy with the Arokalya wheels (which survive) the felloe would have been a U-sectioned sheathing *c.* 5 cm deep for a wooden felloe projecting beyond it, and so perhaps adding 10 cm to the original diameter. The nave perforation was *c.* 5.6 cm diameter and the four tapering spokes had expanded spade-shapes junctions with the felloe-sheath, which was decorated with groups of five rivets, or more likely decorative bosses, at the junction-points. The expanded spoke-ends have good parallels in the bronze sheathing of the wooden spokes of the Ha A1 vehicle from Hart-an-der-Alz described later, and a fragment of comparable bronze felloe-sheathing from Kemnitz in north-east Germany has paired bosses as ornament.

62 The pair of wheels from Arokalya, Szolnok-Doboka (formerly Hungary, now Romania) was again an unassociated find made in the last century (Hampel 1890, pl. LIX, 2a, b, c). They are most elegant and accomplished pieces of bronze-work, four-spoked and with a diameter of *c.* 60 cm to the edge of the felloe-sheathing. The nave and spokes appear to have been cast as a unit, the nave being of double-reel shape, *c.* 21 cm overall, its internal perforation *c.* 6.6 cm diameter, the expanded nave-ends *c.* 10.8 cm diameter. The spokes taper from the nave, with a slight expansion at their junction with the felloe, and have ribbed mouldings, arranged differently on all spokes of the two wheels in a curious and deliberate asymmetry. Each expanded outer end of the spokes has a rivet-hole. The felloe-sheathing, probably of beaten sheet bronze (as at Kemnitz), is formed of eight U-section segments *c.* 3.0 cm wide and deep, each with three transverse rivet-holes, and the wooden

felloe may have projected for 5 cm or so beyond the metal, giving an overall diameter of the wheels of 70 cm or so. A similar combination of bronze and wood in the felloe construction is seen on a more massive scale in the 'Rhône-Rhine' wheels of Late Urnfield date described later in this chapter. The minimum thickness of the spokes is no more than *c.* 2.5 cm, but the riveting of their far ends must presuppose some wooden interior component, even if not for their full length.

63 Segment of bronze felloe-sheathing, early 1st millennium BC, *from Kemnitz, Ostprignitz, Germany*

One additional find may be taken in connection with the Arokalya wheels, the fragment of bronze felloe-sheathing from Kemnitz, Ostprignitz (K. H. Jacob-Friesen 1927, 173). This was an unassociated find and is of 63 beaten bronze, U-shaped in section, 6 cm deep and 2.5 cm wide, with transverse ribbing, two pairs of large bosses, and four rivet-holes near its outer edge. It appears to be one-sixth of a felloe *c.* 46 cm internal diameter, to which its depth and a wooden felloe projecting beyond this could add at least 20 cm, giving a minimum outer diameter 65 cm, with six rather than four spokes.

The foregoing evidence, which with its beaten sheet-bronze technology implies a date within the later, rather than the Earlier Bronze Age, shows an accomplished tradition of fine wheelwright's craft in wood and metal. It also raises the problem which we shall encounter again as to what function the vehicles to which they belonged had in practical terms. The Arokalya and Abos paired wheels certainly suggest two-wheeled chariots of some pretension and opulence, but leave the question open as to whether such vehicles were actually used for ceremonial human transport, or as cult vehicles in a shrine, a function which as we shall see shortly is the likely explanation of a large series of models of varying dimensions in Urnfield contexts, with parallels in, for instance, classical antiquity. The felloes held in the bronze sheathing could have single or multiple segments of bent or worked wood, and that the segmental wooden felloe of 'modern' type was current by Urnfield times is shown by the find of the felloe-segment of a ten-spoked wheel from the Barnstorfer Moor in Lower Saxony, with a radiocarbon date of *c.* 880 bc (1100 BC) (Hayen 1978). This is not of bent wood but is cut from the solid plank, *c.* 4 cm deep, with mortices for two spokes on the inner side and at each end for jointing to its neighbours, The whole wheel would have had a diameter of *c.* 90 cm. In none of these wheels is there evidence for tyres, but these could (on Egyptian analogies) have been of perishable materials such as raw-hide.

The wheels of Arokalya type would have presented a superficial appearance similar to the double-felloe chariot wheels of later second-millennium Egypt, or that shown on the pleasure chariot in the Tiryns wall painting (Littauer & Crouwel 1979, 78, fig. 47; Wiesner 1968, fig. 8). The slight expansion of the spokes at the felloe-junction in Arokalya, and to a more marked degree in Abos, is paralleled in the bronze spoke-sheathings in the Hart wheels of Early Urnfield date already referred to (Müller-Karpe 1956, fig. 7) or in the model bronze *Kesselwagen* from Orăştie, Hunedova, Romania (formerly Szásvárosszék in Hungary). Analogous expansions or spoke-braces appear on some Mycenaean chariot representations (Wiesner 1968, figs 1e; 10b; 11) and on many Greek vase paintings from Geometric times onwards, but as we saw, the thesis of Mycenaean contacts with Central Europe is today hard to maintain.

Still in the Carpathian basin, an interesting group of bronzes, including nave-caps or sheathing, was noted by Moszolics (1956) as distinguished by decoration with openwork triangles in a manner recalling Caucasian bronzework of the later second millennium—as indeed in that associated with the Lchashen vehicle burials already described. The possibility of such contacts in Late Urnfield times is touched on in a subsequent section. Where dating is possible, the pieces in question fall into the Moszolics periods IV and V, approximately equivalent to Bz D and Ha A of the more westerly Urnfields sequence, and they all appear to be of cast rather than beaten bronze. The most informative is that from Tarcal, *c.* 16 cm in 64 length, in the form of a truncated cone with bold mouldings and openwork ornament, covering the outer end of the nave of a four-spoked wheel, *c.* 9 cm in diameter at its open outer end and *c.* 11.5 cm at its inner, where it is broken but preserves arched cusps indicated the positions of the spokes. A comparable piece, from Szilágysomlyó-Perecsény, is complete and shorter (*c.* 12 cm) with an opening *c.* 8.5 cm diameter, openwork ornament and a dentated inner edge ending clear of the spokes. A third, from the Sajóvámos hoard of Opályi (Period IV) date is a shorter cap, for a nave *c.* 10 cm diameter and dentated inner edge, comparable with fittings from miniature vehicles from Swiss Late Bronze Age lakeside sites (G. Jacob-Friesen 1969, fig. 11). Another Hungarian find in the openwork style may be significant here, the bird-headed terminal from the Zsujta, Abauj, hoard with Liptov swords, of Ha A1 date, which Hampel interpreted as a pro- 65 bable terminal mounting of a chariot-pole (Hampel 1890, pl LVII). In all, these finds imply the existence of vehicles, probably chariots, with elaborate metal mountings, in Early Urnfield times.

Additional evidence of such bronze fittings, but not in the openwork style, is afforded by the pair of axle-caps with linchpins from the Komjatna hoard in Slovakia (formerly Komjath in Hungary: Hampel 1890, pls LVI, CXX). As at Zsuyta, this hoard has Liptov swords, and a *Posamentarie* fibula, placing it in Ha A1 (Müller-Karpe 1961, 24). The caps are 'hat-shaped', to fit an axle-end *c.* 6.5 cm diameter, with an external

66 spike and an inner flange, *c.* 11 cm diameter, behind which the nave of the wheel would rotate: the nave-ends of the Arokalya wheels were slightly under this diameter. The linchpin which passes through perforations in the cap is *c.* 14 cm long, with a head formed of three bird protomes, the outer of which has a pendant ring at its base. A fragment of a similar linchpin, rather simpler in execution, comes from an unlocated Hungarian source (Hampel 1890, pl. LVI, 2). Flanged bronze axle-caps of the Komjatna type were not only used in Urn-

75 field times (as in the contemporary Hart-an-der-Alz vehicle) but into the Iron Age on vehicles described in Chapter 5, as for in-

100 stance Wijchen (Ha C) or in Ha D wheels as at Vix. The latter find confirms the inference that their attachment to the end of the axle was not more than a 'knock-on fit', and that they were necessarily removable, with the linchpin, before the wheel could be taken off, for in the dismantled wagon at Vix, the four axle-caps with their linchpins were found, not on the axle-ends, but separately stacked together (Joffroy 1958, 107). As the two Komjatna axle-caps come from a hoard, they could belong equally to a chariot or to two out of the four wheels of a

64 Bronze nave-sheathing, Early Urnfields, from Tarcal, Hungary

65 Bronze pole-tip from Early Urnfields hoard, Zsujta, Hungary

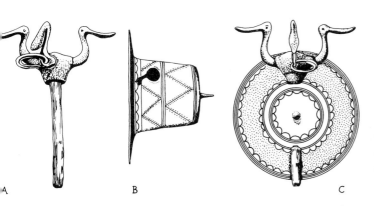

66 Bronze axle-cap and linchpin, from Early Urnfields hoard, Komjatna, Slovakia

A   B   C

113

wagon, though the former may be thought slightly more probable. At all events, the sum of the evidence from the Carpathian region certainly implies vehicles elaborately decorated, and in part constructed of bronze, among which seem likely to be chariots carrying on the tradition of the slightly earlier chariotry explicitly depicted on the Vel'ke Raškovce pot of the Piliny culture, probably of the fourteenth century BC and continuing, in Periods IV and V or B2 D and Ha A1, into the thirteenth and earlier twelfth centuries. This evidence is confirmed by that from the west and north, to which we now may turn.

Bronze models

In dealing with this material it will be convenient to take it in categories rather than combining all forms of evidence in as strict a chronological sequence as the circumstances permit. It is proposed to start with the

67 Bronze cremation urn on wheels (Kesselwagen), *Early Urnfields, from grave at Milavec, Bohemia*

bronze models, basically those carrying vessels of the *Kesselwagen* type, and to follow the earlier of these into south Scandinavia, where their presence must be assumed to presuppose the contemporary existence of full-sized spoked-wheel vehicles, and so take us to the representations of chariots and wagons on the rock carvings. The relationships between the Danubian-Carpathian region and Scandinavia expressed in other bronze types since the Early Bronze Age support such an approach (Lomborg 1959). Of the wheeled bowls the earliest is that found containing a cremation in a grave at Milavec in Bohemia, with a Riegsee type *67* sword of Bz D (Powell 1963, 216; Müller-Karpe 1959, 113; Holste 1953, 28). Moving north we enter the area of the Bronze Age divisions of Montelius, and Thrane, in a review of the earliest Danish beaten bronze vessels, has shown that Bz D must contain not only the final phase of Montelius Period II (II bc) but the early part of Period III as well, though it extends into Ha A1 and A2 (Thrane 1962a). To Period II b, or II bc, belong two important Danish finds, the bronze model wheel from Tobøl near Ribe, and the famous six-wheeled model from Trundholm carrying a horse drawing a gilded 'sun-disc'. Both finds should then date from the thirteenth century BC and be roughly contemporary with Milavec.

The Milavec model (Neustupny *et al.* 960, pl. 50; Powell 1980, fig. 15) shows a four-wheeled vehicle with four-spoked wheels: the felloe is relatively broad and the spokes expand slightly at their junction, and the nave is simple and tubular. The wheel-model from Tobøl is more informative (Thrane 1962b). It was from an inhumation grave and seems to have been lying on the body, probably that of a woman. It is an accomplished casting, 17.7 cm diameter and with incised decoration on both faces. The nave is simple and cylindrical and the felloe relatively broad (one-quarter of the radius), with the decoration in two annular zones, the in-

114

68 Bronze wheeled carriage with horse and gilded disc, Montelius II, from Trundholm, Denmark.

ner decorated, as if to represent a bronze and wood wheel of the type suggested above for Arokalya and Abos, with ornament on the bronze-work. At the four points of the slightly expanded spoke-junctions, arc-shaped motifs again give the impression of decorative bronze attachments in the assumed original. Unless the decoration is purely ornamental, with no reference to a full-sized original, it seems as if we may see in this model an indication of contemporary bronze-sheathed decorated wheels in actual use.

The Trundholm model (Müller 1903; Sandars 1968, 185; Coles & Harding 1979, *68* pl. 12a, etc.) cannot, of course, be taken literally as the representation of a six-wheeled wagon; as at Dupljaja with its three wheels, the object was to make the whole model mobile. The wheels are represented with narrow felloes, very thin spokes, and a long nave with expanded ends in the Arokalya manner (Thrane 1962b, fig. 18). With this should be taken a find from Tågaborg, Hälsingborg, Sweden, where a hoard of Period II consisted of spear-heads, pal-

staves, and a pair of model horses with inlaid amber eyes, all now surviving, and the 'now lost parts of a small car and of a round plate of bronze' (Montelius 1922, no. 980; Althin 1945, 190, fig. 101). This would appear to have represented something similar to Trundholm, but with paired rather than single horse draught. Sandars looked to the middle Danube or Czechslovakia for the origin of the Trundholm four-spoked wheels, and compared the star-shaped treatment of the horse's eyes with the Kličevac figure's breast-stars, and, as we have seen, such comparisons are reinforced by the Čaka and Ducové corselets as well as by the evidence of wheels themselves, and the bowl-on-wheels bronze models. Following on the Milavec example, further finds include those of Peckatel in Mecklenburg, Skallerup in Zealand (with bird-ornaments) and the carriage only from Hedeskoga, Ystad, in Sweden, all of Period III and so within the Early Urnfield phase, with simple four-spoked, narrow-felloe wheels in the general Trundholm manner, perhaps now a standard convention (Thrane 1962a). These, and

115

other examples, are further discussed in a later section where the larger question of the nature of model vehicles in Urnfield times is considered, but it is here convenient to take account of depictions on pottery and to pursue the Scandinavian evidence as amplified by rock carvings.

A single representation of a vehicle on a Late Urnfield vessel has been published from a burial in the smaller of two barrows at Sublaines, near Tours in western France. It forms a panel in a frieze of otherwise geometrical patterns on the shoulder of a handled urn, executed in polychrome and tinfoil inlay, and shows a vehicle with four four-spoked wheels drawn by a pair of stylized horses back-to-back. The wagon is schematically shown with a rectangular body on a Y-shaped perch undercarriage, and a draught-pole with bifurcated articulation in the manner of the rock carvings to be described below. The ends of the straight bar representing the yoke are very slightly expanded with a central dot as if to indicate some terminal feature.

The urn, assigned to the French Bronze Final III, *c.* 800–750 BC, is a western outlier of a class of vessels with tin inlay and schematic figural ornament of human figures and occasionally horses, with a distribution in eastern and southern France, and in Switzerland (Sandars 1957, 312–20, map XIII). The technique of tin inlay is probably of Swiss origin, and occurs again on contemporary Etruscan pottery (Stjernquist 1958). Two further, and roughly contemporary, vehicle depictions are recorded from France, a schematic four-wheeled wagon on a sherd from Camp Redon, Lansargues, Hérault and a two-wheeled chariot on another from Morras-en-Valloire. Drôme (Cordier 1975).

Rock carvings

It has already been noted, at the beginning of this chapter, how the horse preponderates in the representations of draught-animals in the Scandinavian depictions of vehicles with spoked wheels, and the possibility of the augmentation of imported stock from local wild populations has earlier been touched on. The great majority of vehicles are of two-wheeled chariots, as we may continue to call them, of which over thirty have been published from Sweden and Norway.

69 Horse-drawn wagon, in tin inlay on burial urn, Late Urnfields, from Sublaines, France

In the past, 'Mycenaean' origins have been sought or hinted at (e.g. Marstrander 1963, 451), but they belong to a tradition deriving from Early Bronze Age origins in east-central Europe, as exemplified by the Vel'ke Raškovce depictions, which as we have seen do not necessarily demand explanation in these terms. The most famous of these chariot representations is not on a natural rock surface, but on one of the wall-slabs of the unique Swedish stone chambered tomb under a huge cairn of Kivik in Skåne, as part of a series of eight decorated stones depicting cult scenes and objects (Grinsell 1942; Althin 1945, 68, pls 74–77; Hasslerot & Ohlmarks 1966, 252). The tomb was broken into and cleared out in 1748, and its date is uncertain, but on the grounds of the representations of large ceremonial bronze axes on one of the stones it was assigned to Late Period III by Althin. A subsequent find of such an axe, previously only known from stray finds, in a context of Period V, has suggested a wider range of date, within the Ha A–B phases from the eleventh to the eighth centuries BC (Halbert 1955). The chariot representation is on Stone 7, accom-

70 *Rock carvings of horse-drawn chariots, early 1st millennium* BC, *Top, Begby, Norway; above, Frånnarp, Sweden*

panied by two further horses and enigmatic groups of human figures, and is exceptional in showing a profile rather than the plan view normal to rock carvings. The wheels are shown as four-spoked, and the body is reduced to a minimal arched outline on a draught-pole which curves upward to a

117

straight yoke joining two draught-horses. The driver stands in the chariot, holding reins or guide-cords which run as single lines to the noses of the horses. Four horses are shown on another stone (3) and crossed discs which may represent wheels on Stones 4 and 6. While the complex cult significance of the whole range of eight engraved wall-slabs will always elude us, horses and charioteering clearly played an important part among the themes involved. The spirited profile chariot scene has a single parallel, that incised on a detached stone slab, probably from a massive cist grave, found in the early nineteenth century at Villfara, Skåne, in southern Sweden. This is closely allied to Kivik in style and in details, and the stone has also cup-marks and ship depictions. Althin (1945, 96) dismissed the chariot scene (and at least one of the ships) as a complete forgery, copied from the Kivik stone between its publication in 1780 and that of the Villfara slab in 1830, but this view has not received unanimous support, and we may provisionally regard Villfara as a second example of the 'Kivik style' of chariot representation. In neither are the horses represented as stallions.

Of the remaining representations of chariots, 'pecked' on natural rock surfaces in the technique common to all ancient rock art, the most remarkable is a group of seventeen at Frännarp, again in Skåne, (but over 60 km north of Kivik and the main south-eastern concentration) accompanied by fifteen detached wheel-symbols. Here, as with the remaining rock carvings under review, the artistic conventions are those already discussed in the instance of the Armenian vehicle carvings in Chapter 3. (Althin 1945, 102, pls 69–73; Hasselrot & Ohlmarks 1966, 238–43.) The carvings are at a large scale (up to 0.70 m overall) and the most detailed chariot depictions in Western European rock art. The wheels are uniformly four-spoked, and the bodies either minimal or large, D- or omega-shaped in plan, with the

axle set either at the rear or at mid-point in about equal proportions; the draught-pole continues through, and so under the body to join the axle. At least eleven chariots are shown with a pair of entire stallions, indifferently back-to-back or feet-to-feet, others are without horses, or flaked away and uncertain. In five examples the draught-pole is shown flanked by lines indicating reins or guide-cords, and in one of these the flanking lines are doubled, clearly denoting reins. The yokes appear as simple bars joining the horses' necks except in two unharnessed chariots, where inexplicable symmetrical waving 'ribbons' spring from the pole-tip, possibly indicating elaborate curved yokes, and some of the chariot bodies have internal patterning difficult to explain in structural terms; one pair of horses has a double projection at the withers, well behind the yoke on their necks.

The remaining Scandinavian chariots in the rock art do not contribute greatly to our knowledge of the technological details of the vehicles, though they indicate a widespread knowledge of the chariot as a cult symbol, and presumably of its actual use in the same area. At Simris not far south of Kivik, among a large group of carvings probably of relatively early date in the Later Bronze Age sequence, two poorly preserved little two-wheeled vehicles are sketchily shown (Althin 1945, pl. 8), and the same may be said of that at Hjulatorp in Småland (Althin 1945, 21). In Bohuslän, one of the great areas of prehistoric rock art of which no complete corpus has been published, a dozen or so chariot representations have been recorded (Baltzer 1881, pls 5, 6, 47, 48; Berg 1935, pl. XV; Marstrander 1963, figs 45, 46; Fredsjö 1971, 241, 245; 1975, 396, 334, 364). Wheels are shown as four-spoked or occasionally as open circles; the chariot bodies as small, D-shaped or circular, and usually crossed by the draught-pole, and the draught pair, when shown, are invariably roughly depicted horses. A couple from

Østergotland (Anati 1960, fig. 15) show no recognizable bodies, but drivers with reins or guide-cords leading to the horses' noses. In Østfold in Norway, south-east of Oslo, eight or nine chariots have been published by Marstrander, only one of which shows spoked wheels depicted as such, and all are very schematic, with small circular bodies crossed by the draught-pole. The chariots at Begby in Østfold are all depicted with un-spoked wheels (Hagen 1967, pl. 38; Littauer 1977, fig. 24). The isolated wheels which appear to be of cross-bar type from Solbjerg and Nörrköping have already been mentioned in Chapter 3 (Marstrander 1963, 282; Hayen, 1973, fig. 13).

Three chariot representations among the Val Camonica group in north Italy may be considered here (Anati 1960; Berg-Oster-rieth 1972). Two, on Rock 94 at Naquane, have spoked wheels with the axle set well forward under a narrow D-shaped body; that at Campanine no spokes and the axle forms the base of a triangular body. Anati made much of the 'Mycenaean' origin of these chariots, and derived them direct from Greece to north Italy, with a subsequent diffusion from Val Camonica west to Iberia and north to Scandinavia (1960, map, fig. 18) but, as we have seen, such a derivation is now highly improbable, and the chronological position of the Italian carvings need not differ significantly from those of northern Europe just discussed.

The half-dozen or so representations of four-wheeled vehicles which have been published from the Scandinavian rock art series, and rather more from Val Camonica, raise problems of dating and of the possibility of making significant and precise chronological distinctions between a 'Late Bronze Age' (up to *c.* 700 BC) and an 'Early Iron Age' from that date onwards, in terms of vehicle typology. It is clear that many of the north Italian wagon depictions must fall into the second chronological category, and those of Scandinavia are harder to date or to separate

from the chariots or carts just described. It is proposed here to deal with the features they share relatively briefly, and to return to them when in the next chapter we deal with similar constructional details shown on vehicle depictions certainly of later date, within the seventh and sixth centuries BC. Of the Scandinavian wagons as we saw, two are ox-drawn, one (Bottna, Bohuslän: Fredsjö 1975, 364) has a pair of horses and one at Långön the rear end of another probable pair (Berg 1935, pl. XXVII, 2), and the others, at Svenneby in Bohuslän and Solb-jerg in Østfold, are unharnessed (Fredsjö 1971, 232; Marstrander 1963, fig. 44). The anomalous oval outline between four discs at Vårangstad may not be a vehicle at all (Fredsjö 1975, 332). Excluding this last all the rest, including the ox-drawn wagon from Rished, Askum (Berg 1935, pl. XXVII, 1), show variations on the same pictorial convention and constructional feature, in which the undercarriage of the vehicle rather than its superstructure is shown, and the wagon is framed on a Y-shaped perch joining rear to front axle, which also carries a Y-shaped draught-pole articulated for vertical movement, producing a characteristic 'double Y' plan view. Wheels are shown as four-spoked. The Val Camonica wagons are more elongated, show the Y-perch undercarriage in greater detail, wheels with up to six spokes, and as a whole are reasonably dated as at earliest around the eighth century BC, continuing into later centuries, conventionally within the iron-using phases of Ha C and D (Berg-Osterrieth 1972). At this time the Y-perch construction is well attested by other depictions on pottery and elsewhere, and by the last centuries BC the famous Dejbjerg vehicles show it in actuality: it incidentally survived in English farm wagons until the beginning of this century (Jenkins 1961; Arnold 1974). The whole of this prehistoric wagon evidence is therefore better discussed in the next chapter.

Wheeled bowls and cult models

The remaining evidence for wheeled vehicles in the Later Bronze Age consists of a series of bronze models and of wooden wagons and chariots of which bronze fittings alone have survived, either as chance finds or as the less destructable elements escaping to some degree the burning involved in the cremation rite in burials where not merely the human cremated bones were interred, but the debris of the funeral pyre, distinguished therefore by Powell as 'pyre-graves' (1963) and characteristic of the beginnings of the Urnfield period at the end of Bz D and early Ha A1. A problem arises in the use of the term 'model'. Clearly the *Kesselwagen* carrying bronze vessels on wheels are models in the sense that they are miniatures, as are the *Deichselwagen* with birds on wheels at the end of a tubular socket, but neither are models in the sense of reduced versions of the full-sized practical wagons or carts current at the time of their manufacture, and the information they convey on vehicle technology is confined to wheel types (if indeed these are not conventions hallowed by what may well be religious traditions). Difficulties arise with the metal fragments from the pyre graves or chance finds: do these represent 'real' full-size vehicles (as seems to have been the case at Hart and Mengen, shortly to be described) or do they come from 'models' in varying degrees of diminution, especially associated with the funeral rite? At the very end of our period in the eighth century BC comes a well-known series of large metal-sheathed wheels up to an original diameter of *c*. 60 cm, four of which were found in one instance with a large bronze situla and are reasonably assumed to have carried it on a four-wheeled platform and so to have constituted a very large *Kesselwagen;* so large indeed that it could have been at least 2 m long and proportionately wide, a respectable size for a functional wagon. And

71 Bronze Kesselwagen, *Early Urnfields, from grave at Acholshausen, Germany*

72 Bronze vessel with bird protomes on wheels, Late Urnfields, Szászvárosszék, now Orastie, Romania

73 Bronze bird carriage on wheels, Late Urnfields, from Burg-in-Spreewald, Germany

120

throughout the prehistoric vehicle burials we have already described, and the increasingly numerous examples yet to be discussed in the Iron Age, there runs an element of doubt as to whether we are seeing everyday vehicles buried with the dead, or hearses made for funerary purposes.

To return to the small bronze bowls-on-wheels of the *Kesselwagen* class, we have already instanced, from Bz D into Ha A1, those from Milavec in Bohemia, Peckatol in Mecklenburg, Skallerup in Denmark, and a bowl-carriage from Ystad in Sweden, all of bronze, and a pottery counterpart from Kánya in Hungary, to which may be added another rough pottery version from Dergischow in Brandenburg (Forrer 1932, fig. 14, 5). At the very end of our period is the extraordinary little bronze model, but with iron axles, denoting its late date, from Orăştie, Romania, better known from its originally Hungarian place-name of Szásvárosszék (Hampel 1890, pl. LVIII, 2). A recent *71* find, from a richly furnished grave at Acholshausen in south Germany, of the Ha A2–B1 transition, has been made the subject of a penetrating study of the whole group (Pescheck 1972) which illuminates some of the problems outlined in the preceding paragraph. In the first place, where associations are known, the models all come from male graves, in several instances with swords denoting warriors (Milavec, Peckatel, Acholshausen). Further, at Skallerup and Acholshausen the bowl carriage is adorned with two pairs of bird protomes and at Orăş- *72* tie these run riot, to a total of twelve; two pairs on the carriage, four on the bowl and four on its cover (Hampel 1890, pl. LVIII, 2; Angeli *et al.* 1980, no. 3. 30). A direct link is thus established with the water-bird cult of Urnfield, and Iron Age times we have already encountered associated with vehicle models, as well as with the *Deichselwagen* models, eight finds from six sites and all unassociated come from a restricted area birds on wheels. These extraordinary bronze

models, eight finds from six sites and all unassociated, come from a restricted area roughly within a triangle Berlin–Dresden–Wrocław, and consist of a tubular socket as if for a small wooden pole or rod, attached to an axle carrying a pair of four-spoked wheels, or three in line, and numerous models of what appear to be water-birds, though some have a pair of bull's horns (Forrer 1932, figs 17, 29; G. Jacob-Friesen 1968 with map, fig. 8; Coles & Harding *73* 1979, pl. 16b). They are usually assigned to a chronological position in Ha B, probably in its last phase (B3: eighth century BC) and must denote some local cult, and in some sense be related, so far as shared symbolism goes, to the Italian 'bull-birds' on wheels, going back to the eighth-ninth century at Tarquinia, or that from Sokolac (Glasinac) in Bosnia, with iron axles and probably seventh-fifth century BC (Woytowitsch 1978, 62–65; Seewald 1939). The *Kesselwagen* has since the last century been compared with

74 Coin of the city of Cranon in Thessaly showing ritual vessel on wheeled stand, 3rd century BC

the sacred wheeled vessel of bronze in the temple of the city of Cranon in Thessaly which was associated with a pair of holy ravens and was adopted as the city's emblem, depicted on its coins (the vehicle, incidentally, with crossbar wheels) and described in the third century BC by Antigonus *74* of Carystus as being used in a rain-making ceremony in times of drought: Antigonus was himself a bronze-worker and one wonders whether his interest was partly that of the craftsman. Pescheck, commenting on this and the presence of the wheeled bowl models in men's graves, has hinted at chieftain-priests as a possibility (Forrer 1932, 106; Pescheck 1972). As to rain-making ceremonies, these may be thought more appropriate in Thessaly than in Late Bronze Age continental Europe at the time when the climatic deterioration marking the transition from Sub-Boreal to Sub-Atlantic conditions was, in fact, setting in. So far as vehicle technology is concerned, the *Kesselwagen* and *Deichselwagen* series tell us little except that in the models the felloes of the wheels of the first group are shown as narrow and in the latter as broad, a feature shared by the Italian bull-birds. If not wholly artistic convention, this may hint at the co-existence of narrow bentwood felloes with deeper plank-cut forms; an inherently likely situation.

Pyre graves and related finds

From the ambiguities of rock carvings and bronze cult models we may turn to the more direct, if not always simple, evidence provided by the bronze fragments of wheel-sheathing and body decoration from vehi-

cles otherwise of wood and consigned to the funeral pyre, and later gathered together for burial in the distinctive manner which led Powell to distinguish them as 'pyre-graves'. Of these the most important is that at Hart-an-der-Alz in Bavaria (Müller-Karpe 1956), a richly furnished 'chieftain's grave' like Acholshausen, with a sword and the set of bronze vessels—situla, cup and strainer—which may be associated with the serving of *75* wine (cf. Piggott 1959). The grave is firmly dated to Ha A1, twelfth or eleventh century BC in the Müller-Karpe or Sandars absolute chronologies already referred to. The bronze fragments that could plausibly be attributed to a vehicle, partly melted by heat, include several tubular fitments and studs, some with 'duck' figures or protomes, presumably body decoration, and other pieces which can be directly related to the wheels and axles. Of these the diagnostic fittings are three axle-caps with linchpins, many fragments of cast ribbed nave-sheathing, and seven of a presumably original eight tubular coverings, with expanded ends, for the outer parts of the spokes: we may therefore infer a vehicle with four four-spoked wheels. The axle-caps are recognizable versions of those in the contemporary *66* Komjatna hoard already described; the axle-bush itself 5.6 cm diameter, and the flange 12.5 cm overall, with simple linchpins with moulded disc heads. The ribbed sheathing indicates a nave in the form of a truncated bicone, *c.* 25 cm in overall length and *c.* 10 cm diameter at the ends. The outer sheathing is cusped on its inner edge to accommodate the spoke-mortices and held with conical ribbed studs between them. The spoke-sheaths, 9.6 cm long and oval in cross-section, have terminal expansions recalling those shown in the Abos wheels and, in model form, on the Orǎştie model. *72* Müller-Karpe's reconstruction of the wheel (1956, fig. 7), with short spokes sheathed for half their length, and a felloe some 10 cm broad, for a wheel about 0.75 m outer dia-

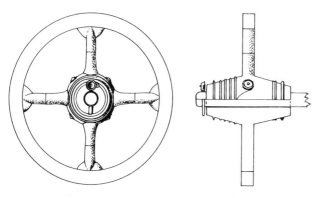

75 Reconstruction of wheel with bronze nave and spoke sheathing, axle-cap and linchpin, from Early Urnfields wagon burial, Hart-an-der-Alz, Germany

meter, seems indefensible in the light of the evidence from Abos, Arokalya, and the models, and longer spokes with a narrow felloe of about the same diameter would be preferable. Nothing survived at Hart to indicate felloe-sheathing or tyres. The inferred size and proportions of the wheels show us that we are dealing with a full-sized vehicle, not a model, but we are left uncertain as to the social or ritual function of such an elaborately constructed and decorated carriage, hearse or processional car. It is here that a point made earlier on in respect of carts and chariots bears repetition. In English usage 'wagon' shares the overtones of farm-yard and harvest-field with 'cart', and is hardly appropriate to a prestige vehicle of limited functions which, however uncertain, must have been in a context of an aristocratic funeral ritual comparable with those depicted in Greek Geometric vase-paintings, and as such, 'carriage' for four-wheeled and 'chariot' for two-wheeled ceremonial vehicles seems preferable usage.

The Hart burial has partial counterparts in a number of broadly contemporary Bz D and Ha A1 graves in Switzerland and south Germany which contain fragments of bronze attributable to full-size or model vehicle fittings, and in some instances bronze horse-bits; other graves have bits alone (Müller-Karpe 1950; Drack 1960; Schiek 1962; Thrane 1963; G. Jacob-Friesen 1969; Balkwill 1973). Although of the greatest interest and importance in the general history of paired-horse draught from Early Urnfield times in Western as well as Eastern Europe, and for the understanding of the further developments in the Early Iron Age of Ha C and D, the surviving metal fragments from these finds are disappointingly uninformative for the technology of vehicle construction over this period. In the first place, with the exception of Hart, already described, and Mengen, with bronze nave-sheathing (Schiek 1962; Paret 1935, pl. XI, 11–16), all the bronze fitments

in question seem to come from relatively small model vehicles, presumably mainly of wood, and many of them can only be interpreted as decorations on body-work of quite unknown form. Information on wheels is confined to a few finds of miniature axle-caps with counterparts in Swiss 'lake-dwelling' material, and in no instance do we know whether these represent two- or four-wheeled vehicles. Socketed protomes of horned birds (as in the Danish finds of Skjerne and Egemose) have been interpreted by Jacob-Friesen as draught-pole tips with Central European prototypes such as that in the Zsujta hoard of Ha A1 already 65 mentioned, but here, though chariot rather than carriage models might be hinted at, the evidence is wholly ambiguous. Among the types of bronze body fitments he isolates are ribbed tubes about 2 cm in diameter which seem to have covered wooden cores; some are curved, and their slightness suggests that in models at least slender bent-wood framing was employed. At Egemose there were fragments which he interpreted in the light of two complete examples from the Swiss Late Bronze Age settlements of Autavaux and Chervoux as hand-grips, again with attachments at right angles which would only take slender woodwork. If these are seen as hand-holds to use when stepping into a vehicle, this can hardly be other than a chariot entered from the open rear end. Though necessarily inconclusive the evidence is at least suggestive and in accord with that for two-wheeled light vehicles of the chariot type in Later Bronze Age Europe and again, as we shall see, in the Ha C phase which follows it. In parenthesis it is worth while observing that the bronze-decorated vehicles implied by the foregoing evidence may well have been further ornamented in carved wood, leather or colour, as a large range of pigments from black and white to brown, red, yellow, green and blue could be available from earth and mineral sources, including copper salts as a by-pro-

123

duct of metalworking. In the Ha D burial at Vix, remains of red and bright blue colouring matter were found in the area of the dismantled funerary wagon, either from its textile or leather covering, or from the vehicle itself (Joffroy 1958, 113). Vegetable dyes had been used for patterned fabrics since Neolithic times (cf. Vogt 1947).

Cult vehicles in the Final Bronze Age: the Rhône-Rhine wheels

A remarkably uniform group of heavily metal-sheathed wheels belonging to the end of Urnfield times, in Ha B3, have long been known as the 'Stade' type, from a north German find-spot of a set of four such wheels (K. H. Jacob-Friesen 1927). The distribution of a dozen find-spots of such wheels and fragments shows, however, that Stade is a far northern outlier of a pattern centering on the Rivers Rhône and Rhine, from the Mediterranean to the Main, and it is suggested that a name should now replace 'Stade' to indicate this fact. To anticipate, the wheels can hardly be other than the products of a single workshop over a short period of time, and with southern technological contacts (Chapotat 1963; Hundt & Ankner 1969; G. Jacob-Friesen 1969). In two instances four wheels were found together, and in another two a pair, and three finds were isolated, while fragments of the cast bronze naves characteristic of the type were present as scrap metal in hoards of Ha B3, within the eighth century BC. With one of the finds of four wheels (Côte-St André, Isère) were a bronze bowl and a huge situla, over 64 cm high and 80 litres in capacity, and this led Déchelette and others following him to interpret this, and the Stade find, also of four wheels, as the remains of large cult vehicles. The situla is of Ha C rather than B type, but if we are dealing with cult vehicles such as that of Cranon already referred to, a difference of a century or so between the vehicles and the vessel it eventually carried need

76

76 *Bronze sheathed wheel, one of four found together, Late Urnfields, Stade, Germany*

occasion no surprise, and we can accept the workshop which produced the Rhône-Rhine wheels as functioning in Ha B.

While closely allied in design and craftsmanship, the wheels show differences of detail, notably in the number of spokes: four at Cortaillod and Stade, five at Hassloch, Nîmes and Fa, and six at Côte-St André. They are particularly fine pieces of bronze-working, both in their initial casting and subsequent finishing, and we will return to some of the technological details later. They all average about 50–60 cm in diameter, and consist of a nave cast as a massive ribbed bronze tube 35–40 cm long, with tubular spokes, sometimes with mouldings, and a U-section felloe about 5–6 cm deep, taking a segmental wooden plank felloe, fragments of which (of oak) survived at Stade and Côte-St André. The wheels from the latter find have cast bronze rings on the inner edge of the felloe between the spokes, presumably for rattling attachments. This bronze and wood construction recalls the Arokalya and Abos wheels already described and tentatively given an Early Urnfields date. At Stade the overlapping of two wheels as they were eventually found left traces of differential patina which showed that the wooden part of the felloe (in four segments) projected *c.* 5 cm beyond the bronze sheathing, giving a total wheel diameter of *c.* 68 cm, and two large-headed nails from the same find could have been part of studding in lieu of a tyre, but the assumption that close nail-studding was normal (as in the reconstruction of the Cortaillod wheel in Drack 1959, pl. 17, 8; Kossack 1971, fig. 30, 1; Sauter 1976, fig. 39) seems unjustified. At Côte-St André, Chapotat estimated a projection of the wood of not more than 3 cm, and here the surviving fragments, of planks from a mature oak-tree, have the surfaces roughened by axe-cuts to afford better adhesion for whatever form of glue was used to fix them to the bronze U-sheathing. The function of these four-wheeled finds as cult vehicles carrying some form of a recipient for liquid—*Kesselwagen* on a large scale—seems probable enough in view of the associated situla and bowl at Côte-St André, and the unattached vessel carried could have been of any type and, as has been suggested, not necessarily strictly contemporary with its wheeled platform. The two instances of a pair of wheels having been found together (Hassloch in the Palatinate and Fa in Aude) are, again like Arokalya and Abos, suggestive of the co-existence of large ritual or processional chariots drawing on the same workshop for fine wheels. If the East European bronze wheels do, in fact, belong to the Early Urnfield period to which we have tentatively assigned them, we have an interesting technological continuity from about the twelfth to the eighth century BC; alternatively they might be thought of as easterly counterparts of the Rhône-Rhine series and of equivalent date. A link between the latter and the wheels

from Hart, unquestionably Ha A1, has been pointed out by Hundt, as the inner mouldings of the wholly bronze naves on some (e.g. Stade and Hassloch) echo in their wavy outlines those of the partial cast bronze sheathing on the wooden naves of Hart.

Hundt also drew attention to the bronze-working technique of repairing flaws in the original castings, during the final processes of cleaning-up and finishing by the application of bronze patches by the 'casting-on' technique, observable at Stade, Hassloch and Côte-St André. Together with the exceptionally accomplished technology of the wheels as a whole, he pointed out the finishing of such a cast-on repair could hardly have been possible without the use of hard iron or steel tools such as chisels and files, and for such features looked to contact with the already iron-using technology of the Mediterranean world, via the Rhône. While this may indeed have been the case, it is worth while remembering that while a full iron-using economy in Europe north of the Alps only emerges in seventh-century Ha C, there was iron-working in East Europe at such sites as Babadag in at least the eighth century BC (Sandars 1971, 23; Powell 1976: Pleiner 1980, 380) and iron objects appearing in Western and Northern Europe to a recently recorded total of some sixty finds in Ha B3 and its chronological equivalent M V period (Kimmig 1964a; Piggott 1964; Snodgrass 1965). An unpublished technical study of the engraving of a series of decorated objects of this latter date has shown that steel tools must have been used for the purpose (R. Savage, pers. comm.). There is then a possibility that the tools requisite for finishing the Rhône-Rhine wheels were locally available, or indeed that other forms of abrasive were used.

Horse harness

In the previous chapter we discussed the bone and antler components of horse harness, mainly in the form of the cheek-pieces or side-bars of 'soft' bits, that could be assigned to the Earlier Bronze Age. The most important type, frequent in the Carpathian Basin, was that made from the curved browtine of a red deer antler, and it was pointed out that this form continued in use in Later Bronze Age times in Central Europe, Switzerland and probably north Italy. The Swiss evidence is of the greatest importance to our consideration of the Urnfield material from B2 D to Ha B, for it is in this context that we first encounter bits made not of organic and potentially perishable materials, but of bronze. Recent studies over the last decade, from Thrane (1963) to Balkwill (1973), have clarified this phenomenon, which is crucial not only in itself but in its antecedent relationship to the problems presented by the horse equipment of Ha C in the seventh century BC (cf. Kossack 1953b). The replacement of bits made of rope, wood, leather and antler by those of metal, in the context of a flourishing bronze industry, comes in itself as no surprise, and this fusion of two technologies, the harnessed horse and developed bronze-working, is usually seen as having taken place in the ancient Near East, perhaps in western Persia or northern Mesopotamia, in the fifteenth and fourteenth centuries BC (Moorey 1971, 104–5; Littauer & Crouwel 1979, 89).

In Later Bronze Age Western Europe we must be seeing the same phenomenon reflected in the Swiss evidence, with the first bronze horse-bits occurring in Bz D/Ha 1 contexts (e.g. Mengen), on Müller-Karpe's chronology about 1200 BC, with curved cheek-pieces echoing antler prototypes.

In Balkwill's classification, the earliest bits have a single-piece smooth or twisted mouthpiece—a bar canon. (For technical terminology Littauer 1969; Moorey 1971, 105; Littauer & Crouwel 1979, 3–7). This type persists but in Ha B, and perhaps towards its end, a new type of mouth-piece appears, formed of two smooth or twisted canons; cheek-pieces are straight or curved but no longer betray antler prototypes. At this point 'it becomes impossible to discuss western Urnfield horse gear without reference to eastern influences' (Balkwill 1973, 441) and we look towards the Carpathians and Romania, where jointed bits with two canons seem already current in Early Urnfield times, and by Ha C these easterly connections can be extended firther, to the Pontic region and Transcaucasia (Terenozhkin 1980), and were at one time used as evidence for 'Thraco-Cimmerian horsemen' invading Europe from the steppe. The bit with two jointed canons of twisted wire or thin rod bronze had been developed in the Near East by the later second millennium BC (Littauer & Crouwel 1979, 88, 119), where, as in prehistoric Europe, it became a predominant type and, while independent translation from organic to metal bits in two unconnected areas is a reasonable assumption, the appearance of the two-canon bit in Late Ha B, with its East European counterparts, must be considered in connection with the ancient East in its widest sense. The use of jointed wire canons introduced a new modification of control: they 'are more severe than the plain bars, and the joint in the centre produces a "nutcracker" effect on the corners of the horse's mouth when both reins are pulled. If the canons are very

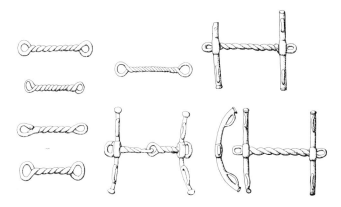

77 *Bronze bits, Earlier Urnfields, from sites in Switzerland*

long—as they often are—this joint could press painfully against the roof of the horse's mouth' (Littauer & Crouwel 1979, 119). Oriental bits of this type were sporadically reaching the west in the second half of the second millennium, as the well-known bronze example from a Late Mycenaean hoard on the Mycenae acropolis shows (Potratz 1966, fig. 45g; Müller-Karpe 1962, fig. 36), and also in Greece a pair of iron bits of the same type were in the 'warrior's grave' XXVII of the Athenian Agora of *c.* 900 BC (Snodgrass 1971a, fig. 84; Müller-Karpe 1962, fig. 28). In Italy one–and two–canon bronze bits are abundant as part of the orientalizing phenomenon of the late eighth-early seventh century BC (Von Hase 1969), a separate episode discussed later in this chapter, but Balkwill is wisely cautious in accepting a Villanovan origin for the Swiss developments (1973, 439–40). By Ha C, as we saw above, the Pontic region suggests itself as a geographically intermediate region linking Europe and the Transcaucasian Orient (with iron technology perhaps also involved), and here we may recall that Moszolics looked for Caucasian connections in the openwork bronze-work of the nave-sheathings such as Tarcal, of Early Urnfield date. If the oriental and European developments in two-canon horse-bits are in some way connected, the possibility of contacts both to the north and to the south of the Black Sea must be given equal weight.

Balkwill made an interesting and important point in relating the width of the mouthpiece of the bit to the size of the horse it controlled. For the European Urnfield series under consideration the range of width between the side-bars or side-rings was *c.* 8–10 cm; for a modern horse the standard size is 15.3 cm. For the Italian bits catalogued by Von Hase the range is *c.* 12–16 cm, and we can add, as a conveniently accessible Oriental group, *c.* 11.0–15.5 cm for fifteen Luristan bits. (Moorey 1971, Chap. III), and *c.* 8–10 cm for eight from the

eighth-century cemetery B at Tepe Sialk (Ghirshman 1939), suggesting significant variations in ancient Persia itself. Later examples become enormous, and must have projected well beyond the horses' lips, exercizing a powerfully cruel lever action: 18.4 to 30.4 cm in eighth-seventh century BC Cyprus (Littauer 1969) and *c.* 20 cm in the early ninth century Athenian Agora grave. At all events, the continental European examples of *c.* 1200–700 BC denote the smaller 'western' rather than the 'eastern' type of horse distinguished by Bökönyi, with a withers height of *c.* 126 cm (1968, 36).

One further class of bronze object must be mentioned here, that of the domed discs with central loop attachment known as 'phalerae' (Latin, from Greek: disc-shaped ornaments and specifically those on horse—harness, as in Aeneid V. 310—*primus equum phaleris insignem victor habeto.*) Such phalerae are a type widespread in Urnfield Europe, with a marked concentration in the Swiss lakeside settlements and found occasionally with horse gear (as at Mengen, or a possible variant at Hart, in A1). In Ha C, as we shall see in Chapter 5, this association is frequent (Von Merhart 1956). Though usually taken to be a part of harness decoration it should not be assumed that all phalerae inevitably performed this function, and shield-bosses, belt ornaments or even cymbals have been proffered as possible alternatives in certain instances (Snodgrass 1973), while a further possibility might be as the metal elements of body-armour of the *kardiophylakes* type (cf. Stary 1979a, pl. 22), with Etruscan counterparts.

The Mediterranean and the Near East: Greece and Italy

Chronologically and conventionally, as we saw at the beginning of the chapter, the Later Bronze Age of Europe north of the Alps is brought to an end by the Hallstatt B3 phase of Müller-Karpe with a terminal date of *c.* 700 BC, a fully iron-using technology characterizing the successive stage of Ha C, thereafter to remain dominant throughout later prehistory. The Aegean world, with mainland Greece, had moved into an 'Iron Age', however, in the eleventh century (Snodgrass 1980); Italy with what was becoming the Etruscan world, by the ninth. These metallurgical factors are a part only of a general pattern of cultural evolution which rapidly led to an increasing dichotomy between the literate civilizations of Greece and Italy and the non-literate, and so prehistoric, communities of the European hinterland with which our enquiry into vehicle technology is basically concerned. The Mediterranean and Aegean present their own problems demanding individual treatment, some of great complexity, but cannot wholly be ignored here, especially as a background to the understanding of the Iberian Late Bronze Age evidence for wheeled vehicles. Two monographs on the Italian evidence are available, that of Woytowitsch on the Bronze and Iron Age vehicles themselves (1978) and Von Hase on the horse-bits (1969); Crouwel's study of Aegean Bronze Age chariots is awaited (cf. Crouwel 1978) while the post-Mycenaean evidence from the Greek Iron Age is inextricably bound up with the much debated question of the status of chariot warfare in Homer, and its tangible archaeology is exiguous apart from vase paintings, from Geometric times onwards (Snodgrass 1971; Greenhalgh 1973). It must, however, be referred to in passing, and again in the chapter following.

The critical factor which engages us is that of mercantile and cultural contacts between the Mediterranean and the Near East often referred to as an 'orientalizing' phase. It owes its origin to a series of historical and political events whereby trade could be established between the east Mediterranean coast and oriental civilizations whose centres of power lay inland, to the Zagros and the Caucasus, but who successively extended their territories westwards to what is now north Syria. In the first quarter of the eighth century BC this was achieved by the Transcaucasian kingdom of Urartu under Argishti I and consolidated by the mid-century by Sarduri II; in 743 Urartian rule over this territory was replaced by that of Assyria under Tiglath-pileser III, and by about 700 had been extended beyond Syria to Cilicia, Palestine and Cyprus. The way was open to tap the resources of trade-routes penetrating deeply into Western Asia by Levantine entrepreneurs, of whom the Phoenicians were the most successful in extending seaborne trade and colonization westwards in Mediterranean waters, while the Greeks were establishing Syrian emporia such as Al Mina by 800 BC if not slightly before. From the late eighth century Greece was acquiring pieces of oriental craftsmanship, or copies, and adopting stylistic traits from a wide range of sources, which have been much discussed (e.g. Boardman 1980, Chap. 3), but here we are concerned with the question as to whether oriental influence can be invoked in the instance of wheeled vehicles and more particularly by the chariot. This must further be discussed in the context of Iron Age Europe as a whole in a subsequent chapter, but one point can be made here.

In Greece there is no archaeological evidence for wheeled vehicles of any kind between the last iconographic and textual evidence of Mycenaean chariotry about 1200 BC, and the first appearance of chariots and funeral carriages as part of the iconography of Geometric vase-painting in the second third of the eighth century, (Snodgrass 1971a; Greenhalgh 1973), and the iron nave bands of vehicles from graves 13 and 58 of the Kerameikos cemetery (Snodgrass 1971, 432; Müller-Karpe 1962a). The bronze horse-bit from Mycenae and the iron pair from the Athenian Agora grave of the beginning of the ninth century inform us of equine control, but not directly of vehicles.

The eighth-century literary evidence of Hesiod implies spoked-wheel farm vehicles, either carts or more probably wagons (Richardson & Piggott, forthcoming); the Homeric references plunge us into a world of intricate scholarly controversy which cannot, however, be ignored by European archaeologists in the present context. Briefly, two alternatives have been offered for the evidence for chariotry in eighth-century Greece; undocumented survival from Mycenean times, or abandonment followed by reintroduction from Near Eastern sources as part of the orientalizing episode under discussion. The issue is further complicated by distinctions of function (an integral part of the Homeric kernel of the problem) as between war-chariots and their use in battle on the one hand, and chariots as prestige vehicles for processions and racing, the latter, with its ritual and religious associations, being after all a peculiar form of competitive procession. Crouwel (1978) represents the case for continuity, assuming a military function for some chariots depicted on Geometric Attic vases and the seventh-century bronze and pottery models from Olympia (before *c.* 590; Snodgrass 1964, 162), and taking them all to 'derive from Aegean Bronze Age prototypes and not from contemporary oriental ones'. A middle position, emphasizing distinction of function, is that of Greenhalgh: 'one-man racing chariots were familiar to Dark Age bards' but 'the *war*-chariot with its two occupants has no place in the history of Geometric Age warfare outside Cyprus' (1973, 38). Finally, Snodgrass sees the evidence as pointing in a direction contrary to that of Crouwel: 'we have not found any material evidence for the use of war-chariots during the dark age ... the chariot was reintroduced to Greece for a strictly limited range of purposes—racing and processions, but not warfare' (1971, 433). He had discussed this in an earlier study, arguing for 'a late eighth century date for the arrival of the

78 Cross-bar wheels on mule-drawn carts, mid 7th century BC, *on Attic black figure vases*

canonical form of racing-chariot in Greece (and, we may add, in Etruria; the agency of a common oriental source, though hardly Egypt itself, may perhaps be detected here); it need not conflict with the suspension of chariot *warfare* which is generally postulated at this time' (1964, 162). Archaeology cannot help us here, in the absence of vehicle remains, partly owing to religious traditions that did not include vehicle-burial as a funerary practice, as they did in contemporary Cyprus and in Italy, to which we must now turn. (Since this book was written J.H. Crouwel, *Chariots and other means of land* transport in Bronze Age Greece (Amsterdam 1981) has been published.)

In contrast to Greece, Italy, as we have seen, provides archaeological evidence of the use of wheeled transport from the second millennium BC onwards, and if this comes from the north of the country owing to the favourable conditions for wood preservation in waterlogged sites in that area, one can hardly suppose that the rest of the peninsula remained unaffected by this development of transport technology, with the use of tripartite disc, cross-bar and spoked wheels well established by the time of the first oriental, and Greek, contacts of the eighth century. The antler cheek-pieces of soft bits further demonstrate the control of horses, presumably as traction-animals, perhaps in the earlier and certainly in the Later Bronze Age. The orientalizing phase, as in Greece, is a phenomenon of the end of the eighth and the seventh centuries and has been the subject of many studies, the most recent being those of Rathje (1979) and Buchner (1979), while as we saw, vehicles and bronze horse-bits have been the subject of monographs by Woytowitsch (1978) and Von Hase (1969). Stary (1979a) has made a full study of the foreign influences in arms and armour of the period. As a mainly seventh-century affair within a technologically iron-using civilization, the question of Etruscan vehicles, and especially chariots, must be postponed until a later chapter, but here we have oriental influences accepted and adopted with far greater avidity than the archaeological evidence suggests for contemporary Greece. While perhaps impinging on an already indigenous tradition of light, horse-drawn, two-wheeled vehicles of prestige, the chariot for parade and procession in the Etruscan world seems reasonably acceptable as an oriental contribution to an emergent culture eagerly and eclectically able to assimilate outside contributions, especially from the eastern Mediterranean and the Aegean (Stary 1979a,

190). With the chariot remains from tombs go a large and individual series of horse-bits, some with elaborate cheek-pieces in the form of birds or double bird protomes in the continental Urnfield tradition, and also the well-known group with figures of horses, often claimed as of 'Luristan' derivation: 'a number of tendentious parallels have been drawn between Luristan and Etruscan horse harness' (Moorey 1971, 27), and a more generalized oriental inspiration from the Levant in the circumstances described above, and mainly in the seventh century, seems preferable for animal cheek-pieces in general. For the horses themselves, the Ridgways have drawn attention not only to the seventh-century Ha C model horses in the Hallstatt cemetery itself, but to stylistic parallels between the horses on Von Hase's eighth century 'Veii' type of bits and those on contemporary Greek vases from Pithekoussai on Ischia: these must 'demonstrate knowledge of the sprightly Greek horse —"probably with Levantine blood": cross-breeding at Al Mina?—and western capacity to produce it in bronze by the middle of the eighth century BC' (Ridgway & Ridgway 1976, 150). With Balkwill's reminder of the large size of mouthpiece of the Italian bits (c. 12–16 cm as against the c. 8–10 cm of continental Urnfield) this knowledge may be thought to have been at first hand and denote imported stock or stallions. As the history of British bloodstock in the eighteenth century shows, an improved breed can be obtained from very few original sires.

In summing up the evidence of Etruscan oriental imports, Rathje (1979, 179) has pointed out that in the matter of trade exchange both the Euboean Greek colonists, already at Pithekoussai in the early eighth century, and the Phoenicians 'were interested in iron, tin and other metals'. Imports broadly from north Syria could have been brought to Italy 'at the western end of a maritime trade-route starting from the

Greek emporia on the coasts of Cilicia and Syria'. But, as she goes on to say, 'this does not exclude the Phoenicians' either from the Levant or from Carthage (traditionally founded in 814 BC). If we are to associate Etruscan chariotry with the introduction of new ideas from outside, and eastward of, the Tyrrhenian Sea, the Phoenicians are obviously, in view of what we have seen of the mainland Greek situations, the better claimants, when we move westwards to the Iberian Peninsula and there encounter evidence of chariotry, and other east Mediterranean contacts associated at the close of the eighth century, we have the problem again presented to us, but in a different form.

The Iberian stelae

In the south-west of the Iberian peninsula is a remarkable group of monuments, datable as we shall see to about the close of the eighth century BC and presenting us with evidence of wheeled vehicles in use among the native Late Bronze Age communities.

These are stone stelae evidently erected as memorials, and incised with what is in effect a pictorial inventory of what might in other circumstances of burial ritual have been buried or consigned to the funeral pyre with the deceased man (Almagro 1966; Powell 1976a). They were warriors' memorials; the dead man is often represented, together with shield and spear, and other weapons such as sword or bow-and-arrow, sometimes helmets and fibulae, and on seven of the twenty-five stones of this type known, a chariot. In addition, a related stele, with a shield, spear, a pair of water-birds and three four-spoked wheels, comes from Substantion (Montpellier) in the south of France. Only one stele has been found in an archaeological context, that from Antegua (Cordoba), found at the foot of the defensive wall surrounding a settlement of the eighth to seventh century BC, producing local pottery and imported Phoenician amphora sherds; the remainder are unassociated finds, south of the Tagus and in the provinces of Sevilla, Badajoz and Carcares.

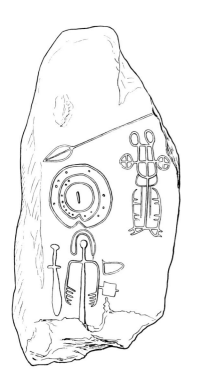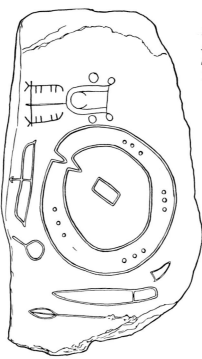

79 Grave-slabs with warrior, sword, spear, shield and chariot, 8th–7th century BC, from Cabeza de Buey and Torrejón del Rubio, Spain

79

The chariots, with paired draught-horses, are shown in the common conventions of rock art already discussed, and are, for instance, very comparable with those at Frånnarp in south Sweden, taken here to be of earlier date. The bodies are D-shaped with central axle and in several instances the draught-pole running through to the rear, and the ambiguity of wheel depictions noted in Scandinavia again applies, as in only two instances, Antegua and Cabeza de Buey, are four spokes shown, the remainder being plain discs. At the rear of the chariot body, in several examples, loops or hand-grips for mounting are shown. The vehicle at Solana de Cabanas is anomalous and appears at first sight to be four-wheeled, but Powell argued cogently for a chariot like the rest being intended, though with an 'initial error in execution of the design' impossible to correct once made.

The associated armament and other objects depicted with the chariots on the stelae are informative and consistent. The spears are indeterminate, and while the published drawing of that on the Antegua stone suggests that it has throwing-loops, the photographs suggest a flaw in the stone at this point. Throwing-loops might suggest the eastern Mediterranean or Aegean, where they became a feature of the Greek hoplite javelin from the seventh century BC (Snodgrass 1967, 57, pl. 34; Yadin 1963, 355), and this feature is also shown on the well-known bronze belt-plate from Vače in Yugoslavia, a piece of 'situla art' of the sixth century BC (Powell 1971a, fig. 2). The swords include the 'carp's-tongue' type characteristic of the Atlantic Bronze Age of Late Urnfield date (Coles & Harding 1979, 474) and the simplified helmet representations could be brought into the general West European Urnfield series (Hencken 1971, 76–77). The fibulae, on the other hand, taken in conjunction with actual Spanish finds of similar types, are exotics. These belong to the 'elbowed fibula' class of the Mediterranean,

variants of which, between the tenth and eighth century BC, stretch from Etruria and Sicily to Cyprus and the Levant (Maxwell-Hyslop 1956; Birmingham 1963). Birmingham, quoting only two Spanish finds which 'if they were of the Cypro-Levantine series, should be dated on typological grounds to the early or mid-8th century', preferred to see the elbowed fibula in both west and east Mediterranean as deriving from the eleventh to tenth-century Sicilian series, with subsequent individual developments. Almagro, however, in several papers summarized in his corpus of the Iberian stelae (1966) has put up a convincing case for an east Mediterranean and probably Cypriot origin. The shields take us further in this direction. Most of those on the stelae are of the individual and characteristic circular (or slightly oval) V-notch type, long recognized as of eastern Mediterranean affinities and sometimes associated with early Greek trade to the west (the Substantion shield is incidentally also of this type). A recent exhaustive review of the evidence concludes that this type of shield 'first appeared in the Southeastern Aegean and Cypriote area. It is not a special Greek phenomenon, it need not be of Phoenician origin either; but it seems obvious that the diffusion of the type in the Mediterranean is connected with Phoenician and oriental activities during the 8th and 7th centuries BC' (Gräslund 1967, 71; cf. Stary 1979a, 189). With this supporting evidence, we can, with Powell (1976a) look to the same eastern Mediterranean source for the chariots themselves, and recognize an 'orientalizing' episode in a restricted area of the Atlantic Bronze Age, among 'the "Tartessian" chieftains of the Peninsular southwest', while recognizing that 'it cannot be assumed, on the evidence of memorial stelae alone, that in or about the eighth and seventh centuries BC chariots were integral to native culture or functional for other than funerary and ceremonial purposes'. Chariots need not mean chariot warfare in the Greece of Homer's time or in the far-off western Bronze Age, but ceremony, parade and display are another matter. Buchner (1979) has drawn attention to the legend, recorded by Athenaeus, of how the early Greek colonists of Cumae 'went native'; 'continually wore gold ornaments and adopted gaily-coloured clothes, and rode into the country with their wives in two-horse chariots'. This charming vignette must have been typical of much of later European prehistory.

The final phase: the British Isles

Apart from the exotic Iberian episode, there is, aside from the items of horse harness already described, very little evidence to contribute to our knowledge of wheeled transport during the final eighth century stage of the Later Bronze Age in Western Europe, unless some of the rock carvings belong to this period. Odd finds of bronze model wheels in hoards, such as those from Kunersdorf, Kr. Westernberg, of northern Period V (K. H. Jacob-Friesen 1927, 185) or in the west, in the carp's tongue sword complex at Venat in Charente or Longueville in Calvados (Briard 1965, 229), tell us little: the pair of four-spoked wheels in the latter hoard closely resemble those of the earlier models such as Trundholm or the *Kesselwagen*. But it is in the late eighth century BC that we have the first evidence of wheeled vehicles and horse harnessing in Britain, exiguous enough and only to be slightly amplified during the subsequent seventh-century phase of contact with the continental Hallstatt C culture.

Bronze items of harness equipment, including bit cheek-pieces, have been recognized in two Cambridgeshire hoards, those of Wilburton and Isleham (Coombs 1975, fig. 10), and only indirectly point to horse-drawn wheeled transport. We saw in the last chapter that domesticated horses had been known in Britain and Ireland since the Beaker cultures at the opening of the second

millennium BC, and horse bones make a sporadic appearance in settlement sites of later date such as those of the Deverel-Rimbury culture (and contemporary Dutch sites) early in the second half of the millennium. Of the few finds of antler cheek-pieces from Late Bronze Age Britain, those from the cave site at Heathery Burn, referred to below, should belong to the late eighth century, and those from the riverside settlement at Runnymede Bridge rather earlier; the remainder probably to the seventh century (Needham & Longley 1980; Britnell 1976). A distinctive and curious form of *80* kidney-shaped 'rattle-pendant' which dangled from a bit, links the Baltic area to France, Wales and Ireland (Thrane 1958; Rynne 1962). It seems to have its origins in the northern period V, broadly eighth century, though surviving into the seventh, and a south Scandinavian source for the finds in the British Isles seems likely, even if the type has ultimate Urnfield origins. A Swedish find from Svartarp, Åsle, of Period V, in- *81* cludes a tubular bronze mounting with a pair of horses' heads modelled in the round, with miniature pendants of this type, confirming their association with paired horse draught: Thrane suggested that it might be part of the decoration of a vehicle, such as a pole-tip (1958, 224).

The bronze phalerae of Urnfield Europe which later, in Ha C, are frequently associated with harness trappings, have already been mentioned. Those from the British Isles—over forty from eight find-spots in Britain and one of a probable pair from Ireland—have recently been listed and discussed (O'Connor 1975). Their range of date, on the grounds of continental analogies, and British associations where they occur, is from Ha B into Ha C, and for our purposes the presence of six in the Heathery Burn find, with the nave-bands of a wagon and cheek-pieces of bits, is significant, and they are plentiful in the Isleham hoard. Seventh-century (Ha C) contexts, such as in

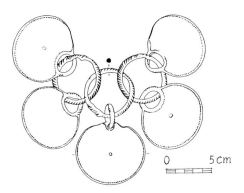

80 Bronze 'rattle pendants' from horse harness, 8th century BC, from Parc-y-Meirch, Wales

the Llyn Fawr hoard, are discussed in Chapter 5.

A group of bronze mountings deserve mention here as having a claim to be connected with the driving of harnessed horses, the so-called 'flesh-forks' known in the British Isles and France as stray finds or in 'carp's-tongue sword' contexts of Late Ha B. They consist of tubular bronze mounts for rods of uncertain length, with a two-pronged curved fork at one end, a knobbed ferrule at the other, and on occasion surviv-

ing intermediate tubes with pendant rings and, in the best known, that from Dunaverney Bog in Antrim, bird figures in the Urnfield tradition. The writer (Piggott 1953), following an earlier suggestion by Mariën (see later Mariën 1958, 115–17) in respect of a Ha C iron example of rather different type, suggested that the 'flesh-hooks' could be better interpreted as goads, used as alternatives to whips for horse-driving, in the manner of the Greek kentron or Roman stimulus. Comparison can now further be made with the bronze 'bidents' from Transcaucasia mentioned in the previous chapter, or with the straight pointed goads from Etruscan contexts (Woytowitsch 1978, 108).

The one piece of direct evidence of wheeled transport in Britain comes from the Heathery Burn cave in Co. Durham already mentioned (Britton & Longworth 1968). Here was a large deposit of objects, domestic or votive, including antler cheek-pieces, phalerae and four pairs of ribbed bronze nave-bands, c. 10 cm diameter. A four-wheeled vehicle is clearly indicated, and the cheek-pieces imply horse draught, but the

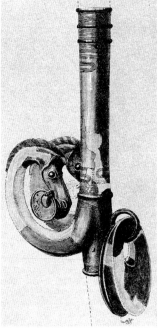

81 Bronze horse heads with 'rattle pendants' and reconstruction, 8th century BC, from Svartarp, Åsle, Sweden

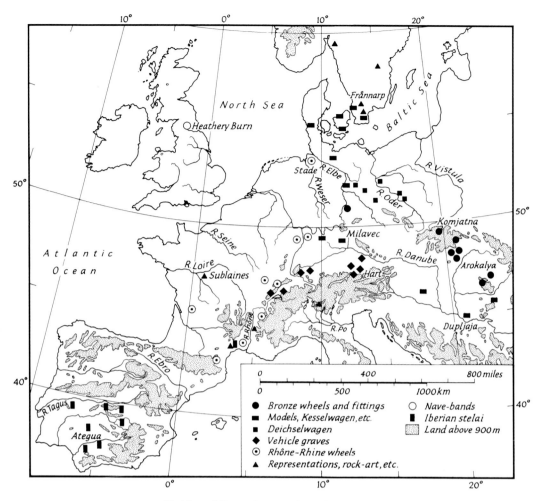

82 *Map of Later Bronze Age vehicle remains*

presence with the other finds of a bronze situla recalls the Côte-St André association, and a large cult vehicle remains an alternative possibility to a wholly secular wagon or carriage. Some other finds of horse-harness which are in the Late Urnfield tradition are discussed with similar finds of the seventh century in the next chapter.

Transcaucasian epilogue

It remains to complete our survey, to take a final glance at Transcaucasia in the Later Bronze Age, between the Lake Sevan vehicle burials described in Chapter 3 and the rise of the kingdom of Urartu in the ninth

century BC, when the region became a part of the literate Near Eastern world of later antiquity. A number of finds with representations of vehicles can in general reasonably be regarded as pre-Urartian, within the Armenian and Georgian Late Bronze Age, notably the bronze belt-plates engraved with scenes of persons and animals, probably mythological but embodying hunting, warfare and chariotry. Martirosyan (1964, 132) assigns them to his Late Bronze Age III, *c.* eleventh to tenth century BC, in Armenia, and the comparable Georgian pieces are presumably of the same date (cf. Lang 1966, fig. 14; pl. 18 for illustrations). An early find of such a belt-plate is that from

135

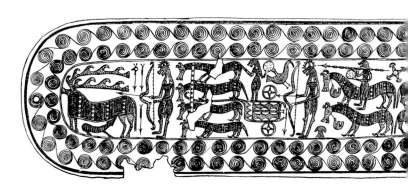

83 Engraved belt-plate with quadrigae, 8th–9th century BC, *from Astkhi-blur, Armenian SSR*

the Lalvar cemetery, Akthala, with a figure standing in a chariot drawn by two horses, with four-spoked wheels and a draught-pole turning up at the end with a flower-shaped terminal. No yoke is shown (de Morgan 1889, 141, fig. 145). Another chariot is shown on one of three belt-plates found un-associated near Lchashen (Esayan 1978). A most remarkable plate from Burial 14 at Astkhi-blur, dated to the ninth to eighth century BC, has an elaborate scene with stags, bow-men, horses (some with riders armed with spear and shield) and three char-iots, each carrying a man, two holding slings (Esayan 1967). The chariots are shown in some detail, with four-spoked wheels, an elongated D-shaped body, and what must, on analogy with the Lchashen bronze chariot-models already described, and another from Khaketia shortly to be mentioned, be a medial lengthwise division. The curved draught-poles have a crook-shaped terminal and the most surprising fea-ture is that the chariots are quadrigae, each drawn by four horses under a single straight yoke. Such harnessing is not attested until the eighth century BC in Assyria, Cyprus and the Levant (Littauer & Crouwel 1979, 113–15). The fragmentary belt-plate from Stepanavan (Martirosyan 1964, fig. 65; Pig-gott 1968, fig. 10) has already been men-tioned in connection with the horse-drawn A-frame cart it depicts, but it also shows a horse-drawn chariot with six-spoked wheels and two schematized men as occupants,

shown standing within the body of the char-iot on each side of a transverse division. The style of drawing is cruder and more schema-tic than that of the other belt-plates and one might suggest that a longitudinal division is, in fact, intended, but made to appear transverse so that the men can be drawn upright (and so alive) and not horizontal (and dead). The longitudinal dividing bar is present on the unassociated bronze chariot model from Gokhebi in Kakhetia in eastern Georgian SSR, variously dated between the second millennium and the eighth century BC, but on the whole more likely to be ninth century (Anon. 1977; M. A. Littauer *in litt.* 1978). It has openwork sides and is of D-shaped plan, with a long slightly curved draught-pole and yoke, with two entire stal-lions as draught. The wheels have relatively broad felloes and ten spokes, with the axle set centrally to the body. The remaining chariot representations in Later Bronze Age Transcaucasia comprise a small schematic design on a pot from Dilijan in Armenian SSR, dated by Martirosyan to his Late Bronze I, thirteenth to twelfth century BC (1964, 99, fig. 46), and an engraved stone slab of a burial cist at Berekey, Daghestan ASSR, in a Late Bronze Age cemetery probably contemporary with the later Lchashen graves (Littauer 1977; Hančar 1956; Chernykh 1976b, 146). Here the char-iot, with four-spoked wheels, resembles the Syunik rock carvings; it has no draught ani-mals but is accompanied by two detached

83

136

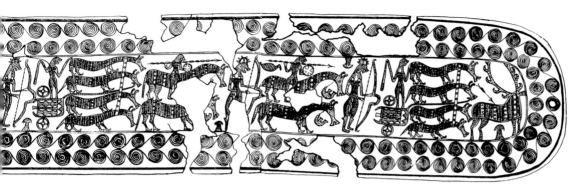

wheel symbols, one with four and the other with eight spokes. All the foregoing examples seem to antedate the consolidation of the Urartian kingdom in the eighth century BC, when a court or palace style of art sharing features with that of Assyria and the Levant was developed. Even if some of the belt-plates should prove in strict chronology to be contemporary, they represent a provincial art style as against the variants on the international oriental conventions adopted by the Urartian artists working under royal patronage.

The evidence for European wheeled transport set out above brings us in chronological terms to the end of the eighth century BC; in technological nomenclature to the end of the Later Bronze Age north of the Alps. The now ancient disc-wheel tradition can be seen to persist, though by accident the evidence is slight, as archaeology has involved few sites favourable to the preservation of wood and throughout the period we have had to rely increasingly on the metal fitments of vehicles of prestige or cult, or of their models, surviving in hoards, as votive deposits, or as grave-goods. Here we can see two factors operating, the dominance of the horse as a traction animal, and the development of spoked-wheeled vehicles of relatively light construction, whether four-wheeled wagons or carriages, or two-wheeled carts or chariots. The latter cannot be demonstrated as war-chariots, thought certainly vehicles of prestige in

which chieftains in a socially stratified order of things paraded or 'rode into the country with their wives', and represent a continental European tradition of earlier ancestry in which it is unnecessary to seek new oriental contacts, even if some modification in horse-bit types may look in this direction. In the Mediterranean the situation is otherwise, with strong orientalizing contacts affecting both the nascent civilizations of Greece and Etruria, and a small group of barbarians in the Iberian peninsula. By the late eighth century too the first evidence of wheeled vehicles and horse control in the continental Urnfield manner begins to appear in Britain and Ireland.

From about 700 BC we move into a technologically fully iron-working temperate Europe and the first Iron Age period of Hallstatt C, where the ancient tradition of vehicle burial is again displayed in the changed circumstances of an inhumation rather than a cremation burial rite, to continue, through Hallstatt D and La Tène, until the last prehistoric centuries before the Roman Conquest. For the prestige vehicles at least, which played a part in such funerary practices, the evidence becomes almost embarrassingly abundant.

5 The Early Iron Age: vehicles in Hallstatt Europe and beyond

Cultures and chronology

We saw in the last chapter how conventional archaeological nomenclature has divided the original Hallstatt A–D scheme into two, on technological and chronological grounds, with Ha A and B constituting a Late Bronze Age, with bronze as the standard metal for weapons and edge-tools, giving way in the late eighth century to the technological adoption of iron for these purposes and, with the inception of iron working, the formation of the assemblages of material culture classed within two Early Iron phases of Ha C and D, between them covering rather over two centuries, from a generation or so before 700 to a roughly similar period of time after 500 BC. Convenience and convention allow of the useful shortened terminology 'Hallstatt' to apply specifically to these two iron-using phases while relegating A and B to a Late Bronze Age. The eventual Hallstatt continuum, in this sense, extends from the Rhône and Upper Seine eastwards, north of the Alps to the Danube at Budapest, but stretching northwards from the head of the Adriatic to the southern edge of the North European Plain (Angeli *et al* 1970 with map, fig 1; Filip 1962). The complex circumstances of the beginning of iron-using economies in Europe north of the Alps and Balkans have been discussed by Powell (1976 b), and Pleiner (1980), making *inter alia* points of direct relevance to our present enquiry to which we will return. The aspects of continuity and discontinuity between Final Urnfield and earlier Hallstatt traditions have been much discussed, but there does seem

some sort of break in settlement type and occupation, as well as the appearance from Ha C of a new type of inhumation burial rite, not infrequently including vehicle burial in the time-honoured mode of earlier antiquity, over an area from France to Bohemia and providing the fundamental material for our study in the seventh and sixth centuries BC. On the other hand, it is possible to see, in vehicle technology, clear links between the Bronze Age and Iron Age traditions, and indeed we saw in the last chapter that vehicles accompanied their owners to the funeral pyre from Ha A1 onwards, and with the change-over to the inhumation rite the custom was continued in a form demanding greater labour in the preparation of a suitable grave to contain a complete and unburnt vehicle, but fortuitously and happily providing greater archaeological evidence for its nature and technology.

The changes in social structure between Urnfield and Hallstatt populations implied by the archaeological evidence from graves, fortifications and settlements in the period *c.* 700–500 BC have been a matter of active consideration and discussion, admirably set out by Härke (1979) and Wells (1980). The broad outlines of a development pattern can safely be inferred, in which societes relatively unstratified in Urnfield times (though, as the vehicle evidence alone suggest, not without some differentiation of status) becomes increasingly, from the seventh to the sixth century, hierarchical in structure, with a rise to power of some form of ruling class of chieftains with status and resources able to command wealth in natural resources, goods (including costly imports) and labour. The German archaeologists, foremost in this debate, have shown the open hill-top settlements, or the hillforts, of Late Urnfield (Ha B) times from *c.* 1000–750/700 BC, appear to have had minimal occupation in earlier Ha C, though this increases before *c.* 600 BC and the changes marking the inception of Ha D. At this time too begin the first of what are interpreted as 'chieftains' seats' *(Herrensitze)*, themselves developing into distinctive defended centres of authority and of territorial units as the *Fürstensitze* or 'princely residences': the terminology is sometimes further complicated by the introduction of *Adelsitze* or 'noblemen's seats', as an intermediate term. With the *Fürstensitze* go the *Fürstengräber* or 'princely graves' and indeed these inferred aspects of the Hallstatt nobility in life and death slip into circular arguments whereby each is defined by the presence of the other; *Prunkgräber* ('ostentatious graves') also make an appearance, and it is in such types of grave, however designated, that the vehicle burials with which we are directly concerned occur. This terminology, with its overtones of nineteenth-century German aristocratic society and political patterns, can better be replaced by a broader model in which the Hallstatt phenomena are seen as the expression of a situation widespread in other ancient and recent societies, that of the appearance of a 'prestige goods economy' maintained by the acquisition and redistribution of material goods denoting and maintaining status by a variety of socially accepted means including the type of gift exchange typified by the *keimelia* of the ancient Greek world (Frankenstein & Rowlands 1978; F. Fischer 1973). The rich graves belong to the type originally classed as 'Royal Tombs' by Childe, recurrent throughout prehistory and history in the social contexts of the prestige economies just described (Piggott 1978b), and the vehicles take their place as objects denoting

status and prestige in Hallstatt (and later, in La Tène) societies as much as those already described among earlier peoples from Transcaucasia and south Russia.

The chronological framework of Ha C and D is provided by the well-known phenomenon, itself an integral part of the prestige economy just described, of the importation into the Hallstatt world north of the Alps of luxury products from the Etruscan and Greek worlds of the Mediterranean. An early zone of contact was that between the Golasecca and Este cultures of northeast Italy, and the head of the Adriatic; after the foundation of the Greek colony of Massalia around 600 BC the Rhône gave access to the western Hallstatt area. Unfortunately many of the bronzes, which form our main sources of information, are difficult to date with precision in their homeland, but a consistent pattern can be seen (Dehn & Frey 1979; D. Ridgway 1979; F. R. Ridgway 1979; Shefton 1979, Wells 1980). The earliest pieces, in Ha C, include the Etruscan pyxis from Kastenwald in Alsace, and the ribbed bowls from the same grave and from a grave with a bronze-studded yoke and an early bronze situla in the Stadtwald, Frankfurt-am-Main, all of the seventh century BC (U. Fischer 1979); at the end of the Ha C and beginning of Ha D come the so-called 'Rhodian' flagons, Greek products of the late seventh to early sixth century BC, two of which come from the vehicle graves of Kappel and Vilsingen. The bowls with beaded rims, as from the vehicle graves of Hohmichele 6 and Hradenin 28, are again of probable Etruscan origin but not more closely dated than the sixth century BC. But with graves containing Attic painted cups and fine Greek bronzes as at Vix, more precision becomes possible, and a date just before 500 BC can be proposed for this final Hallstatt vehicle grave, which also contained a beaked flagon *(Schnabelkanne)*, a type which continued into the early fifth century, when it affords a point of departure for the chronology of the La Tène culture with its vehicle graves, dealt with in Chapter 6. The Hallstatt D chronology is reinforced by the presence of Attic black-figure pottery of the second half of the sixth century in many fortified hill-settlements of the *Fürstensitze* type already mentioned, and the period itself can be subdivided.

Graves and vehicles

The circumstances of archaeology, with an absence of identified or excavated settlement sites with soil conditions favourable to the preservation of wood, render the Ha C and D periods a blank so far as knowledge of utilitarian and agrarian vehicles presumptively disc-wheeled and ox-drawn goes: a few isolated finds of such wheels with radiocarbon dates could possibly be placed late in Ha D but are better considered in the next chapter with comparable later finds. All our evidence, with the exception of a few, but informative, representations, comes from 'chieftains' graves' of the type already mentioned, in the form of horse-drawn vehicles, mainly four-wheeled wagons or carriages, but a few two-wheeled carts or chariots, deposited with the dead. These, which with their yokes and harness were clearly often of considerable elaboration, were therefore ceremonial vehicles that in the form in which we encounter them were funerary by deposition or design and, as Powell reminded us (1976 a, 168), likely to be associated with rites akin to the *ekphora* and *prothesis* of the early Greek world. Strictly speaking, the evidence would allow us the alternatives of seeing the vehicles as for parade and ceremony in life and eventually used as a demonstration of the status of their owner in the grave, or as hearses made with a funerary purpose alone in mind. In the chariot burials of the eight and seventh centuries BC at Cypriot Salamis, certain simple two-wheeled vehicles were distinguished as funerary hearses by Karageoghis (1969, 32),

in contradistinction to the elaborate chariots for parade and warfare which were also buried. In the instance of the Hallstatt vehicles under discussion the likelihood that we are seeing ceremonial vehicles used in life before they were deposited in the grave could be supported by two considerations. In the first place, other valuable objects in the grave-weapons, ornaments, bronze vessels—were clearly some of the status symbols that demonstrated the standing and distinction of the deceased while alive, and in the second, the placing of a vehicle in the grave together with such other attributes of social status implies prestige of a carriage or chariot in life. Vehicle burials in themselves have meaning only in a society which places a value on ceremonial wheeled transport as an accepted attribute among the ranks of society entitled to 'ostentatious graves'. It is therefore proposed here to proceed on the assumption that the vehicles in the graves of Hallstatt and La Tène were, by and large, constructed and used in life before reaching their eventual destination in the grave, or were directly representative of such carriages and chariots.

Within the seventh and sixth centuries BC in Hallstatt Europe the most distinctive technological feature in vehicle construction is the increasing use of metal, and particularly of iron tyres and iron or bronze nave-sheathing on the wheels. These and other features are more fully discussed in a subsequent section, but are of importance here *84* because the higher survival value of iron as against wood has enabled the identification of vehicle graves, even under poor excavation conditions, from such remanent fragments as that of iron tyres alone. Records of the recovery or excavation of metal fittings *85* (usually tyre fragments) go back to such finds as those of de Bonstetten in Switzerland in 1848, and at a conservative estimate we have for Ha C and D, a current total of some 150 graves with widely varying standards of record, excavation and publication.

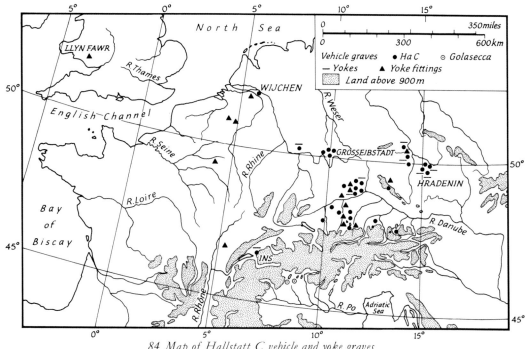

84 *Map of Hallstatt C vehicle and yoke graves*

These were listed and mapped by Schiek (1954), and later detailed surveys include those of Drack (1958) for Switzerland, Joffroy (1958) for France, and particularly the studies for southern Germany by Kossack (1953a; 1959, 1970, 1971) and for the final Hunsrück-Eifel phase by Haffner (1976). Koutecký (1968) made a general survey of the graves of the Czech Bylany culture, and there have been important individual excavations reported from Bohemia and Germany, notably the Hohmichele (Riek & Hundt 1962) and the Grosseibstadt cemetery (Kossack 1970). The most recent and one of the most important finds is that of the unplundered Ha D burial chamber at Hoch-

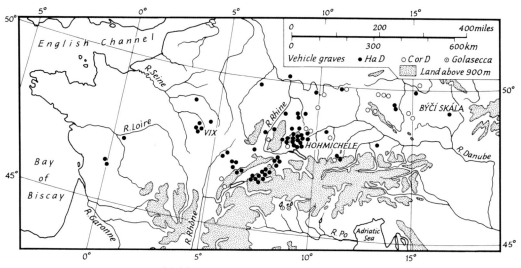

85 *Map of Hallstatt D vehicle graves*

dorf excavated in 1978–79 (Biel, 1981; Bittel 1981). This last discovery draws attention to a potentially limiting factor, the plundering in prehistoric antiquity of the wooden chamber graves of Ha D, with the consequent removal of valuable objects: in the central primary grave of the Hohmichele barrow, the tomb-robbers had so thoroughly cleared the chamber that they even removed a four-wheeled vehicle, leaving its former presence to be marked only by the rust-stains of its iron tyres on the floor.

The distribution of Hallstatt vehicle graves as mapped by Schiek in 1954 has on the whole been intensified rather than drastically altered by more recent discoveries, north of the Alps from the upper Seine and the Marne to Bohemia, with new finds slightly extending its northern limits to just beyond the River Main in Franconia. Outliers occur in little enclaves of Hallstatt culture as the Poitiers region (Ha D) or the Wijchen burial in the southern Netherlands (Ha C). The feature brought out by Schiek, that within this western Halstatt province, the earlier, seventh century, Ha C vehicle burials lay to the east, with the later, sixth century, Ha D graves to the west, with a dividing line roughly from Innsbruck to Nuremburg, remains substantially unaltered. The well-known votive (and perhaps burial) deposits in the Býči Skála cave in Moravia, excavated in the 1870s, have never been published in detail or to modern standard, but they contained the remains of at least three wagons represented by iron tyres and other fittings (Angeli *et al* 1970, 99–150; 1980, 230, pl. 11. 2; Bittel 1981, fig. 17).

Vehicle burial quite obviously demands the provision of a grave of adequate size to accommodate the carriage or chariot it is designed to contain, and one of the features distinguishing at least the aristocratic burial rite of Hallstatt times is the appearance not only of inhumation burial as against the traditional cremation rite of Urnfield times, not exclusively, but predominantly, but also the adoption of a plank- or log-built burial chamber, set in a grave-pit or free-standing at ground level until incorporated in a barrow or cairn. Alternatively a pit grave with a log or plank roof may be used, though it is impossible to tell in how many instances the circumstances of decay, or of incompetent excavation, may have led to a failure to recognize original timbering. Powell (1976b) drew attention to the timbered pit graves at Pécs-Jakabhegy in Hungary, one with a cremation burial and horse-gear of the general type associated with eighth-seventh century 'Thraco-Cimmerian' series, and commented on the typological affinities of these and the Hallstatt examples with other tombs widespread in time and space, from the second-millennium Timber Graves in south Russia and by way of Transcaucasia to the eighth-seventh centuries tombs of Gordion. Koutecký (1968) and Kossack (1970) both attempted interpretations of the exceptionally large graves of Ha C in socio-economic terms of hierarchy, status and lineage, but so far as the vehicle burials are concerned we may make an alternative approach based on certain practical necessities, and the technological considerations of wood and metal construction.

All the graves under consideration were chamber tombs of one kind or another, whether a pit with horizontal roofing of wooden baulks or stout planks in the manner of the south Russian Pit and Timber Graves described in Chapter 2, or a wholly wooden construction in a pit or on the old ground surface under a barrow. The remains of roofing of the first type were noted, for instance in the Ha C graves at Hradenin, or, better recorded, at Lovosice; the Ha D wooden chambers survived almost intact on occasion, as at the Hohmichele. In such circumstances the burial deposit, including the vehicle, would have survived intact in an air-space until the rotting of the timbers caused the the roof to collapse, bringing with it the super-incumbent earth and stones to crush the burial and grave-goods and fill the chamber. It also follows that the minimum height of the roof above the floor would have to be sufficient to allow clearance of the highest object placed with the dead: if a vehicle, its wheels and any superstructure above the undercarriage at axle height. If decay took place within the chamber while the roof was still bearing up, a wagon would be especially vulnerable at four points, the junctions of the cylindrical axle-arms (normally *c*. 5.0–6.0 cm diameter) with the square-section axle-bed, allowing the wheels to fall flat or at any angle if they did not remain vertical to be crushed in the final collapse of the roof. This can be well seen in the photographs of the excavated graves at Hradenin (Dvořák 1938, figs 19, 34, 41) but does not mean the wheels were dismantled at the time of burial: in Grave 28 one wheel was flat, the remainder crushed upright. In fact, the only certain evidence for dismantling wheels is at Vix, exceptional in this as in so many other respects. Sufficient excavation records have been published for us to compare the sizes of vehicles and their graves, though many are less reliable than would appear at first sight: the well-known 'plans' of the Hradenin graves, for instance, are sketch-diagrams, not to scale and schematic in detail (Dvořák 1938; Koutecký 1968) and few approach the excellence of the Grosseibstadt or Frankfurt presentations (Kossack 1970; U. Fischer 1979). A problem facing those conducting an inhumation burial ritual involving the deposition of a vehicle in the grave would be the minimum size of pit or timber housing to allow of such a deposit, and the degree of completeness of the vehicle itself as prescribed by rite and custom. The evidence implies that the latter factor was one of retaining or dispensing with the vehicle's draught-pole, as earlier examples described in Chapter III demonstrated, with the composite draught-pole hacked off to accommodate the wagon in the small grave-

pit of Trialeti 5 but retained in the ampler Lchashen tombs. We shall encounter the same circumstances in some of the fifth and fourth-century chariot burials described in the next chapter, where the dismantling of the wheels of the vehicle from the body may sometimes also be inferred.

To return to the Ha C and D vehicles, the four-wheeled carriages, as will be demonstrated in a later section, are built to singularly uniform proportions. Wheel diameters (determined by their surviving iron tyres) average *c*. 80 cm, carried on fore and rear axles and axle-beds about 1.80 m apart, and with a gauge of *c*. 1.30–1.40 m; the overall length of the vehicle being *c*. 2.60 m. Where it can be estimated, the body and the undercarriage supporting it, were narrow, between about 1 m and 0.70 m across. The length of a draught-pole depends on the interval to be allowed for the draught-pair between their yoke at the base of the neck or on the withers, and the front of the vehicle they draw. As no horses were buried with the Hallstatt vehicles (save in one ill-recorded old excavation, with no details of the skeletons, at Lhotka in Czechoslovakia: Dvořák 1938) we must rely on the average size, confirmed by the width of the bit mouth-piece which, as in the Late Bronze Age examples quoted in the previous chapter, represents a horse about 125–30 cm at the withers. As very approximately withers height can be equivalent to the withers to rump length, this would imply the clearance between yoke and vehicle for an animal of such proportions of *c*. 1.50–2.0 m. The overall length of a four-wheeled vehicle with its draught-pole would therefore be of the order of 4.0–4.5 m. Since the draught-pole of a four-wheeler is, as we have seen, necessarily mounted to permit vertical movement, its removal at the point of articulation can be readily effected, leaving only the body of the vehicle with its two pairs of wheels, a little over 2.50 m long. On these figures alone, if ritual demanded the

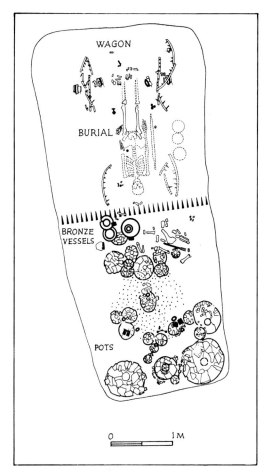

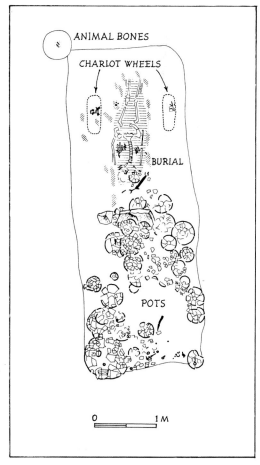

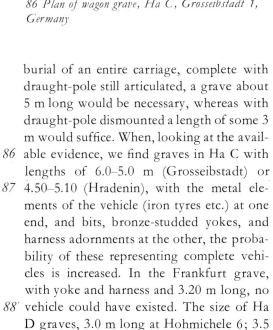

86 *Plan of wagon grave, Ha C, Grosseibstadt 1, Germany*

87 *Plan of chariot grave, Ha C, Grosseibstadt 4, Germany*

burial of an entire carriage, complete with draught-pole still articulated, a grave about 5 m long would be necessary, whereas with draught-pole dismounted a length of some 3 m would suffice. When, looking at the available evidence, we find graves in Ha C with lengths of 6.0–5.0 m (Grosseibstadt) or 4.50–5.10 (Hradenin), with the metal elements of the vehicle (iron tyres etc.) at one end, and bits, bronze-studded yokes, and harness adornments at the other, the probability of these representing complete vehicles is increased. In the Frankfurt grave, with yoke and harness and 3.20 m long, no vehicle could have existed. The size of Ha D graves, 3.0 m long at Hohmichele 6; 3.5 m at Offenbach–Rumpenheim, 3.0 m at Vix, with the iron-tyred vehicles in the former two only just fitted into the available space, goes to confirm the removal of the draught-pole in these instances. At Vix, exceptional as in so many other respects, the whole vehicle was dismantled, the wheels removed, and the draught-pole laid alongside the carriage-box which functioned as a bier. On the other hand, the Ha D burial at Hochdorf included a carriage with its draught-pole in a chamber 4.80 m long, and the large wooden chambers, robbed in antiquity, in the Magdalenenberg (Villingen) 7.65 m long, or the primary Grave 1 in the Hohmichele (5.80 m long) could have had similar complete vehi-

86

87

88'

89

90

144

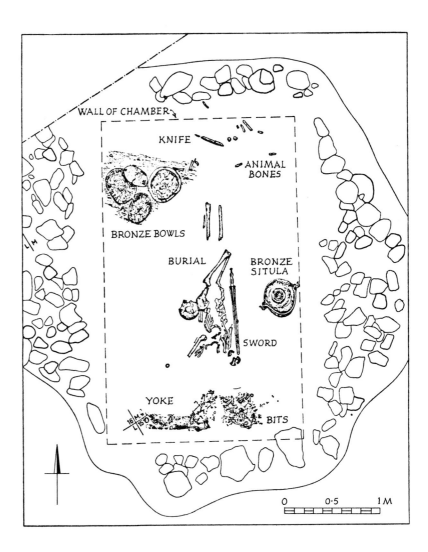

WALL OF CHAMBER

KNIFE

ANIMAL BONES

BRONZE BOWLS

BURIAL

BRONZE SITULA

SWORD

YOKE

BITS

0   0·5   1 M

88 Plan of yoke grave, Ha C,
Frankfurter Stadtwald, Germany

cles (cf. Kimmig and Rest 1953 on large burial chambers in Ha D). Practice was clearly not consistent.

The matter can be taken a stage further by applying to the Ha C vehicle graves the technological considerations of the varying degrees in which metal may be used in the construction of the vehicles themselves. Kossack (1959, 16–24) made an elaborate classification of all Ha C graves containing weapons, vehicles and harness (*Kombinationsgruppe* A) in three main groups broken into eleven sub-divisions, and the scheme about to be put forward is not offered as an alternative, but as a suggested technological counterpart. The adoption and modification

of iron tyres beginning in Ha C and continuing through Ha D and La Tène is discussed in a later section, but as a novelty in the seventh century it allows for the continuance of wheels without metal tyres from Urnfield times. Similarly metal sheathing or banding of wheel-naves need not be an invariable practice, nor the use of metal linchpins. The basic assumption is that graves of around 5.0 m in length, and containing metal bits and harness mounts, and some instances metal-studded yokes as well at one end, contained a complete wooden vehicle, normally a four-wheeler, with a greater or less use of metal fitments, at the other, even when, if wholly of wood, its traces were not

145

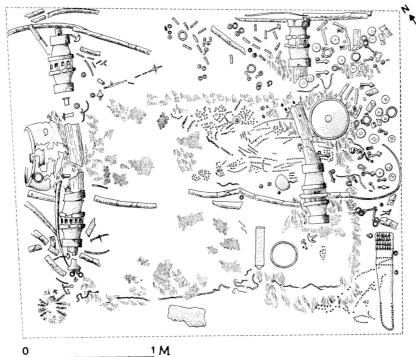

89 *Plan of wagon grave, Ha D,*
Hohmichele Grave 6, Germany

0   1 M

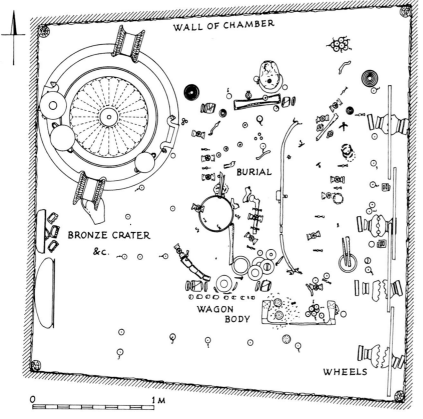

WALL OF CHAMBER

BURIAL

BRONZE CRATER
&c.

WAGON
BODY

WHEELS

90 *Plan of wagon grave, Ha D,*
Vix, France

0   1 M

perceptible on excavation. The sequence would then run as follows:

Type 1 Vehicles with full complement of metal as tyres, nave-sheathing and linchpins. Examples: Hradenin 24, 46.

Type 1a As above, but with no metal linchpins. Examples: Grosseibstadt; Leipheim.

Type 2 Vehicle with iron nave-bands, but no metal tyres or linchpins. Examples: Grosseibstadt 4 (two-wheeler).

Type 2a Vehicles with metal linchpins but no tyres or nave-sheathing. Examples: Hradenin 30, 33 (two-wheeler), 58; Rvenice (two four-wheelers).

Type 3 Assumed wholly wooden vehicle in grave of adequate size, with metal bits, harness-mounts and sometimes bronze-studded yokes. Examples: Lovosice 2, 3; Grosseibstadt 2, 3, 5, 7.

Type 4 Yoke and harness in grave, but presence of vehicle excluded. Examples: Frankfurt, Plaňany.

Such a sequence seems to make technological sense, and in particular offers an explanation of the otherwise disproportionate size of such 'yoke-graves' as Lovosice 3 (Pleiner 1959; Pleiner & Moucha 1966), where, in fact, the traces of planks recovered under the burial could now be seen as the flooring of the vehicle, as in the Type 1 graves of Hradenin 24 and 46 (Dvořák 1938). Similarly, the iron linchpins of Type 2a, would be better explained in the positions in which they were found, as having been originally *in situ* as the only metal fittings of an all-wood vehicle (Koutecký 1968, figs 2b, 2d, 10; 1966). In the Grosseibstadt cemetery, Kossack (1970) devised a chronological (and dynastic) scheme with the seven graves spanning the seventh century, the iron-tyred carriage of Grave 1 being demonstrably the earliest, at the beginning of Ha C, and Grave 4, with the chariot, the latest. In between come the graves which would now fall into place not as merely having *pars pro toto* deposits of harness, but all-wood vehicles as

well. The foregoing explanation, allowing for vehicles with minimum metal fittings, would permit of the inclusion as vehicle burials such cremation-graves as the famous rich grave, no. 507, of the Hallstatt cemetery, with its four bronze linchpins, as well as others less remarkable (Kromer 1959, 118, pl. 98, 2; Barth 1973).

As far as technological sophistication is concerned, the foregoing proposed scheme could be taken in reverse, with an increasing amount of iron-work as one moves from Type 3 back to Type 1, but this is fairly certainly too simplistic a reading of the evidence. The seventh-century Ha C phase saw the adoption, and by its end the almost universal usage, of the nailed-on iron hoop tyre, at least in the vehicles deposited in graves on which our knowledge relies, and the circumstances and details of this important technological innovation are discussed in a subsequent section. But the substitution of iron for an earlier tradition of a softer and archaeologically perishable bearing surface —the wooden felloe alone or a tyre of shrunk-on rawhide or similar material— would not be a simple unilinear or universal process among the craftsmen concerned, the wheelwrights and smiths, but one of variable and tentative experiment over three or four generations and a wide geographical area, dependent also, at least in its earlier stages, on adequate supplies of iron outside the areas of the natural occurrence of suitable ore. The long tradition of building wholly wooden vehicles in prehistoric Europe has been demonstrated in earlier chapters, and in making the suggestion that 'empty' spaces in large graves might have contained vanished vehicles of perishable materials, the similar circumstances in the burials of paired animals in late third-millennium BC graves described in Chapter 2 have not been forgotten. A German archaeologist describing a Final Ha D grave in which only four wheel-pits indicated the former existence of a vehicle nicely characterized it as a *Phantomwagengrab* (Driehaus 1966b, 38), but ghosts once had bodies, and these we may reasonably infer from the archaeological evidence.

The practice of vehicle burial, once established early in the seventh century in the Ha C phase, continued and was developed in the following century in what can now, in Ha D, be recognized as a western Hallstatt province. Four-wheeled prestige wagons or carriages predominate, in what became stereotyped forms already established in Ha C, but a significant number of burials with two-wheeled vehicles or chariots appear, again well documented in Ha C in Grosseibstadt Grave 4 and in less well recorded instances. In the Final Hunsrück-Eifel phase, which in that part of Germany bridges the conventional Late Hallstatt and Early La Tène division, such vehicles predominate, in a new tradition of vehicle burial which then persists throughout the pre-Roman centuries, in certain areas of La Tène culture.

In view of this essential continuity, it is proposed to deal with the material from Ha C and D graves as a whole, supplemented by the important contemporary information to be derived from representations in rock art, and a few on pottery or other artifacts, even though these may be outside the bounds of the area of strictly Hallstatt culture. One must also assume a major technological development in the increased use of iron, and more especially the use of iron hoop tyres, on vehicles which in the earliest stages could be wholly of wood. The archaeological evidence from Ha C graves, set out above, favours such a sequence of indigenous development in continental Europe north of the Alps, although for reasons which are discussed below, the origins of the iron tyre seem likely to lie rather in the adoption of this crucial innovation from Mediterranean sources, as part of the traffic between northern Italy and the Hallstatt world.

Technology and typology

An outstanding factor in vehicle construction as summarized in the last chapter is the beginning of what was to become a dominant feature, the interdependence of two types of craftsmanship, the woodworking skills of carpenter and wheelwright on the one hand, and the expertise of the smith in bronze or iron on the other. Examples are apparent at least from Early Urnfield times, in the bronze-sheathed wheels from the Hart vehicle grave, or those from Arokalya and Abos, and by the eighth century BC the Rhône-Rhine series. As we move into ironworking and especially from Ha C onwards, with the constant use not only of nave-binding and sheathing, but especially of the hoop-tyre, shrunk on to the wheel when red-hot, the close co-operation of wheelwright and smith moves from convenience to necessity. The importance of this combination of skills, and its social implications for the later Celtic world were pointed out by Sandars some years ago (Sandars 1962), and her conclusions can equally well be applied to the earlier period under review. One factor, which may well have applied throughout the prehistory of wheeled transport, is that of the use of seasoned and dry rather than freshly cut 'green' wood, even if the initial splitting of riven planks is best carried out on newly-thrown timber. If not for one-piece disc wheels, then certainly for tripartite discs, timber which will not warp and distort in drying is desirable, but with spoked wheels it is a necessity, and the seasoning of timber implies permanence of residence and suitable storage conditions in a social structure in which these can be regarded as normal circumstances available to the craftsman. The Hallstatt evidence certainly implies that by the seventh century vehicle building had reached a degree of competence which can with justice be termed professional, and this not only in the woodwork but the iron-work involved, and it is by no means irrelevant and anachronistic, but directly pertinent, to interpret our archaeological evidence in terms of the combined wheelwrights' and smiths' skills employed in building English farm wagons until the early decades of this century (Sturt 1923; Hennell 1934; Jenkins 1961; T. Arnold 1974). Here (as elsewhere in pre-industrial Europe), the workshop of the wagon-builder and wheelwright, and the iron-worker's smithy, were not only in close co-operation, but physically contiguous, above all to achieve the 'split-second timing' of exactly fitting a red-hot hoop tyre on to the fully made wooden wheel (on a wagon-wheel of 1.5 m diameter the expansion allowed by heating was only of the order of about 5 cm in the tyre's circumference—'anxious, impatient, gasping moments' as Sturt recalled from his personal experience.

In Hesiod's Boeotian village of the eighth century BC there were a resident carpenter and a smith, and while the farmer might be able to make an ard-type plough for himself, he took his own selected and presumably seasoned wood to the carpenter to make into a vehicle. This may well also have obtained in Hallstatt Europe, though as we have seen, we are ignorant of vehicles at this level of society and our evidence is from graves of the equivalent of Hesiod's infamous *basileis,* the chieftains who have been seen, by Ha D at least, as the occupants of seats of power and prestige. It is here that we should look for the settled craftsmen responsible for the construction of fine vehicles and, if this were the case, regional variants of style and workshop practice might be expected to develop, in the manner to be inferred in La Tène from fine metalwork. Distinctive forms of linchpins or metal-studded yokes in Ha C, or of nave-sheathing in Ha D, would be explicable in these terms. The concept of the itinerant craftsman in European antiquity, much used by Gordon Childe, has received serious criticism in recent years,

and is, as we have seen, very unlikely to be applicable to the twin crafts brought into such close relationship by the development of the iron-tyred vehicle of prestige, whether carriage or chariot, which eventually found its way into the grave.

It should perhaps be made clear at this point that what follows in this chapter is not a study of the vehicle graves of Hallstatt C and D, but of the technology and typology of the vehicles which formed a part of the funerary deposition, in so far as this can be inferred from the surviving remains. A complete corpus and detailed analysis of the graves in question, while obviously highly desirable, is a major piece of research to be left for full treatment by others. The evidence of the surviving vehicle remains can be taken into connection with, and illuminated by a few representations within the canonical area of Hallstatt culture, and others likely to be broadly contemporary outside this central area, in north Italy, the North European plain, and south Scandinavia. The vehicles depicted in 'Situla Art' form the subject of separate study later in the chapter.

Vehicle representations: pottery, rock art and models

The relatively scanty, but nevertheless valuable vehicle representations attributable to the seventh and sixth centuries BC, can be listed here, and their features dealt with later in conjunction with the grave-finds. A well-known find is that of the scene on one of the large urns with figural representations from the Ha C cremation-cemetery adjacent to the Burgstall hillfort at Sopron in Hungary, in Barrow 80. Here a processional scene of a horse-drawn, four-wheeled carriage with eight-spoked wheels and carrying a conical-topped object is shown, followed by a figure who may represent the driver, and preceded by a man on horseback. A more roughly sketched version of a similar scene appears

91

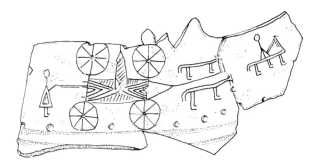

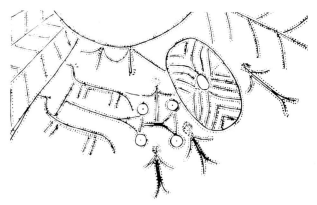

91 Representation of wagon and horses on cinerary urn, Ha C/D, Sopron, Hungary

92 Representation of wagon and horses, men and long shield, on cinerary urn, 6th–5th century BC, *Elsenau, Germany*

93 Rock carving of wagon and horses, 6th–5th century BC, *rock 62, Naquane, Val Camonica, Italy*

150

on a vessel from Barrow 28, accompanied by dancing women and two lyre-players, and again a horseman ahead of the carriage (Gallus 1934, pls II; VI–VII). Similar scenes again appear, outside the Hallstatt area, on the North European plain, in the large series of anthropomorphic cinerary urns of the so-called 'Pommeranian Face-Urn' class, and apparently only on those urns purporting to portray males, lacking the earrings, necklaces and neck-rings of those apparently for female burials. Wagons or carriages, with paired horse draught and sometimes drivers, are shown on a number of vessels in various degrees of stylization or sketchiness, the most informative being those from Grabau, Elsenau and Darslub. La Baume, in his *92* detailed studies of these urns, assigns the wagon representations to his Middle Phase, approximately equated with Ha D (La Baume 1928, 1963). Both the Sopron and the Pommeranian scenes strongly suggest funeral ritual akin in concept and spirit to that lying behind the earlier and more sophisticated friezes on Greek Geometric funeral pottery. A single isolated find of a wagon with paired horse draught incised on bone is that of a knife-handle of the Platenice culture, lying somewhere within Ha C/D, from Dobřcice in Moravia (Neustupny *95* 1961, 133; pl. 60) and in the Hochdorf grave of Ha D, wagon representations adorn the bronze sheathing of the funeral couch (Bit- *94* tel 1981, fig. 87).

Two wheeled models in metal must be mentioned here. The first is the unique and extraordinary cult-scene of figures in the round, set on a wheeled base, from Strettweg near Graz (Schmid 1934; Piggott 1965, pls XXVIb; XXXI; Sandars 1968. pl, 218). It was found with a cremation burial under a barrow, the ashes with three iron horse-bits contained in a large bronze vessel on an elaborate open-work stand, and the wheeled platform, 35 by 18 cm, carries a group of figures of riders on horseback, with helmets and oval shields, other figures, and stags, grouped round a central slender 'goddess', carrying a shallow bowl on her head, 22.5 cm high. The four wheels are shown with very broad felloes and eight thin spokes, and long reel-shaped naves, the whole resembling the four-spoked wheels of the Urnfield *Deichselwagen* models described in Chapter 4. The find is certainly Ha C in date: the bits and a fragment of a harness-ornament from the crest of a head-stall relate it to such graves as Mindelheim 7 and 11 (Kossack 1953 a, b) and as Sandars pointed out (1968, 214) the central 'goddess' figure is stylistically 'unthinkable apart from Greek Geometric bronzes', with a good seventh-century parallel from Olympia. Ultimately the idea of a wheeled stand with figures goes back to oriental and Cypriot prototypes (cf. Catling 1964, 207) and their counterparts in Italy such as those from Bisenzio, Lucera, etc. (Woytowitsch nos 125, 127, 132), but the strong Greek feeling in the Strettweg figures gives the group a numinous dignity quite alien to the scampering mannikins of Bisenzio. The second model is that in lead (from the local deposits) at Frög in Carinthia (Modrijan 1950; Pittioni 1954, 631), which has not been published in the detail it deserves. It has four ten-spoked wheels with relatively massive felloes and long naves, and a rectangular undercarriage with four cross-members, with four outward-curving uprights at the corners: more than this cannot be said from the published photograph. The Frög finds included many more lead figures including riders on horseback, and the whole, with its cross-handle-attachment situla of Von Merhart's Type B 2b, would date within Ha C.

The wagon representations in Scandinavian rock art have already been noticed in Chapter 4, where their salient features, including the structural detail of joining the fore and rear axle-beds by a Y-shaped longitudinal perch, have been referred to and the same applies to the Val Camonica series of similar construction. This same feature is

94 Representation of wagon and horses on bronze sheathing of couch, from wagon grave, Ha D, Hochdorf, Germany

95 Representation of wagon and horses incised on bone, Ha C/D, Dobřcice, Moravia

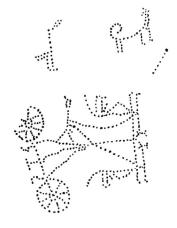

96 Representation of chariot, horses, driver and dog on pot, Ha C/D, Rabensburg, Austria

151

depicted at Hochdorf, Dobřice, and in, Pommerania; less clearly at Sopron and as we shall see, it helps us materially in the interpretation of the four-wheeled vehicles from the graves. Before turning to these, there remains for mention a final vehicle representation, that of a chariot, executed in pointillé on a sherd of a Ha C/D pot of Pittioni's Bernhardsthal type from Rabensburg in Lower Austria. It shows a vehicle of roughly triangular plan with two eight-spoked wheels (one in fact appears to have nine spokes) in which stands a driver with single reins leading each to a pair of entire horses yoked to the draught-pole (Felgenhauer 1962). As two-wheeled vehicles or chariots form a minority of the known vehicles of Ha C and D, it is convenient to deal with them after the four-wheeled wagons or carriages to which we can now turn in detail, combining the evidence from the representations with that of the surviving archaeological material from the graves.

96

Four-wheeled vehicles: undercarriage and body

The fortuitous circumstances of survival, and the inequality of excavation standards, leaves our knowledge of the essential wooden structure of Hallstatt vehicles extremely imperfect, though fortunately much can be inferred from the position of metal fitments recorded *in situ*. The evidence from the graves in no way conflicts with the contemporary representations just enumerated and in a few fortunate instances goes to confirm them, and provides us with enough measurements to show we are dealing with remarkably standardized constructional traditions. The 'double-Y' construction, especially as shown in the best known Scandinavian rock carvings, has long been held to demonstrate a theoretical derivation of the wagon in prehistory from a pair of linked carts of Y- or A-frame type (Haudricourt 1948; Haudricourt & Delamarre 1955,

162–64) and the writer at one time also inclined to this view (Piggott 1968, 295). A more detailed review of the evidence as set out in the previous chapters, however, leads one to question such a sequence, since A-frame carts are hardly known in antiquity outside Transcaucasia where, although they must lie behind the peculiar form of draught-pole in that area, the wagon undercarriages as at Trialeti and Lchashen, are of rectangular plan owing nothing to A- or Y-form construction. Furthermore in earlier prehistoric Europe carts of any kind are remarkably elusive in the archaeological record, and the four-wheeled wagon with a rectangular body (as at Bronocice or by inference at Zürich) is the earliest type of wheeled vehicle known, and must therefore be regarded as an independent and primary form.

The Hallstatt wagons or carriages under consideration, however, were clearly of a standard constructional form whereby the fore and rear axle-trees are joined by a central bar (technically a 'perch') which was evidently bifurcated or trident-shaped at its rear junction; a form of wagon undercarriage construction found again in the Dejbjerg carriages from Denmark of the first century BC and which continued in pre-industrial Europe until modern times and is particularly well documented and described for England in Jenkins 1961 and Arnold 1974. This is the undercarriage form shown in the wagon representations in Italy and Scandinavia in rock art, and in the other similar depictions. Hochdorf is particularly explicit, with Y-shaped perch and draught-pole attachment, long narrow body, and a pair of entire draught-horses (Bittel 1981, fig. 87). Some form of square-sectioned longitudinal perch was recorded in the Ha C vehicle in Hradenin Grave 46 (Dvořák 1938) and the iron reinforcements of a Y-shaped perch survived in the Early Hunsrück-Eifel culture carriage at Bell (Rest 1948; Driehaus 1966 a); at Vix an analogous

98

94

97

straight iron reinforcement of the perch was found (Joffroy 1958). In the representations Y-perches are normally shown, but trident forms are characteristic of the Val Camonica series at Naquane Rocks 57, 62 and 23 *93* (Berg-Osterrieth 1972) and appear on a Pommeranian 'Face-Urn' from Elsenau (La *92* Baume 1928, pl. 10): this is incidentally the standard form in English farm wagons. The surviving woodwork of the carriage in Hohmichele Grave 6 presents peculiar problems, as the oak axle-trees alone survived, but no trace of a perch, nor mortices to take *89* one. The fore axle-bed, however, carries a vertical central iron bolt or king-pin, which could have held the end of a perch, and the rear axle-bed is decayed and broken two points, about 40 cm apart, where lay a pair of tubular bronze mounts of wooden rods or pegs, 6.8 cm long and *c.* 2.5 cm in diameter which could similarly have held the bifurcated rear of a perch (Riek & Hundt 1962). Differential decay of wood other than oak might be more reasonable to assume than removal of the perch, and the matter is further discussed below in the context of the vexed question of a pivoted front axle in vehicles of this type is prehistory. The sizes to which such undercarriages were built were remarkably constant, as can be seen from measurements of sufficiently well-re-

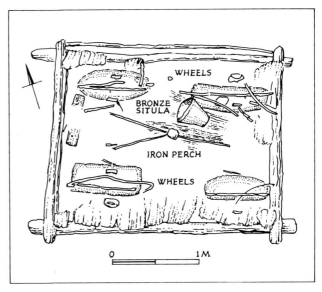

97 *Plan of wagon grave, Hunsrück-Eifel culture, Bell, Germany*

corded excavations. In Ha C, utilizing four vehicles (Grosseibstadt 1, Hradenin 24 and 46, Rvenice), the wheel-base (centre-to-centre of axles) ranges from *c.* 1.60 to 2.0 m; the gauge (centre-to-centre of wheels) from *c.* 1.10 to 1.30 m. In Ha D, with eight vehicles (Bad Cannstatt, Apremont, Hohmichele 6, Hradenin 28, Offenbach-Rumpenheim, Bell, Les Jogasses, Vix) the variations are even less, with wheel-bases ranging from *c.* 1.50 to 1.80 m, with five *c.* 1.80 m, and the

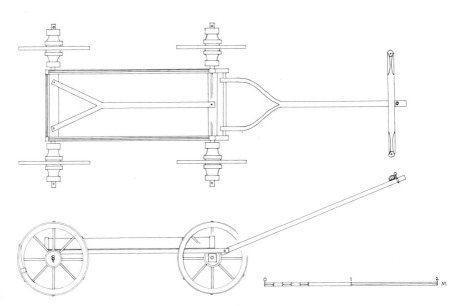

98 *Constructional diagram of Ha D wagon undercarriage*

gauge from *c*. 1.10 to 1.30 m, with seven at *c*. 1.20 to 1.30 m. The wheels will be discussed in detail in a later section, but so far as overall measurements of the vehicles go, the naves in Ha C and especially Ha D, have a fairly constant length of 40–45 cm, thus, allowing for linchpins and in some instances axle-caps, giving an overall axle-tree length of *c*. 1.85–2.0 m. The Val Camonica vehicles are shown with two longitudinal members parallel with a central perch forming an undercarriage of elongated rectangular plan rather narrower and longer than the measurements from the graves imply, if we assume their proportions approximate to reality, and Berg-Osterrieth has pertinently compared them with the long narrow wagons in use in the region today.

We next turn to the problem of reconstructing whatever superstructure in the form of body or box the undercarriage may have supported. Traces of plank flooring, 2.0 to 3.0 cm thick, were recorded in the Ha C vehicles in Graves 24 and 46 at Hradenin and (if we are correct in assuming an all-wood vehicle) in Lovosice 3, and again in the Ha D vehicle in Hohmichele Grave 6. Above floor-level, all the evidence implies that we are dealing with a very shallow structure indeed. Dvořák pointed out that one controlling factor in estimating this, quite apart from any surviving remains, would be the clearance available between the floor of the grave on which the vehicle stood and the flat timber roof: at Hradenin the grave-pits were between 1.0 and 1.5 m deep, and the vehicles had wheels *c*. 80 cm diameter, so limiting the height of any wagon-body on an axle-bed 40 cm above ground level to the remaining space under the roof-timbers and in Grave 46 he estimated the possible height of the sides of the body as 50–60 cm but, in fact, in the remaining graves suggested far less (Dvořák 1938). In Ha D we have direct evidence from the bronze or iron decoration on the sides of a vehicle which confirms this view of a very

shallow box or platform with decorative edging. At Bad Cannstatt, where ornamental sheathing, corner knobs and traces of actual wood survived, the box was under 15 cm high and the ornamental sheathing of the Ha C vehicle at Ins in Switzerland and the closely similar decoration on the Ha D vehicle of Ohnenheim in Alsace is only 10–12 cm high, and the iron-decorated sides of the recently excavated Ha D wagon at Hochdorf were 8.5 cm in height (Paret 1935; Drack 1958; Joffroy 1958; Biel 1981). In the reconstruction of the Bell carriage, further to be discussed, the actual body was made to be 14 cm high but could well have been less, and in that of the Vix vehicle 27 cm was allowed for elaborate open bronze-work and decorated wood sides (Rest 1948; Joffroy 1958). The published reconstruction of these vehicles, and that of Ohnenheim, are all related to that of the first-century carriage from Dejbjerg already referred to, and depend on certain assumptions relating to pivoted front axles which, as we shall see, are not in fact necessary, but led to raising the reconstructed body high above the axles on a hypothetical complex undercarriage. Excluding these, which alternative reconstructions would bring into conformity with the evidence from other examples just quoted, we seem to have no more than a plank platform with a decorated flanged edge as constituting the superstructure carried by the axle-and-perch undercarriage. With such a construction, the total height of the vehicles deposited in the graves would not be greater than the diameter of their wheels, and a clearance of about one metre between floor and roof would therefore be adequate. The width of the body would have been no greater than of the square-sectioned axle-bed between the cylindrical arms forming the axles on which rotated the 40–45 cm long naves of the wheels and so recoverable from a knowledge of the gauge afforded, for instance, by surviving metal nave-sheating, tyres, or even linchpins *in situ.* This averages under one metre and the width of the surviving superstructure at Bad Cannstatt, *c.* 70 m, seems a sound figure. The length is less easy to estimate but it seems normally not greatly to have exceeded the wheel-base average of 1.80 m, and perhaps *c.* 2.0 m might be a fair enough figure.

The decorative features in bronze or iron applied to these shallow bodies include corner-knobs: a characteristic form with an iron rod topped by a vase-moulded bronze head links the vehicles in the Ha D graves of Bell, Kappel, Hennweiler and Hatten (Driehaus 1966 b), and presumed corner-knobs were already present in the Ha C vehicle remains at Ins, and again, in Ha D, at Birmenstorf in Switzerland (Drack 1958). The hollow knobs on a rectangular base-plate, bronze at Bad Cannstatt and Býčískála and iron at Apremont, were found *in situ* at the corners of the box of the carriage in each instance and the close similarity of the wheels in the two graves may indicate the products of the same workshop. In the Hohmichele Grave 6, at the front of the vehicle were two sets of four bronze-knobbed mounts on ash stems, the width of the body and 40 cm apart, presumably from some sort of a decorative openwork front. The metal appliqué to the sides was of closely comparable openwork in both the Ha C vehicle at Ins, and the Ha D example at Ohnenheim, and at Vix turned bronze baluster-pins alternated with openwork plaques along the sides, with the front of the box framed in bronze. The relief bronze sheeting along the sides at Bad Cannstatt had rows of figures of little horses and men, in the style of a well-known Ha D series of bronze belt-plates with figural motifs (Maier 1958, 167–71) with ultimate Etruscan prototypes such as the bronze sheathing of a model chariot-body from Vulci, of the early seventh century BC (Woytowitsch 1978, no. 34).

At Ohnenheim and Birmenstorf were found the bronze mountings of what, taken with a wholly wooden example from Dejb-

jerg, can only be regarded as ceremonial chairs or 'thrones' standing on the floor of the vehicle and presumably performing some processional or ritual function. At Vix, the position in the grave of four ribbed bronze discs at the rear of the vehicle led France-Lanord, in his reconstructed model, to assume the presence of two uprights which would give the appearance of a throne-like structure at this point, again presumably ceremonial in intention (Joffroy 1958, figs 23, 24).

The draught-pole

Very little evidence of the nature and construction of the draught-pole of the vehicles of Ha C and D under review is available, and we await the detailed description of the elaborately iron-sheathed example from Hochdorf. Lengths, as we saw, are approximately known or estimated, and in the published evidence available there seems an almost complete absence of metal fittings which might enlighten us, save at Vix, to which we will return. The antecedent evidence from prehistory, and the contemporary vehicle representations, combine to make one fact least clear, that the articulation with the body of the vehicles would have been in some way bifurcated, with two points of attachment capable of vertical mobility. The first-century BC Dejbjerg carriage has its draught-pole complete, with decorative metal sheathing, and the pole bifurcates at its inner end to join a bar carrying two parallel arms about 50 cm long which presumably functioned for the articulation itself (Petersen 1888; Klindt-Jensen 1949, 87–108). The detailed reconstruction of the Dejbjerg vehicles is problematical and is further discussed in Chapter 6. In the two Ha C vehicle graves at Hradenin, 24 and 46, parallel woods traces were recorded forward of the vehicle towards the yoke, and in the former they are given as 64 cm long. These then might, in fact, be the arms of a draught-pole of approximately the type implied by Dejbjerg, but the absence of further details or of scale plans debars us from pursuing the matter. An alternative interpretation is that they represent, on analogy with modern English wagon construction, the two prolongations of the undercarriage known as 'hounds' which carry the transverse draught bar on which the shafts pivot to obtain the vertical mobility necessary in all four-wheeled draught to adjust to the height of the harnessed animals, the shafts in modern harnessing replacing the bifurcated draught-pole of antiquity (Jenkins 1961, 83–87; Arnold 1974, 10–11). Whatever the details of construction, some sort of arrangement would have been necessary in the Hallstatt vehicles, as the draught-pole's vertical movement of adjustment could not be obtained by attachment to the immobile square-section axle-bed, as Hohmichele 6 shows, and Rest's well-known reconstruction of the Bell wagon is erroneous here as well as in other details commented on below (Rest 1948, figs 9, 10). Vix supplies part at least of the answer. Here, when the wheels were removed and stacked against one side of the burial-chamber, it was necessary to free the naves from the axle-arms by removing the linchpins and the decorative bronze axle-caps through which they ran, and these were piled in a heap at the fore end of the body of the vehicle, on its metal-sheathed axle-trees, which was removed to form a bier for the corpse. Immediately forward of the front axle a row of bronze tubular mounts 3 cm in diameter were identified by Joffroy as the decoration of a draught-bar taking the arms of a bifurcated draught-pole, the bar like the axles being provided with miniature bronze caps and linchpins, which had similarly been removed to slide out the bar and remove the pole, and had been placed together on one side. No other details of the Vix draught-pole were recoverable except for two bronze fittings, a small knob-terminal and a knobbed yoke-

pin through a half-band for a pole 6 cm diameter (Joffroy 1958, 103–14). While there is a danger in reading too much into schematic rock art representations, the vehicle on Rock 62 at Naquane in the Val Camonica does seem to have a bifurcated draught-pole articulated to the front axle but independent of the undercarriage (Berg-Osterrieth 1972, figs 22, 23).

The problem of the pivoted front axle

In all discussions of ancient four-wheeled transport technology, one problem has been under continuous debate, the time and place in prehistory or history of the first use of the front axle moving on a pivot to facilitate the turning of the vehicle in motion. In early, and largely conjectural, studies, what appeared to be common sense (and an ignorance of primitive vehicles) denied the possibility of wagons with a pair of fixed axles, and this view was reinforced by the early publication of the Dejbjerg carriages (Petersen 1888) of 'double-Y' perch and pole construction, which in turn could be related to the Scandinavian rock art representations and those on the Pommeranian Face Urns. This led to hypotheses involving an 'evolution' of the wagon from linked carts, the front turning on the perch of the rear: judiciously summarized by Gösta Berg (1935), these theoretical views hardened into dogma (e.g. Haudricourt 1948) though as Jope pointed out (1956), Dejbjerg remained the one piece of tangible archaeological evidence. When the question of the use of a pivoted front axle in medieval Europe came under debate a few years later (Boyer 1960; Hall 1961) it was possible to write confidently that 'four-wheeled vehicles with the pivoted front axle came in at least by the time of the Hallstatt iron age' (Boyer 1960, 128), a view doubtless influenced not only by Dejbjerg but the reconstructions based on it by Forrer (1921) of the Ohnenheim carriages and by Rest (1948) of that from Bell. There then followed a period of inaction in the discussion, and a certain reaction as the early fixed-axle vehicles of Mesopotamia and Transcaucasia were described (cf. Piggott 1968). The problem must now be considered afresh in the light of the evidence of the vehicle from the Hohmichele Grave 6, and less clearly that of Vix.

If the front wheels of a four-wheeled vehicle are to turn, to however limited a degree, the centre of the axle-bed must be provided with a pivot holding it to the undercarriage. In the Hohmichele wagon, as we saw, the oak axle-trees alone survived of the structure of the vehicle, and the fore axle-tree, a 10 cm square piece of timber, had a central perforation holding an eyed iron pin or bolt, 1.4 cm diameter and 20.8 cm long, which can only be interpreted as a king-pin which would also pass through the front end of a missing perch, which it has been suggested might have been Y-shaped and vanished by reason of differential decay: it could hardly have been removed as traces of its superincumbent flooring planks were recorded. A pivoted front axle must in this instance be assumed. At Vix the documentation is unfortunately less precise but the excavator assumed a pivot, with a presumed wooden king-pin, from the remains (?wooden) of 'une plage de rotation de 16 cm de diamètre, cerclée d'une bande de bronze de 1 cm 5 de haut' at the front of the dismantled body of the vehicle (Joffroy 1958, 110). At Bell, a wooden king-pin was also assumed by Rest in his well-known reconstruction (1948, figs 9, 10) as at Ohnenheim in Forrer's restoration, which was closely modelled on Dejbjerg. In this last vehicle, further discussed in the next chapter, Klindt-Jensen in his careful consideration of its structure, assumed a missing iron pivot (1949, 87–100).

In the four reconstructions of Dejbjerg, Ohnenheim, Bell and Vix the archaeologists concerned all felt is necessary to make

allowance for the front pair of wheels wholly or partially to pass under the body of the vehicle in turning, which therefore had to be supported on a high undercarriage for which no evidence survived. As we have seen, the admittedly imperfect evidence from other Ha C and D graves consistently implies a very shallow body resting directly on the undercarriage and axle-beds. A consideration of recent English farm wagons suggests, however, that a pivoted front axle can achieve a satisfactory turning-circle for a vehicle of such construction, without invoking any hypothetical substructures. The degree to which the fore-wheels can turn is technically known as their 'lock', and traditional wagon-building recognized four possibilities, from quarter-lock to full lock, which could be embodied in wagon design, and what concerns us is the potential of the first, the 'quarter-lock in which the movement of the wheels is limited by the straight sides of the wagon floor'. This involves several constructional factors, notably the width of the body (if straight-sided, equivalent to that of the axle-beds and undercarriage), the gauge or wheel-track in relation to this and the length of the naves, the wheel diameters, and the wheel-base (axle-to-axle length). If these measurements are available it is easy to draw a diagram which gives one the angle of turn, or wheel-lock, and from this the diameter of the turning circle of the vehicle (Hennel 1934, 28; Jenkins 1961, 101–2, 112). To take the Hohmichele vehicle, it has as we saw the approximately 'standard' Hallstatt dimension of body width of 70 cm; gauge of 1.20–1.30 m; wheel diameters *c.* 80 cm; nave length 45 cm; wheelbase 1.80 m. As a quarter-lock vehicle it has an angle of turn of 20 degrees and a turning circle of about 8.5 m. Nineteenth-century English quarter-lock wagons (though, of course, larger vehicles) had angles of turn between 21 and 27 degrees and turning circles with diameters of between 9.0 and 11.5 m. (Cf. table in Jenkins 1961, 102). Inter-

157

preted in these terms a credible (and creditable) performance could have been achieved without the necessity of assuming elaborate undercarriage structures, and accepting the evidence for a shallow body only slightly raised above the level of the axle-beds. Nor need all Hallstatt four-wheeled vehicles have had pivoted front axles, though it is clearly a feature to be looked for in future excavations: the single iron bolt recorded without further comment between one of the pairs of wheels in the wagon-grave at Les Jogasses in the Marne (probably Late Ha D) is a case in point (Favret 1936, fig. 19).

Two-wheeled vehicles: carts or chariots

On several occasions in the early excavation reports of the Ha C and D graves two-wheeled rather than four-wheeled vehicles were claimed to have been found, usually on the grounds of the number of identifiable iron tyres. Later, and probably as a result of the recognition that the normal Hallstatt burial-rite involved a four-wheeled wagon or carriage and that of the succeeding La Tène phase a two-wheeled cart or chariot, these early records tended to be ignored or attributed to imperfect excavation techniques, though it was seen that in the context of the Hunsrück-Eifel culture at least, the latter form of vehicle burial was becoming dominant in Final Hallstatt times. The impeccable excavation of an undoubted Late Ha C burial with a two-wheeled vehicle in *87* Grave 4 at Grosseibstadt, however, renders a reconsideration of the situation necessary, and although several of the old excavation records may indeed be faulty, they cannot be totally dismissed, even if they may contribute little to our knowledge of the vehicles themselves. They constitute ten or a dozen graves in the west Hallstatt province and to them may be added the chariot represen-*96* tation of Ha C/D from Rabensburg already

referred to, and beyond the west Hallstatt area, probably both the vehicle graves at Sesto Calende within the Golasecca cultural zone of north Italy, and one of those from Atenica in west Serbia, of the late sixth or early fifth century BC, to be described at a later stage when the Hallstatt material has been dealt with.

Grosseibstadt Grave 4 has already been mentioned in connection with vehicles having a minimum of metal fittings, here confined to simple iron nave-bands. The wheels, which were set in a pair of slots in the floor of the grave, giving a gauge of 1.30 *87* m, had no tyres and were estimated by Kossack (1970) to be no more than 40 cm in diameter. The provision of pits or slots for the wheels looks back to the mid second-millen- *47* nium chariot graves in the Urals (Chap. 3) and forward to those of La Tène from the fifth century BC to be described in the next chapter, and the same wheel-track of *c*.1.30 m is common to all. The Grosseibstadt chariot was accompanied by a pair of iron bits at the far end of the grave, of anomalous three-link form with the very large overall width of 21 cm in contrast to the 12–13 cm mouthpieces of the bits ·from the other graves in the cemetery, implying a different method of control if not a larger breed of horse in this instance. Of the remaining graves of Ha C which could have contained two-wheelers, that in Hradenin 33 is inferred from a single pair of metal linchpins and would have been otherwise wholly of wood; that at Straškov was associated with a highly decorated yoke of 'Bohemian' type described below and iron nave-sheathing fragments; and the Lhotka (Welhotta) grave contained iron tyres 88 cm diameter and iron nave-sheathing together with bronze bits and harness trappings, together with, apparently, two horse skeletons (Koutecký 1968, fig. 2d; Dvořák 1938). At the end of Ha D in Czechoslovakia graves such as Sedlec-Hůrka 44 and Maňetin-Hradek 196 (with wheel-pits and a gauge of *c*. 1.30 m) look forward to Early La Tène (Soudská 1976). In Ha D Germany, the two graves of Uffing 1 and 11 were thought by Naue (1887) and Paret (1935) to contain two-wheeled vehicles though others (e.g. Schiek 1954) have been uncertain; in Switzerland the Ha D1 and D2 graves of Adiswil, Ins '1848–VIII', Jegenstorf and Payenne were similarly classed as having two-wheelers by Drack (1958), and in France, Grandvillars and Ste-Colombe La Garenne again have claims, the latter grave containing also the famous imported Greek griffin-cauldron and tripod (Joffroy 1958). The Hunsrück-Eifel chariot burials already referred to, and overlapping with Early La Tène in the fifth century BC, are considered at the beginning of the next chapter, as are those of comparable date in the Plzen region of Bohemia. Haffner (1976, 33–34) has remarked on their standardization in measurements and details, clearly traditionally established in Ha D and continued into La Tène, and has looked to individual workshops with recognizable products.

The foregoing examples, with the Rabensburg representation, may fairly be taken to demonstrate the continuance of the central European chariot tradition, going back at least to the Piliny Culture in the second millennium BC, through Urnfield and into Hallstatt C and D times. Grosseibstadt 4 certainly, and Hradenin 33 probably, show that metal hoop tyres did not necessarily form an inevitable concomitant from the first, even if, as with the four-wheeled vehicles, they were soon to be adopted in Ha C. With such a pedigree there is no need to seek anything but a local and indigenous origin for the chariot which has come to be so closely associated with the later world of Celtic Europe. In both the four-wheeled and two-wheeled vehicles the most significant elements are the wheels themselves, which, as we move from Ha C into Ha D, become increasingly provided with metal components in the form of sheathing for naves

and sometimes spokes, with iron tyres becoming constant; with these go in certain instances the use of metal axle-caps and linchpins. These, with a high survival value, enable us to make a typological approach denied to us by the perishable nature of the basic wooden structures to which they belonged.

The wheels: sizes and spokes

The presence of iron tyres enables us, where the record is adequate and sufficiently long segments of tyres survive, to see that the range of diameters is relatively restricted. For five Ha C examples the range is from *c.* 83 cm (Grosseibstadt 1) to 90 cm (Hradenin 24); for about twenty-five Ha D finds, the overall variation is *c.* 90 cm to *c.* 70 cm, with most clustered around *c.* 80 cm: Bell is *c.* 95 cm; Bad Cannstatt and Apremont *c.* 90 cm; Offenbach-Rumpenheim, Les Jogasses, Hohmichele 6, Hradenin 28, Burrenhof and Winterlingen lie between *c.* 85 cm and *c.* 80 cm; the Swiss examples between *c.* 80 cm and *c.* 75 cm; Vix is *c.* 74 cm and Augsburg *c.* 70 cm. The number of spokes, where this can be ascertained, is more variable. In Ha C, Hradenin 24 had eight spokes, and the same number is given by the excavator to the wheels in Grave 46, but Kossack, observing traces of the spoke-ends in the corrosion of the inner surface of the tyre, estimated fourteen, and sixteen to the wheels in Grosseibstadt 1 (Dvořák 1938; Kossack 1971). The fuller Ha D evidence aided by metal sheathing of the nave or of the spokes themselves, varies from ten (Hohmichele 6, Vix, Býčí Skála, probably Bad Cannstatt); eight (Ludwigsburg, Ste-Colombe La Butte, Sulz, Hügelsheim); six (Vilsingen Uffing 6, Hradenin 28) to four (Apremont). Metal sheathing of spokes as a whole or in part was either in bronze (Vilsingen, Ludwigsburg, Uffing 6, Hügelsheim, Býčí Skála, and small dentated collars at Vix) or iron (Apremont, Ste-Colombe La

Butte, Hochdorf). At the Magdalenensberg, Villingen, the spokes and half the felloes had leather sheathing, glued on by birch-pitch and finished with bronze nails (Spindler 1971). Leather may have been used to a greater extent than the excavation records, so many of them imperfect, suggest, as the leather covering of the magnificent bronze studded yokes of Ha C, described below, might imply. Little is known of the woods used for spokes, but those of the Hohmichele 6 vehicle were oak *(Quercus)* as in modern wagon-building; apple *(Malus)* was identified at Ins in Switzerland and fruit-wood *(Malus* or *Pyrus)* at Villingen.

Naves, axle-caps and linchpins: Hallstatt C

The wheel naves constitute one of the most informative features at our disposal, and with them go axle-caps and linchpins. For the presumed wholly wooden vehicles in Ha

C graves, for which a case has been put forward above, we have naturally no details of nave types, but the two pairs of iron nave-bands from the chariot-burial of Grosseibstadt 4 show us the beginnings of metal bindings to the wooden naves, which were to become wholly metal-clad in certain Ha D vehicles. But the wagon or carriage in Grosseibstadt 1 shows us the beginnings of a nave type which is long (33 cm overall) and symmetrical, decorated with alternating iron and bronze strips nailed on moulded mushroom-shaped terminals, and in Hradenin 46 the same nave-type with bronze and iron sheathing appears. As we shall see, the peculiar felloe type further links the wheel construction of these two vehicles, and the contemporary vehicle from Hradenin 24 only has simple hoop nave-bands. The Ha C vehicle from Ins (Switzerland) and that from Lhotka (Czechoslovakia) appear to have had conical ribbed iron sheathing on the naves, ultimately reminiscent of that in bronze on the Early Urnfield (Ha A1) vehicle at Hart-an-der-Alz described in Chapter 4.

In the vehicle graves of the Bylany culture in Bohemia metal linchpins, usually iron, were developed as a distinctive decorative feature, with pendant rings dangling from a normally double-looped head, or alternatively, triangular pendants similarly hang from the loops. As we suggested earlier, the record of the position of such linchpins in graves implies that they were the only components of otherwise wholly wooden vehicles but they equally appear with other vehicles having nave-sheathing and iron tyres. No comprehensive study of these linchpins seems to have been made—they were said for instance to have been present in almost all the important burials at Hradenin, a cemetery of sixty-nine graves—but they are recorded in Hradenin Graves 24, 30, 33, 46 and 58; in the two-wagon grave of Rvenice, and at Nymburk, Ohrada and Mirkovice (Dvořak 1938; Koutecký 1966; 1968). Out-

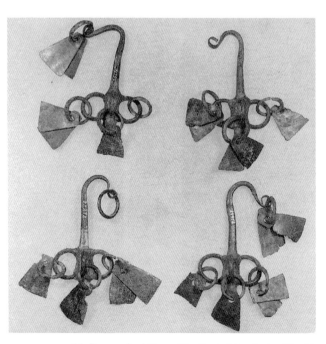

99 Bronze linchpins with attached pendants, Ha C, Grave 507, Hallstatt cemetery, Austria

side Czechoslovakia a couple of Ha C examples are published from vehicle burials with iron tyres at Illkofen, and with tyres and a bronze nave-band at Lengenfeld in Bavaria (Kossack 1953a, nos 4, 5). It is here that the four bronze linchpins with looped heads and tringular pendants in Grave 507 in the Hallstatt cemetery itself belong: it was an extremely richly furnished cremation grave with an iron sword with pommel inlaid with ivory and the linchpins would represent a four-wheeled vehicle otherwise wholly of wood and presumably burnt in the funeral pyre (Kromer 1959, 118; 1963, 46, pl. 24). Fragments of a linchpin of 'Bohemian' type with triangular pendants (as too at Lengenfeld) have been recognized by Spindler (1971) from the main tomb-chamber, robbed in antiquity, in the Magdalenenberg at Villingen of Ha C/D date. But the most remarkable linchpins of this group are the four bronze examples from a grave at Wijchen in the southern Netherlands. This is one of a group representing a Ha C enclave mainly within the bend of the Maas (maps in Mariën 1958, fig. 56; Verwers n.d., 2) and including the graves at Oss, Meerlo and Court-St-Etienne, all with horse-bits and other elements of harness. The Wijchen find, made in 1897, has still not been adequately published (cf. De Laet & Glasbergen 1959, 162, pl. 37; Megaw 1970, no. 16) but seems to have contained, in addition to a pair of bronze bits and harness trappings, iron tyres and four extremely elaborate bronze linchpins and their accompanying axle-caps, together representing an original vehicle-burial. The heads of the linchpins, trident-shaped with a cross-bar, each terminate in three human heads with long braided tresses of hair behind, and carry a profusion of rattling rings; the axle-caps through which they pass are cylindrical and 'hat-shaped', with a flange 10.5 cm overall, and decorated with stamped ring-and-dot ornament. Linchpins and axle-caps alike look back to those in the Komyatna hoard

99

100

100 Bronze axle-caps and linchpins with human heads and attached rings, and pair of bronze bits, from wagon grave at Wijchen, Netherlands

of Ha A1 described in the previous chapter; the caps are closely paralleled in the Early Ha D vehicle from Vilsingen (with a 'Rhodian' flagon of late seventh-early sixth century date). In common with the whole group of Ha C finds in the Low Countries the antecedents of the Wijchen vehicle lie in southern Germany and Bavaria (Kossack 1953a; Mariën 1958; Verwers n.d.) and a good parallel for the trident-and-bar linchpin in simple form is that from the Ha C vehicle grave of Illkofen (Kossack 1953a, no. 5). The combination, in Germany and for instance at Oss or Meerlo, of a warrior's sword and 'selective' harness elements in a grave which may contain no direct evidence of a vehicle is again characteristic, and here it is worth while making a parenthetical comment on a find from the British Isles. The well-known hoard of bronze and iron objects from Llyn Fawr in Wales, further discussed below, includes not only two bronze cauldrons, bronze tools and an iron sickle, but a Ha C iron sword, an iron spear-head, and bronze harness elements including bit cheek-pieces, looped discs ('phalerae') and what, in the continental Ha C contexts under discussion, would be termed a 'yoke-buckle' *(Jochschnalle)*, as Mariën pointed out. This assemblage, in fact, is exactly what might constitute the 'selective' furnishings of the graves under discussion, and one wonders whether the Llyn Fawr hoard includes the results of a successful tomb-robbery (Fox & Hyde 1939; Mariën 1958; Alcock 1961).

Naves, axle-caps and linchpins: Hallstatt D

As we move chronologically towards the end of the seventh century from the Ha C to Ha D phase, which then continues into the early fifth century, our material becomes abundant owing to the development of the technique of sheathing or binding the wooden nave of the wheel with bronze, iron, or a combination of both metals, increasing the chances of survival and opening opportunities for typological classification. Technologically these wood and metal naves must have marked an increasingly close association between wheelwright and smith, demonstrated in parallel as we shall shortly see by the universal adoption of iron hoop-tyres. They also raise explicitly a technological question which has been latent implicitly in all our discussions of spoked wheels from Urnfield times or even before, that of the employment of a lathe of some type for turning the naves with the required precision and indeed for finishing the mouldings of the metal coverings.

The technique of turning cylindrical objects of such substances as wood or ivory seems to have an origin in the ancient Near East and to have been transmitted to Egypt, the Levant and Mycenaean Greece by the later second millennium BC. In all instances rotating the object under manufacture while suitable chisels or gouges were engaged on its surface was the end to be achieved; there is no evidence to suggest that the continuous rotary motion of modern lathe-turning was achieved, but that the reciprocal or intermittent rotation imparted by a strap around the lathe spindle, actuated by ancillary human motion or that provided by a spring controlled by the turner (as in a pole-lathe), was the normal type of mechanism. Its first use in Europe north of the Alps is difficult to pin down, but the use of a lathe in the finish of bronze bowls in Urnfield times has been tentatively suggested (Piggott 1959) as it has in the finishing of the Ha D 'barrel-shaped' bronze armlets, and its use in turning contemporary wooden wheel-naves hinted at (Rieth 1940; 1950; 1955; Rieth & Langenbacher n.d.). It is tempting to look towards Italy, and Greek and Etruscan technology transmitted, as well as tangible objects, across the Alps and from the head of the Adriatic; the elaborately turned wooden cup from a Ha D grave at Uffing in Bavaria must be a Greek import (Naue 1887; Rieth 1940). Hypothetically, if lathe-turning of the nave and finishing of the mouldings of its metal sheathing, were considered as the last stage in the process of manufacture, an otherwise complete wheel (with indeed its iron tyring), mounted in a pole-lathe could soon be made to generate considerable speed and momentum, and, acting as its own fly-wheel, greatly facilitate the action of chisels and other tools brought to bear on the rotating nave. As will be seen, the Vix type of wheel-nave has as its profile a typical Greek enriched ogee-moulding proper to classical wood-turning, and it is difficult to see

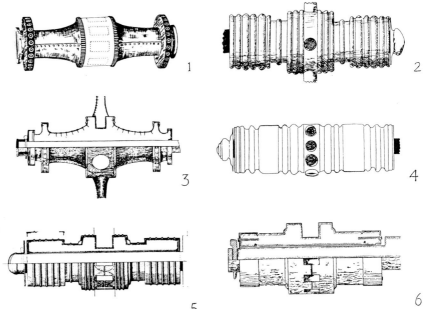

101 *Metal-sheathed naves, Ha D. 1, 3, Type A, from Winterlingen and Vilsingen; 2,, 4, 5, Type B, from Bad Cannstadt, Apremont and St-Colombe; 6, Type D, from Bell*

how the composite wood components of the Offenbach-Rumpenheim naves were achieved without a lathe (Frohberg 1974). Though the two technologies are not inevitably connected it may be worth while remembering that wheel-turned pottery was being imported (from the Este culture-area) and then locally produced by Late Ha D times (Kimmig 1974; Dehn 1963).

A sufficient number of Ha D metal-sheathed naves have been recorded (some thirty finds) to make a classification not only desirable but possible. Paret (1935) recognized the necessity and pointed the way to a typology developed by Drack (1958) which *101* is used here with minor amplifications, resulting in a classification into four groups which, primarily divided by typological and technological considerations, prove also not to be inconsistent with a chronological sequence. All are large and massive, *c.* 40–50 cm long, 12–14 cm in maximum diameter and with an axle perforation of *c.* 5–6 cm diameter.

Type A Biconical, with a bold squared moulding at the outer ends, wholly of bronze or of bronze and iron in combina-tion. In only one instance (Vilsingen) are metal axle-caps and linchpins present (Bur-renhof, Hohmichele 6, Hradenin 28, Hügelsheim, Offenbach-Rumpenheim, Sulz, Uffing 1 and 6, Vilsingen, Winterlingen). *101*

The bronze and iron elements are moulded accurately to the wooden cove of the nave and are sometimes made into decorative features: round knobbed rivets at Sulz, for instance, or decorative bronze inlaid iron studs in a zigzag pattern at Uffing 1. At Vilsingen the axle-caps are closely comparable to the Ha C examples at Wijchen, with ring-and-dot ornament, but with small knob-headed linchpins: the flamboyant 'Bohemian' linchpins of Ha C have gone out of fashion. Except for Burrenhof, which appears to be Ha D2, and both the Uffing finds, imprecisely within Ha D, all finds have been assigned to Ha D1. The Vilsingen grave contained an imported 'Rhodian' flagon, a Greek type already referred to but not more firmly dated than late seventh-early sixth century BC, and Hohmichele 6 and Hradenin 28 are linked by both containing bronze bowls with bossed rims, again sixth-century Italian imports (Dehn 1965; 1971). The distribution of

Type A naves is wide, extending from the Rhine and Upper Danube to Bohemia, and typologically the conical sheathing with bold end-mouldings recalls the examples, in openwork cast bronze, from Tarcal and other east European sites and with a date equivalent to the very beginnings of the Urnfield period, described in Chapter IV, and with the axle-caps of Wijchen in Ha C and Vilsingen in the present Ha D group, argue for continuity of workshop practice over a long period.

Type B Cylindrical with ribbed mouldings and as with Type A, symmetrical on both sides of the spokes. All are of iron, and have bulbous or knob-shaped axle-caps and small *101* iron linchpins. (Allenlüften, Apremont, Bad Cannstatt, Hochdorf, Ludwigsburg, Ste-*102* Colombe la Butte, Býčí Skála).

This is a very interesting small group, of such consistent design as to suggest the products of closely related workshops or even of a single tradition of craftsmanship. All appear to belong to Ha D2; the Hochdorf grave is dated to the Ha D1/D2 transition of *c.* 550 BC by the imported Greek cauldron with lions on its rim, and Ludwigsburg has a handled bronze bowl, again an import probably from Italy, but of uncertain date within the sixth century (Dehn 1971, fig. 3). The distribution is markedly western, with the exception of Býčí Skála, with one find (Allenlüften) in Switzerland and two (Apremont and Ste-Colombe La Butte) in France. The Apremont vehicle shares other features in common with that from Bad Cannstatt and Býčí Skála such as the globular knobs at the corners of the body. The flanged and knobbed axle-caps with bulbous ends have parallels at Vix (Type C) and one, with what may be a fragment of the outer edge of a ribbed Type B nave-sheathing, survives from a cart or chariot burial at Niederweiler, of the Hunsrück-Eifel Final Hallstatt phase (Driehaus 1966 b, 39). The long ribbed cylindrical form of these naves might be

102 Reconstructed wheel of wagon, Ha D, from the Býčí Skálá cave, Moravia

thought to embody a reminiscence of the Rhône-Rhine bronze wheels of Final Urnfield date described in the last chapter.

Type C Wholly bronze-sheathed, with symmetrical ogee-mouldings, with bulbous *103* axle-caps and small linchpins (Vix).

This type is represented by a single Hallstatt find: the four dismantled wheels of the vehicle in the famous Vix grave. While preserving in a sense the features of the Type A nave-sheathings, with prominent outer collars and a narrowing of the profile between these and the central part taking the spokes, their angularity has been replaced by an elegant double curve which is, in fact, the classical *cyma recta* or ogee-moulding. This was to become a standard nave form, wholly in wood or with at most iron nave-bands or bushes, in La Tène, represented by some of the latest examples of native wheel construction in the last centuries before and after the Roman occupation of Britain, as we shall see in the next chapter. The bronze axle-caps, with their small bronze crescent-headed linchpins,

164

stacked in a dismantled heap at Vix, are typologically closely akin to the iron examples of Type B just described. The Vix grave, with its Greek imported bronze crater and painted cups, must date from the very end of the sixth century, just before 500 BC (Dehn & Frey 1979) and it is to Greek sources transmitted through Italy, or via Massalia up the Rhone, that we must look for its classical mouldings. South of the Alps, a good ogee-moulded counterpart is found in the bronze nave-sheathings of the carriage from Ca' Morta, Como (Woytowitsch 1978, no. 112), of the Golasecca IIb phase of the late sixth century BC, and so contemporary with Vix.

Type D Naves mainly of wood, but with outer bushes of iron, often fastened by three or four dome-headed nails. No axle-caps, probably some small crescent-headed linchpins, but otherwise of wood (Bell, Chatonnaye, Grandvillars, Ins/Hermingen, Ivory, *101* Les Jogasses (?), Rances, Ste-Colombe La Garenne, Saraz, Urtenen).

After the elaborate decorative sheathing in bronze or iron of Types A and B, Type D becomes less demonstrative and more utilitarian. Dating is less well defined within the limits of Ha D: Drack assigns Rances to D1 and Chattonaye and Ins/Hermingen to D2. Ste-Colombe La Garenne was associated with the famous imported Greek griffin-cauldron and tripod of the mid sixth century (Boardman 1980, 221), with one griffin protome (no. 2) stylistically different from the rest, which has been thought to be possibly a locally made replacement, and so suggests a rather later date for its deposition in the grave (Joffroy 1960). The Bell grave, of the Early Hunsrück-Eifel culture, would fall late in Ha D, around 500 BC (Driehaus 1966a, b). The distribution of the type is again westerly and especially characteristic of Switzerland and France, and it is possible that the flat iron nave-end from the vehicle burial of Quinçay near Poitiers, one of the

103 Bronze sheathed naves, Ha D, Type C, from Vix, France

two most westerly of all such finds, may be related to this group. The vehicle burial with ribbed iron nave-bands from a robbed grave in the larger of the two barrows at Sublaines (Indre-et-Loire), assigned to a Late Hallstatt date, may be mentioned here (Cordier 1968; 1975).

The foregoing scheme is, of course, applicable only to the surviving metal-sheathed naves, and leaves unanswered (and unanswerable) the questions of the proportion and typology of plain wooden naves that may have co-existed with them. In itself, however, it does show the transmission of earlier technological traditions into Hallstatt C and D times, what must have been a short-lived and localized style of cylindrical iron-sheathed naves in Ha D2, and the anticipation of La Tène styles with the naves of Types C and D. Continuing our survey of Hallstatt wheel technology we finally turn to the construction of the felloes and the use of iron tyres.

Felloes and tyres

The basic discussion of the felloe-construction in Iron Age spoked wheels is that of Kossack (1971). His first conclusion, to which reference has been made in Chapter 1, was that the bent felloe, well attested in La Tène Europe in single-piece construction and known in at least two-piece form in

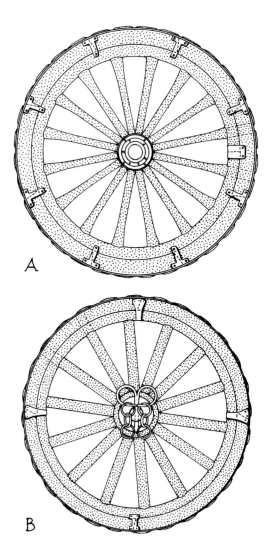

A

B

*104 Reconstructions of wheels, Ha C; A,
Grosseibstadt, Germany; B, Hradenin Grave 46,
Bohemia*

the plank, must be looked for, or multiple-ring felloes which might embody both plank and bent-wood components. Proceeding from this, he was able to study two exceptional sets of Ha C wheels, from the graves of Grosseibstadt 1 and Hradenin 46, already discussed, by a careful examination of the wood grain preserved by corrosion products on the insides of their complex iron or bronze felloe-mountings. From these observations he demonstrated that in both sets the felloes were composite, with an inner bent-wood element (single-piece with a butt-ended joint at Grosseibstadt) and an outer, segmental felloe, again with butt-end (and presumably morticed) joints, in eight pieces at Grosseibstadt and four at Hradenin, the whole held together by T-shaped or expanded-ended clamps and by a nailed-on iron tyre. At Grosseibstadt the inner bent-wood felloe had a U-shaped clamp at the junction of its two ends. In their complex double form these wheels appear to be without parallel in Hallstatt Europe.

Iron joint-braces or felloe-clamps are recorded with Ha D wheels at Bell, Les Jogasses, and Vix. As Kossack pointed out, the thickness of the felloe implied at Bell (8 cm) makes it unlikely that it was a bent-wood construction, and the wood appears to have been oak, hardly appropriate to this technique, but at Les Jogasses the thickness would only be about 5 cm, and so bent-wood is likely. The Vix joint-braces were of openwork bronze, more decorative than functional, but appropriate to a thin bent felloe, and thicknesses of up to 5–6 cm can sometimes be inferred from tyre-nails hammered over to clench them on the inside of the felloe. Nail-lengths (where recorded) of over this length imply plank-wood segmental felloes of some kind. Before considering the possible affinities of these forms of construction outside Hallstatt it is necessary to turn to the question of the iron tyres themselves.

second-millennium Egypt and elsewhere, can only be achieved with wood of a limited thickness, and he suggested a useful rule-of-thumb proportion of 1:10 for felloe thickness to wheel radius. In other words, a single-piece felloe for a wheel of 80–90 cm diameter cannot be bent from a straight piece of wood much more than 5 cm thick. If there is evidence from metal mountings or clamps that the felloe thickness is in excess of this, alternative forms of construction, such as the use of segmental felloes cut from

Our survey up to the present has shown that the spoked-wheeled vehicles with presumptively horse draught from the second millennium BC did not employ metal tyres, and if tyres of any kind were, in fact, used, they were of perishable substances such as rawhide, and that there is good reason to believe that such construction continued into the seventh century and the Ha C phase. It is in this period, however, that we first encounter north of the Alps a technological innovation which was thereafter to remain an essential feature of wheeled vehicles up to and beyond the Industrial Revolution, the provision of a hard bearing-surface on the outer edge of the felloe as an iron band or tyre. This took the form of an iron strip, averaging about 5–6 cm wide and 0.5–0.6 cm thick, bent in a hoop, the ends either welded together or, in two recorded Hallstatt instances, Kappel and Les Jogasses, nailed through an overlapping or scarf joint. The welded hoop-tyre demands considerable technological skill to make and fit to the felloe of the finished wheel, depending as it does on an accurately measured correlation between the circumference of the outer edge of the felloe and that of the closed iron hoop, whereby the latter when raised to a red heat will expand sufficiently to enable it to be fitted over the wheel, and by its subsequent contraction to hold tightly in position and bind the component elements of felloe, spokes and nave together. It has already been mentioned early in this chapter that in recent English wagon-building the tolerance allowed was only about 5 cm in the circumference of a 1.5 m diameter wheel. For the 80–90 cm diameter Hallstatt wheels, the smith would have needed a wrought iron strip, finished to the width and thickness of about 5.0×0.5 cm, of over 2.5 or 2.8 m long respectively, measured accurately to fit the individual felloe, bent to a true circle, provided with punched holes to take the fastening-nails, securely welded and then uniformly heated for its whole circumference to obtain the calculated degree of expansion to enable it to be dropped by tongs over the prone wheel, quickly hammered into place and allowed to cool. It is a technical feat of no mean order, and it is ironical that this prehistoric achievement was forgotten by the European Middle Ages, when wheels were clad with nailed-on iron segments or 'strakes' until the single-piece hoop-tyre was reinvented in the Industrial Revolution for carriages in the late eighteenth century, but was first being used on English farm wagons little over a century ago (Sturt 1923; Jenkins 1961). To return to prehistory, if one-piece wet rawhide tyres were being shrunk on to wheel felloes, the principle of contraction and compression would already be familiar to wheelwrights.

The tyres themselves are on the whole too uniform in pattern to be used typologically, except for those which have a groove wrought on the outer surface to take the nail-heads. It is the nails themselves which are more informative. Cataloguing the vehicle finds from seventh and sixth-century Italy, Woytowitsch drew attention to the fact that the spacing of the frequent nails on the iron tyres there could be roughly divided into two groups, one with close intervals (3.0 to 7.0 cm) and the other with wider spacing (18 to 20 cm), but did not pursue the matter further (1978, 6–8; pl. 59, 1–2). Independently, the writer had noted the same for contemporary Hallstatt Europe and for the Later La Tène tyres, where the fastening-nails become even fewer, and are eventually dispensed with altogether (Piggott 1979). Before turning to the Hallstatt evidence in detail, a distinction can be made between the types of nails, which in part reflects function. The first type may be termed *stud-nails,* with large domed or lentoid heads, set closely and constituting the 'tread' of the wheel rather than the tyre itself, beyond which they project even in the occasional grooved type (e.g. at Lhotka), the second *fastening-nails,* with small heads nearly flush with the tyre face and set at wider intervals, no longer as an armature but as additional security to the shrunk-on tyre. When the approximate spacing and the wheel diameter (and so circumference) is known, an estimate of the number of nails per tyre can be made from fragments, and checked against complete contemporary examples. Expressed in this manner, and on the basis of the published details of about 20 wheels, a division can be made between 'close' spacing of stud-nails, their heads sometimes contiguous, of between *c.* 30 to 40 nails per tyre; 'medium' (and now increasingly fastening-nails), of *c.* 20 to 30; and 'wide', invariably fastening-nails, and with numbers below 20 and sometimes as few as 6. Into the first group would come the tyres of Grosseibstadt 1 (?32), Hradenin 24 (33), and Nymburk (30–40) in Ha C, and Sulz (42), Ebingen (?40) and Villingen (?40+) in Ha D1. The second, medium-spaced, nails would include Hradenin 46 (20) and Ins in Ha C; Bad Cannstatt, Apremont, Les Jogasses, and the Swiss group, all around 24 to 26, in Ha D2. In the wide-spaced series the numbers drop drastically and all are in Ha D2—Offenbach-Rumpenheim 18, Saraz ?12, Vix and Bell 10, Poligny 8 and Hohmichele 6 and Hradenin 28, 6. Although, as might be expected, there is not a simple unilinear progression, a general tendency for the nail numbers to diminish, and for fastening-nails to replace stud-nails, is clearly apparent from the seventh to the late sixth century. While a fuller study is clearly desirable to check and amplify this sequence, it does seem to offer in outline a likely pattern, in which the primary innovation in a technological world of wheels without tyres in the seventh century north of the Alps is the iron tyre with close-set stud-nails, with subsequent developments, varying between workshops and local centres of production, tending towards a later tradition of wider spaced fasteningnails.

The Hallstatt wheels: continuity and innovation

As the length of the foregoing discussion alone indicates, it is the wheels of the Hallstatt vehicles which provide us with the largest body of information in assessing the technological traditions in the horse-drawn transport of Europe north of the Alps in the seventh and sixth centuries BC. Most of these constructional features can be seen to be indigenous and of long standing. Vehicles with one or two pairs of spoked wheels were in use by Early Urnfield times, from the beginning of the first millennium, and with their ancestors in the second. Felloes of bent-wood construction in one or more pieces have a similar pedigree, if only inferentially; segmental felloes cut from the plank are known from the Barnstorfer Moor example and from those slotted into the rims of the bronze wheels of the Rhône-Rhine series of the eighth century. The bronze sheathing of naves and spokes is again a technique which goes back to such vehicles as the Hart-an-der-Alz carriage of the beginning of the first millennium BC, as do the decorative axle-caps and linchpins from this and from contemporary hoards. All these features demonstrate an accomplished vehicle technology already flourishing and easily accommodated by and indeed enhanced and developed with, the new iron technology of Hallstatt C and D.

What stand out as innovations are the complex felloe construction with iron joint-braces and felloe-clamps, and above all the nailed-on iron tyres which form a part of these exceptional wheels from Grosseibstadt and Hradenin and become a standard feature of wheels of simple felloe construction, supplanting the wooden wheels without tyres which we have seen reason to believe were also in use in the seventh century. The question, of course, is whether such iron tyres are a Hallstatt invention, or whether an origin may more plausibly be sought out-

105 Assyrian relief of Asburbanipal (668–626 BC) showing chariot with composite stud-nailed wheel, from Nineveh

side. Nearly a century ago Julius Naue, considering the wheels of the Ha D vehicle he had excavated in Barrow 6 at Uffing, with its iron tyres and close-set stud-nails, looked to the representations of Assyrian chariot-wheels, and reconstructed his Bavarian find in accordance, assuming a 'double-felloe' wheel in six segments (Naue 1887, 38, 144, pl. XXXIX, 2). As we saw, Kossack (1971), using the evidence of the iron fastenings on two sets of Ha C wheels, interpreted these as belonging to complex 'double-felloe' construction, and for parallels looked also to Assyria, and the depictions of wheels with composite felloes, tyres, and prominent close stud-nails. But so far as the tyres go, Woytowitsch has made it apparent in his publication of the Italian material that we have to look no further than across the Alps for counterparts, and it would seem a sequence from close-set stud-nails to wider spaced nails with smaller heads, and we have already noted his classification in these terms. Of the Etruscan iron chariot tyres which can be roughly classified, close-set stud-nails occur in four tombs of seventh-

century and two of seventh-sixth century date; wider spacing in one seventh-century tomb, two of the seventh-sixth century and two of the sixth century BC. In addition the tyres of the two Sesto Calende vehicles, of the Golasecca culture and contemporary with Ha C in the seventh century, have very close-set nailing—48 in one instance and about 42 in the other (Woytowitsch nos 15b, 23, 28, 34, 67, 100; 15a, 86, 95, 98, 89).

But, as we saw at the end of the last chapter, Villanovan, and later, Etruscan Italy was in contact with the Assyrian and Near Eastern world from the eighth century BC at least, and Stary (1979a) has suggested that military innovations from that quarter might be represented by the horse-bits from graves, as a *pars pro toto* symbol of chariotry, though as he goes on to say 'whether there was also introduced, together with these bits, the two-wheeled war-chariot ... cannot be established'. But by the orientalizing period, from the late eighth to the early sixth century BC, Near Eastern types of offensive and defensive weapons were certainly being adopted and 'together with the weapons from the Near East ... war-chariots were also taken over from there'. Stary instances the Assyrian reliefs for contemporary representations, and so far as stud-nailed tyres are concerned, it is, of course, to these that Naue and later Kossack turned. In the Near East, the use of large stud-nails as the armature of tripartite disc wheels without metal tyres goes back well into the third millennium as at Susa, or the Tell Agrab model, and appear represented, with tyres presumably of iron, on reliefs from the time of Sennacherib (705–681) and Ashurbanipal (668–626) and elsewhere (Frankfort 1954, pl. 110; Littauer & Crouwel 1979a, 108; figs 5, 7; Kossack 1971, pl. 20, fig. 35, 3). Stud-nailed iron tyres may indeed be Assyrian in origin, but in Europe their presence must be seen as part of the Etruscan adoption of Near Eastern weaponry together with other aspects of oriental culture. Cyprus may incidentally be excluded as a source for chariotry, as although composite felloes were in use in the eighth and seventh-century chariots there, metal tyres or nail-studding were not a part of their construction.

So far as Hallstatt Europe is concerned, it is difficult to see the appearance of the stud-nailed iron hoop-tyre as other than a technological novelty introduced in the seventh century from Cisalpine Italy, where it was itself an introduction from the Near East. That the sequence from close-set stud-nails to wider spaced fastening-nails seems to have proceeded in parallel in Cisalpine and Transalpine Europe argues that the technology once established, craftsmen on both sides of the Alps maintained contact and exchanged ideas: the adoption of lathe-turning and the ultimate use of classical mouldings on the wheel naves may be involved here. Contact between the two areas has long been recognized at the top level of Hallstatt society in the imported bronzes, Greek painted pottery and other exotica, but Ridgway, commenting on the relationships of the culture-areas of Golasecca, Este

and Hallstatt, saw 'a degree of "contact" between the three areas—two Cisalpine and one Transalpine—at a social level that certainly cannot be defined as uniformly "elevated" or "princely"', and looked to 'a way of life that lent itself as a basis for the free circulation of artisans—primarily from south to north—and ultimately the political alliances that required public recognition in the form of magnificent gifts' (D. Ridgway 1979, 417). Into such a pattern the transmission of technological innovations in vehicle construction in the seventh and sixth centuries BC would find an appropriate place.

The Hallstatt C yokes

Perhaps the best known and most spectacular finds from the Ha C vehicle graves are the elaborately decorated and massive yokes surviving virtually complete from five Bohemian graves and in fragments from at *106* least two burials in Bavaria and one in Switzerland, in addition to that from the Stadtwald, Frankfurt grave. (U. Fischer 1979). None of the earlier complete finds has been adequately published in detail (Dvořák 1933; 1938; Pleiner 1959; Pleiner & Moucha 1966; Mariën 1958, fig. 55a; Piggott 1965, pl. XXXIII; Filip 1966–69, pl. LXXXII). The yokes come from Hradenin Graves 24 and 46 (four-wheelers), and Lovosice Grave 3 (with a possible vanished *107* vehicle), and Straškov with a two-wheeled vehicle, and Frankfurt and Plaňany (with no *108* vehicle). They are a metre or so overall (the size of that from Hradenin 24 is given as 1.26 m and of those from Straškov and Frankfurt about 1.30 m). Constructionally they have a wooden basis (Hornbeam *(Carpinus betulus)* at Lovosice) overlaid with leather (and apparently also fabric at Hradenin 24), and elaborately decorated with closely set bronze studding. Typologically, Hradenin 24 and 46, and Lovosice 3 are very close and suggest the products of one workshop or craftsman, with a broad central part

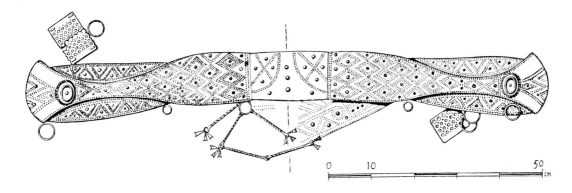

106 *Wooden horse-yoke, leather-covered with bronze studding, bronze rings and yoke-buckles, Ha C, Grave 46, Hradenin, Bohemia*

and a pair of lateral neck-pads over which is modelled a flattened-down version of recurved terminals each with a bronze 'cup' at its tip. Frankfurt is similar, but the 'cups' are moulded and studded, not bronze. At Plaňany there are similar cups, but no mouldings to suggest recurved terminals, and at Straškov neither mouldings nor cups, but a broad scrolled end. Frankfurt, Plaňany and Straškov are also linked by their decoration of close-set square-headed studs in a chequer-board pattern, while the Hradenin and Lovosice yokes have dome-headed studs in elaborate lozenge and zigzag patterns, as have the fragments from Moritzbrunn, Gaisheim and Ins. There are no attached terrets or rein-guides, but at Hradenin rings spaced out along the inner edge, and similar rings at Frankfurt, might have served this purpose in some way.

Apart from the bronze studding, two types of metal fittings from the yokes, characteristic in form and with a higher potential of survival than wood and leather even in a cremation grave, enable us to extend the distribution of the basic yoke type over a large

107 *Part of bronze-studded horse-yoke, Ha C, from grave at Lovosice, Bohemia*

108 *Part of bronze-studded horse-yoke, Ha C, from grave in the Stadtwald, Frankfurt-am-Main, Germany*

area of Ha C Europe. The first is the terminal bronze cup or 'rosette', oval or circular and found in pairs, the second the openwork and bossed plaque or 'yoke-buckle' which in some way seems to have been a strap-fastening associated with the yoke, near which such buckles lie in undisturbed graves such as Hradenin. These have been listed and mapped by Kossack (1953a, map 4) and Mariën (1958, fig. 56) with other elements of harness discussed in the next section, and take the distribution of versions of the 'Bohemian' yoke westwards from Bavaria to the Ha C 'enclave' in the Low Countries already referred to and with new finds, into Champagne and the Jura (Flouest & Stead 1979, 12–15) and eventually to the yoke-buckle at Llyn Fawr in Wales, already mentioned and discussed further below.

One remarkable piece of technical information comes from the fragments of leather surviving from the Baron de Bonstetten's 1848 excavations of the Ins grave in Switzerland. Examined by the distinguished leather chemist Gansser-Burckhardt, this leather proved to have been prepared with the valonea tanning agent, derived from the dried acorn-cups of the oak *Quercus aegilops,* used to the present day. The natural distribution of *Q. aegilops* is Turkey, Syria, the Levant, and the Greek islands and mainland, an area within which the valonea tanning process must have originated, but up until the 1930s a long-standing traditional trade in valonea existed between Smyrna and Trieste, up the Adriatic, to supply the north Italian leather industry in such centres as Milan and Turin (Drack 1958, 14; Reed 1972, and *in litt.* 1980). The subject cries out for further analyses, but in the Swiss instance we are undoubtedly dealing with imported leather, either from the eastern Mediterranean direct or, if the recent trade is a reflection of a long-standing order of things, perhaps more likely from the Atestine region, as yet another transalpine contact in the seventh century BC.

In themselves, the yokes are unparalleled in the elaborate form in which we know them from the complete examples, but this may partly be due to their perishable nature. Of the typology of the wooden yokes of more widespread distribution implied by the bronze fittings we have no knowledge, though the cups or 'rosettes' imply decorated terminals of some kind. The little representation on the Sublaines urn of Late Urnfield date referred to in Chapter 4 may imply some sort of terminal and, as Dvořák (1938) and later Mariën (1958) pointed out, *69* the surviving wholly wooden yokes from La Tène, with emphasized neck-pads and cupped terminals, offer significant comparisons at a much later date, probably the third *137* or second century BC (Vouga 1923, pl. XXV; Jacobsthal 1944, no. 172). Cupped bronze terminals, probably from yokes, appear in La Tène contexts, both on the Continent and in Britain (Mariën 1961; Piggott 1969). Yokes with recurved ends and cupped terminals are again represented on *138* Assyrian reliefs from the ninth to seventh century BC (Littauer 1976a; Littauer & Crouwel 1979a, 84, 113; Yadin 1963, 420; Potratz 1966, pls II. 3, XXIX. 64a, XLIV. 97). What we know of Etruscan yokes shows nothing similar (Littauer 1977a, pl. 19; Woytowitsch nos 85, 168, 280, 169, 169a). If eventual oriental styles are to be involved, they do not seem to have been transmitted through Etruscan intermediaries.

Bits and harness ornaments: Hallstatt C

When turning to the horse-bits and harness of Ha C we encounter once again a very complex set of circumstances, still imperfectly understood, which manifests itself first in Urnfield times as we saw in the previous chapter. Despite a number of studies since that date, the fundamental survey is still that of Kossack thirty years ago (1953a), though some of its assumptions have been rendered invalid by later work, notably that of Thrane (1963) and Balkwill (1973), and new material was assessed by Mariën (1958). It can now be seen that already by Early Urnfield times a translation of 'soft-mouthed' bits with antler cheek-pieces to bronze forms had taken place in the western part of what by Ha C times was to become a very large province of Europe, extending from Switzerland to Hungary and with counterparts in the Pontic region and Transcaucasia (Kossack 1953a, map 5) and including the material from the Carpathian Basin originally ascribed to 'pre-Scythian rider-folk' by Gallus and Horváth (1939). It was this easterly material which came to be associated with the historically documented Cimmerians on the grounds of an ambiguous passage in Herodotus, and a 'Thraco-Cimmerian' origin sought for the whole series (cf. Gimbutas 1965, 479–517). The Pontic and Carpathian material is difficult to date in itself or in correlation with the Western Urnfields—Ha C sequence, but it seems that none of it need be earlier than Late Urnfields (Ha B 3). The nature of the eastward, Transcaucasian, element remains obscure, but as we saw, Moszolics looked in this direction for the openwork ornament of pieces such as the Tarcal nave-sheathing (1956) and some such contacts are involved in some of the precocious iron objects in Late Urnfield times, notably the short swords or daggers of Gamów type (Powell 1976b; Podborsky 1967, with map, cf. Pleiner 1980). So far as the wheeled vehicles under consideration are concerned, their occurrence in graves, associated with the characteristic horse-gear under discussion, extends no further east than Bohemia.

The problem of these bits and associated bronze harness mounts is to assess the relative weight, in the area of the vehicle burials which are our primary concern, of those features which are of clear Urnfield derivation or Ha C modifications of this tradition, and

those which may appear to be innovations. Kossack recognized four main provinces within the general area of distribution, those of Bohemia, Bavaria, West Hungary and Carinthia-Carniola, with vehicles and yokes represented only in the first two. In addition, there are the sporadic westerly outliers, notably the Low Countries 'enclave' including Oss, Court-St-Etienne, Meerlo and Wijchen (Modderman 1964b; Mariën 1958; Verwers n.d.) and isolated finds in eastern France (Flouest & Stead 1979) and even in the south (Taffanel 1958; 1962). The western finds Kossack classified into five groups as follows:

1 Graves with simple harness ornaments, bronze or iron bits (e.g. Lager Lechfeld)
2 As above, with iron bits and a vehicle (e.g. Gehrsricht)
3 Graves with elaborate harness ornaments, bits with variant forms of knobbed bar cheek-pieces, strap-crossing buttons and sometimes phalerae (e.g. Mindelheim 11, Gilgenberg)
4 As above, but with iron bits, yoke and vehicle (e.g. Hradenin 24 and 46)
5 As Group 3 but with yoke-mountings alone and no vehicle (e.g. Thalmässing, Oss, Court-St-Etienne)

Of these, Group 4 is obviously the most significant for our purpose, and the relevant burials have already been discussed. It should also be emphasized that, as with all assemblages of material from graves, we are dealing with a conscious selection of components in accord to rites, customs and beliefs now lost to us, so that our categories are no more than our own rational attempts to impose order on circumstances which may well be the product of mental and emotional states wholly alien to us. The bits are normally found in pairs in the graves in a manner proper to paired horse draught but in certain exceptional graves, for instance Hradenin 24 and 46, Lovosice 3 and Rvenice, three bits are found, interpreted

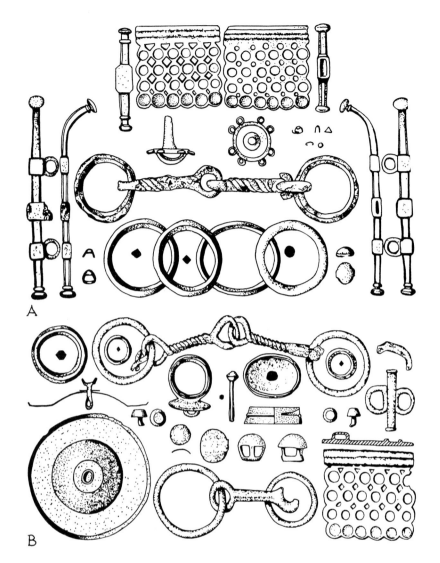

109 Grave-groups with bronze bits, yoke-buckles and harness attachments, Ha C, from A, *Beilngries;* B, *Gernlinden, Germany*

variously as representing two horses for draught and one for riding, or three-horse draught with two animals under the yoke and one with traces. On the evidence available, no certainty can be expressed.

The bits are all of the two-canon type *109* already described as having emerged as a standard form in Urnfield times and the differentation of types has largely been based on the cheek-pieces of bronze or iron: Balkwill (1973) makes an interesting point with regard to the adoption of iron mouthpieces, in that they may have been less likely to produce toxic corrosion-products with the horse's saliva than those of bronze. The

commonest cheek-piece forms are versions of straight knobbed bars deriving from easterly types of Urnfield date (e.g. Kiskőszeg) and continuing with the commonest form, often in iron, with a curved upper end (Thalmässing) or a spatulate lower end (Gilgenberg). A contemporary Ha C variant form is the crescentic form as at Lengenfeld, with its differentiated counterparts among the western outliers such as the long thin crescentic form at Maillac in the south of France (Taffanel 1962) or the distinctive forms of Court-St-Etienne in Belgium or Llyn Fawr in Wales (Mariën 1958, 236). With these various forms of bit are asso-

174

ciated other bronze elements of what must have been elaborately decorated leather harness, especially head-stalls and other ancillary bridlery, attempts to combine which have been made in more than one reconstruction (Kossack 1953a, fig. 1; 1953b, fig. 4; Mariën 1958, fig. 46, whence Piggott 1965, fig. 99). The distribution of these harness ornaments has been mapped by Kossack (1953a) and Mariën (1958) and for our purposes we may draw attention to two main types, both with Urnfield antecedents, the phalerae with internal attachment-loops already referred to in Chapter 4, and devices for carrying two straps at right angles, either as cruciform tubular mounts or, more commonly, buttons with a cruciform attachment at the base.

The phalerae of Ha C (which continue into Ha D) in the reconstructions just cited, have been incorporated in the head-harness of the horse, but may also have been used in other parts of the body-harness. At Plaňany, eight examples of the variant 'tutulus' form, with a central spike, were arranged in two rows across the centre of the decorated yoke already described (Filip 1966–69, pl. LXXXII). The cross-attachment buttons have an extraordinarily wide distribution, with the earlier examples, contemporary with the western Urnfields (Ha A—B) in the Carpathians and beyond, even occurring in the Iranian cemeteries of Giyan and Sialk (Gallus & Horvath 1939) and those of Ha C appearing in outlying westerly contexts such as Maillac (Taffanel 1958; 1962) and Britain, (discussed in the next section). One curious type of bronze head ornament, so specialized as to suggest a passing fashion prompted by an individual craftsman, is what in modern English cart-horse harness was known in the late nineteenth-century trade catalogues as a 'fly terret', a stemmed ring or rings mounted on the poll of the
110 horse between the ears carrying free-swinging discs within the rings (Keegan 1978, 161, figs 35–36). The Ha C examples, from

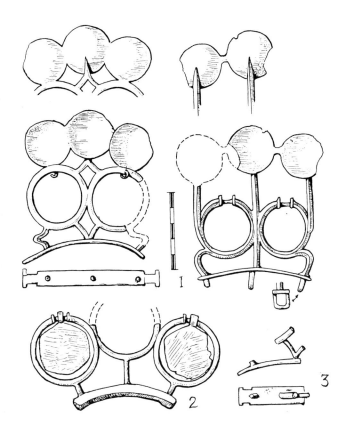

110 Bronze 'fly terrets' from harness, Ha C, from 1, Leibnitz, Austria; 2, Seddin, Germany; 3, Strettweg, Austria

Mindelheim, Leibnitz, Seddin and Strettweg (with the cult-wagon already referred to), Langenthal in Switzerland and Court-St-Etienne in Belgium (Kossack 1953b; Mariën 1958), show how widely distributed individual forms of harness ornament could become within what must have been a quite brief period of popularity. As a possible element of horse control tubular mounts from Hradenín 24 and 46, and Court-St-Etienne, with disc or bulbous terminals, may perhaps have mounts of whip stocks or goads. A Ha D example comes from Ohnenheim.

Indeed, the Ha C phenomenon of elaborate and flamboyant horse-harness and the accessories of driving such as the richly decorated yokes is notable for its sudden flowering and subsequent withering within such a short space of time, perhaps no more than four or five generations from the end

175

of the eighth to the last decade or so of the seventh century. However, its eastern affiliations are interpreted, it was an essentially Bohemian and Bavarian phenomenon even if it quickly adopted some ultimately oriental ideas which as we saw may have been current already in the eighth century, if not before. It was a sudden efflorescence of what might in more sophisticated circumstances have been termed a 'court style' in equitation and the driving of prestige vehicles among a group of chieftains which, owing to economic and social factors that now elude us, had come into being with appropriate status and patronage to provide fine craftsmanship to display and reinforce that prestige. The status remained and perhaps the power was augmented in sixth-century Ha D times, but its particular expression in elaborate horse-gear appears diminished in the grave-archaeology and though the vehicle still symbolizes some potent element in the ritual, harness takes on a diminished aspect.

Bits and harness ornaments: Ha D

When we move in the sixth century to the vehicle graves of Ha D we find, in contrast to those of the previous century, that horse-gear is rarely deposited with the numerous finds of four-wheeled wagons or carriages. As this applies to competent and well-documented excavations such as Vix, Bell or Offenbach-Rumpenheim, none with bits or harness, we are encouraged to accept similar negative evidence from older and less adequately recorded finds. Of the half-dozen or so vehicle graves also containing harness gear, Hradenin 28 stands apart with its profusion of bronze trappings as well as containing three and not two iron bits, all quite in the manner of its earlier counterparts in the same cemetery, nos 24 and 46. In Grave 28 there was no studded yoke, but abundant fittings, rings and four attachments with jangling rings at the far (southern) end of

the grave that could have adorned a yoke; further north lay the bits and more bronze mounts, and in a double row in front of the front wheels of the carriage one large and seven flanking smaller phalerae. In the Hohmichele 6 grave, which like Hradenin 28 contained an imported Etruscan bronze bowl with bossed rim probably of the second half of the sixth century (Dehn 1971), two iron bits and a quantity of bronze harness mounts, including two sets each of nine phalerae, lay forward of the front wheels in the narrow space between them and the wall of the burial chamber. Apart from these two examples, the evidence is restricted to pairs or fragments of iron bits from Uffing 6 and 11, Les Jogasses and Ohnenheim, the last also with bronze studs which may come from harness and surprisingly, what appears to be a single-ring 'fly terret' in the tradition of the Ha C series described above. A possible whip-stock mount from the same grave has been mentioned above.

Horse-harness: British Isles

In Chapter 4 we saw how certain Late Urnfield material associated with horse riding and wheeled vehicles appears in the British Isles in the eighth century BC, notably the Heathery Burn wagon nave-bands and 'rattle pendants' from horse-bits. In the succeeding century Ha C contacts become clear, notably in bronze sword types and in some horse-gear as at Llyn Fawr. There remain certain other objects and three important hoards of bronzes which though in part at least in the Ha B 3 tradition may in chronological terms be put into a position of about 700 BC and are therefore considered here. Of the phalerae already mentioned, a proportion of these may belong to the seventh century, such as O'Connor's nos 1, 2 and 5 from the Thames at London, Sompting, Osgodby, Newark and perhaps the Inis Kaltra find in Ireland (O'Connor

1975). The setting within which the hoards lie has been reviewed in Hawkes & Smith 1957; Savory 1958; Coles 1960 and Burgess 1969; they come from Welby, Leicestershire, in England (Powell 1950), Horsehope in Scotland (Piggott 1953) and Parc-y-Meirch, Abergele, in Wales (Sheppard 1941; Savory 1971a, b; 1976).

The Welby, Horsehope and Parc-y-Meirch hoards are linked by a type of strap-mount in the form of a thick ring with a rectangular loop at its base present in each, and at Welby are 5, and at Parc-y-Meirch 8, ribbed 'cross strap junction' discs of the flat rather than domed type associated with eighth rather than seventh-century contexts on the continent. At Welby were also the 'cross-handle-attachments' of a bronze bowl, again of comparable date, and at Parc-y-Meirch the 'rattle pendants' of Montelius V, which seems to span the late eighth and early seventh centuries, as well as many other presumed harness attachments of unknown function, of which eight objects (Sheppard 1941, nos 24–31; Savory 1976. 44) curiously foreshadow the basic terret forms of Iron Age Britain discussed in Chapter 6. The hoard has been connected with the adjacent hillfort of Dinorben, radiocarbon dates for the earliest phases of which go back to the ninth and eighth centuries BC. The Horsehope (Peeblesshire) hoard is of rather different character. It includes a number of bronze rings, one of Parc-y-Meirch-Welby type; five dish-shaped mounts with tubular sockets, and a pair of concentrically ribbed discs with shallow collars for attachment, 7 cm in diameter; and a number of anomalous mountings of thin bronze all with nail-holes for attachment. The ribbed discs were interpreted as possible axle-caps and the sheet-bronze mountings for attachment to the wooden body of a vehicle, and the dish-shaped mounts to examples from the Hradenin graves. As a whole, the find remains without parallels. The Llyn Fawr hoard from south Wales

has been referred to earlier in this chapter, when discussing the selective assemblages of objects expressive of equitation and driving in graves of Ha C in the Low Countries, and it was pointed out that though incorporated in a hoard, the harness elements present were exactly what might have been deposited in a grave. The hoard (Fox & Hyde 1939; Mariën 1958; Alcock 1961; Savory 1976, 46) includes bronze socketed axes and gouges, two bronze socketed sickles and one of iron, all of British origin, and two bronze cauldrons of 'Atlantic' type, *c.* 700–650 BC (Hawkes & Smith 1957, 176–90). What can be attributed to contemporary Ha C, and more specifically to the Low Countries cultural enclave more than once referred to, are the pair of bronze cheek-pieces, three phalerae and a 'yoke-buckle' which a Ha C iron sword (the only one in Britain), an iron spear-head, a bronze belt-hook and a razor also of Ha C type make up a personal assemblage representative of a warrior-horseman. The yoke-buckle is an individual rendering of the general continental type, but the cheek-pieces have close parallels in those from Court-St-Etienne, as Mariën was the first to point out (1958, 33, figs 3, 12). It is then in this direction that we should look for the immediate antedecents of this group of objects at Llyn Fawr.

There remain two finds which should be mentioned if only to dismiss them from our survey. The burial under a small barrow on Beaulieu Heath in the New Forest of woodwork including a morticed plank, a small bronze ring, and an Iron Age sherd was very tentatively compared by its excavator to continental vehicle graves, but without great conviction, and it does indeed seem difficult to maintain (C. M. Piggott 1953). On King's Weston Hill, Bristol, an iron object apparently associated with Iron Age pottery in a small barrow, was interpreted as the cheek-piece of a bit, but some alternative use seems more probable (Tratman 1925).

Beyond Hallstatt: vehicles in 'Situla Art'

We have already intermittently referred to the Cisalpine contacts of the Hallstatt culture, notably with the two archaeological provinces of Golasecca in the north-west of Italy, and that of Este in the east and at the head of the Adriatic (cf. Peroni *et al.* 1975; F. R. Ridgway 1979). It is the latter area which now concerns us, centred in the Po valley around Este and Bologna, where in the eighth and seventh centuries the Iron Age was already strongly influenced by Etruscan art and technology, but by the seventh and sixth with independent orientalizing and Greek contacts up the Adriatic coast. 'From the end of the sixth century, and especially during the fifth and fourth, Etruscan influence became very marked', and the Po valley an Etruscan province, with Adria and Spina among its coastal townships, the latter also an important Greek trading centre from the sixth century (Pallottino 1975, 97; Boardman 1971; 1980, 228). In this area there was from about 600 BC a remarkable, and as we shall see, probably short-lived, efflorescence of figural art on bronze-work which has come to be known, from the characteristic bronze pails or buckets (probably holding wine or other drinks) ornamented in this style, as 'Situla Art'. Its characteristics have been admirably summed up by Sandars (1968, 223–25), it has a 'common repertoire drawn from a common Mediterranean and Near Eastern stock' and 'it is in essence not of Europe' with 'a gaucherie that betrays the artist working in a way that is uncongenial, too much at variance with the temper of the craftsman and the craft, so that the result is neither civilized *nor* barbarian, but provincial', highly characteristic, however, with its scenes of people with 'funny hats, dumpy bodies and big heads, and individual in the vitality and enjoyment that are its most engaging qualities'. In these engaging

scenes of contemporary Atestine life is a number of representations of vehicles which it is instructive to analyse; some forty-five examples of the situla style have been catalogued, distributed from the Bologna-Este region to Nesazio across the Adriatic, the Tirol, Carniola (the Ljubljana area) and outliers such as Kuffarn on the Danube. The vehicles are presented in lateral profile views in the classical and ultimately oriental tradition, rather than in the indigenous European convention of the rock art or representations on pottery of the types discussed above. (Gabrovec *et al.* 1961; Lucke & Frey 1962; Kastelic 1964; Frey 1969; Boardman 1971; Bonfante 1979 for general surveys; Kastelic 1956 on Vače; Frey 1962 on Kuffarn).

The dating of the style, and consequently its duration, has been a matter of some discussion. Basing a chronology partly on style and partly on associated finds, the situlae and allied decorated bronze-work have usually been spread over a long period, from *c.* 600 to the fourth century BC, inherently difficult to believe of a style with such closely linked mannerisms throughout its range. Viewing the material with the critical eye of a classical archaeologist and art historian, Boardman (1971) offers a more convincing alternative. Stressing the 'extreme homogeneity' of styles, and seeing the bronzes in the practical terms of masters and workshops, with better or worse artists and early and late production over a short span, he sees 'no good reason to think that the Alpine situlae were being made over a long period'. He would suggest that sixteen pieces could have been the work of one master-craftsman, and another twenty-one products of the same atelier; a few others are not so closely related, Kuffarn being 'in a rather wilder style' than others. The Benvenuti situla from Este itself can be dated to *c.* 600 BC by its associations and must come at the head of the series; Greek decorative motifs on situlae are of the early sixth century; the chariot-racing scenes on two of them should date to later in the century, for, as we shall see, they derive from Etruscan prototypes themselves due to the popularity of Attic black-figures vases with such scenes of later sixth-century date imported into Etruria (Bronson 1965). Bonfante (1979, 74) would see the composition in registers on the situlae as deriving from the imported 'Phoenician' bowls referred to below. A few workshops and three or four generations of craftsmen at most could then be responsible for the whole phenomenon of situla art. To quote Sandars again, 'an art that came like a revealing flash to the Greek world was fearfully strange to other Europeans and after struggling with it for a little while they gave it up'. Boardman suggests that the situlae were valued in themselves and preserved as heirlooms for some time in their area of origin, a case much strengthened by Bonfante's fascinating demonstration that examples of situla art were available in Italy for copying (and up-dating) by the artist carving the Late Republican marble Corsini Throne as an act of deliberate archaism (Bonfante 1977). In such circumstances, the situla art found in fourth-century La Tène contexts, as at Moritzing or Kuffarn, would be explicable as part of the transalpine contacts of that period between Italy and Central Europe, with already old pieces looted or otherwise acquired by the northerners (of. Stary 1979c; Kruta 1978).

It is proposed therefore to regard situla art as a sixth-century phenomenon, contemporary with Ha D north of the Alps. But owing to the peculiar nature of the style, in which Etruscan and Greek prototypes may be involved as well as local indigenous scenes to be represented, we must remember an important concept, touched on when discussing the conventions of prehistoric European rock art in Chapter 3, the use by the artist of set schemata to represent standard scenes or objects as so convincingly argued by Gombrich (1960). When we are considering the vehicles which are depicted on these embossed and engraved pieces of metalwork we have in effect to decide whether the artist is adopting (and perhaps adapting) a schema for a carriage or chariot already present in other linear art available to him—copying a picture with its own set of schemata so to say—or whether he is 'drawing from the life' when confronted with problems of representation for which no ready-made schema is available. In the discussion which follows, the individual pieces are for convenience referred to by their numbers in the Lucke-Frey corpus of 1962, using when necessary the names of the find-spots—the Benvenuti situla is no. 7, Kuffarn no. 40.

The chariot-racing scenes on two situlas (3 and 40) offer an introduction to the problem of schemata outlined above. The Etruscans seem to have adopted the sport, and with it the specialized type of light racing-chariot, from the Greeks in the second half of the seventh century (Bronson 1965), and so far as their own artists were concerned, they had an available schema ready to copy on imported Greek vases painted with such scenes. The situla artists could have turned to Etruscan art or, with the independent Greek contacts with the head of the Adriatic, directly to Greek black-figure painting themselves: both situla friezes have the Greek trick of showing one charioteer looking back on his competitors, again copied in Etruscan art. But two modifications have been made. First, the charioteers wear local 'funny hats' in the form of long stocking caps, and on 40 the horses have U-shaped cheek-pieces to their bits of a local type discussed below. But more important, while on 40 the chariots are all more or less of Greek type with four-spoked wheels, on 3 (the Arnoaldi situla) three out of the five have a feature which is neither Greek nor *111* Etruscan, a pole-brace with central knob running obliquely from the breastwork of the chariot box to the draught-pole. To

111 chariots in situla art: A, *racing scene, Arnoaldi situla;* B, *processional chariot, Rovereto*

quite unknown in Etruscan contexts except on two imported Phoenician or Cypro-Phoenician silver vessels from the Bernadini tomb of the seventh century (Woytowitsch nos 224a, 225a; Rathje 1979) where simple pole-braces are schematically shown. Such simple braces, set at an acute angle to the pole, are also shown in Assyrian reliefs (Littauer & Crouwel 1979a, 110) but the knobbed form seems peculiar to situla art. The well-known bronze chariot models among the votive deposits at Olympia, of before about 600 BC, have an inclined strut in front but no body-work behind it (Snodgrass 1964, 162; Greenhalgh 1973, fig. 25). We may therefore suppose the knobbed pole-brace to have been a local structural device, presumably the result of a chariot type developed as a result of independent orientalizing contacts of Adriatic rather than Tyrrhenian origin, present on the parade chariots and, on the Arnoaldi situla, incongruously introduced in the Graeco-Etruscan racing scene on the basically Greek chariots adopted with the sport by Etruscans (Woytowitsch nos 173, 174, 175 for models). In the Assyrian reliefs, the presence of a draught-pole with Y-shaped articulation has been detected where the pole is shown as if it 'impinges on the lower side of the box, as a central pole never would' (Littauer & Crouwel 1977b; 1979a, 109) and the same seems to apply to chariots on situlae 30a and 33, and we shall see another clear example on a two-wheeled vehicle in a moment. The situla parade chariots have wheels with fairly narrow felloes and normally six spokes (7 has five spokes and 32 has seven, probably due to bad drawing) and the driver has a goad in his right hand hels up above the reins in his left: on the racing scenes the goad is aimed at the horses' rumps. No bits are shown on the racing horses except on 40 (Kuffarn), but on all the others, whether driven or ridden (as for instance on Vače belt-plate, 35) large crescentic cheek-pieces are shown, not

appreciate the significance of this we must look at the non-racing parade chariots in situla art.

These appear on 7, 12, 13, 30a, 32 and 33, *111* and are clearly versions of Etruscan parade or war chariots, holding one or two people standing (the racing chariots naturally only hold the charioteer) with a high upright front and curved sides ending in scrolls in front as in the surviving bronze mountings of the Monteleone di Spoleto vehicle (Woytowitsch no. 85). But all have pole-braces, simple in 7 and 13 but in all others with a central bulbous swelling or knob, a feature known from actual finds but presumably derivatives of Von Hase's Bologna-Ronzano type of the eighth to seventh centuries BC (1969, 25–27; map fig. 9B). The cheek-pieces are most clearly shown on six pieces and divide into two types, a wide 'C' (2, 7, 13, 35) and a narrower 'U' (33, 40) form. The former type, nearer to these local proto-types but taking the form of a long curved knobbed rod, recalls those in the Ha C warrior grave at Maillac in the south of France (Taffanel 1962). The latter, U-form, shown on the Vače and Kuffarn situlae, may well represent bits with large U-shaped cheek-pieces such as that from the Tomba di Guer-riero B at Sesto Calende, of the Golasecca II A period, *c.* 600–575 BC (Peroni *et al.* 1975, fig. 71.3; F. R. Ridgway 1979, 466, fig. 37.4). This is discussed further in the subsequent section: the type gives rise to a distinctive Early La Tène form of cheek-piece which Dehn (1966, 146) compared to those in situla art.

A group of two-wheeled vehicles depicted in situla art but not chariots are those on 8e, 15 and 33, all embellished with bird-head protomes and carrying sitting rather than standing passengers. Only 33 is complete and here the artist is faltering for want of a suitable schema. For the shallow vehicle with railed sides Etruscan proto-types would have been available, as in the architectural reliefs at Murlo (Poggio Civi-tate), where two persons sit on a chair under a parasol (Macintosh 1974; Bonfante 1979, fig. 15; Woytowitsch no 251) and the cart has decorative curved ends, but the representation of an undoubted Y-shaped forked draught-pole is ineptly shown askew on 33, the Vače situla. It has six-spoked wheels set on a high undercarriage and carries two men, one driving and with a goad. Fragment 15 shows the rear end of an almost certainly two-wheeled vehicle with a railed shallow body, six-spoked wheels, one sitting man and a probable second, and two bird protomes, one on the body as 33 and

112

112 Vehicles in situla art: A, *Vače situla;* B, *lid from Mechel*

another on a structurally inexplicable strut. On 8e there is a tiny fragment only of the rear of a vehicle with a long-eared animal protome and a male passenger's head. While the railed-sided cart may be common to Etruscan and Atestine Italy, the bird protomes take us back to Urnfield and Hallstatt Europe and the *Kesselwagen* and we might have confidence in seeing here a local vehicle.

There are two representations of four-wheeled vehicles, on 11 and 14. The latter is linked to the two-wheelers 15 and 33 just described, as a fragment with a bird protome at the front of a vehicle, boat-shaped and with eight-spoked wheels, with three, probably originally four, seated male passengers facing forward: the front figure has neither reins nor goad, but the horse has a large crescentic cheek-piece. In depicting the vehicle on 11 the draughtsman completely flounders for lack of an appropriate schema to adopt, showing diagrammatic four-spoked wheels in profile, with a Y-shaped perch in plan view; the Y-attachment draught-pole is half in plan, half profile, and the body appears as an oval with beaded rim on a vertical support running from the perch, and contains probably two seated figures with the bulbous knees of other personages on the piece. A rein leads to the horse, represented without cheek-pieces. Clumsy and inept though the rendering is, it does provide us with evidence of Y-perch four-wheeled vehicles in situla art, and the very incompetence of the depiction suggests a local form for which the ready-made schemata of Etruscan or Greek art were lacking. As a postscript, attention should be drawn to a group of bronze fibulae characteristic of Este IIIB, *c.* 675–575 BC (Peroni *et al.* 1975, fig. 45.4; F. R. Ridgway 1979, 439; Müller-Karpe 1962 b). These are usually interpreted as horse-and-rider figures but each has four large concentrically ribbed discs at the corners of the platform formed by the little schematized horse figures, taken by Müller-Karpe to be phalerae. Similar ribbed discs form the wheels of a bronze cista from Oschiri in Sardinia, of seventh-century date (Woytowitsch no. 147). A possible alternative would then be wheels, and two comparable fibulae, from Magdalenska Gora and St Lucia in Slovenia, certainly appear to represent figures driving two-wheeled chariots.

North Italy: vehicle burials in the Golasecca culture

In the area of the north Italian lakes, within the ambit of the Golasecca culture, are four vehicle burials which stand apart from the Etruscan chariot graves and have extremely interesting connections with the west Hallstatt province north of the Alps. Two of the graves are at Sesto Calende, at the south end of Lake Maggiore, and two at Ca'Morta, Como. Their chronological position has recently been assessed in detail (Peroni *et al.* 1975; F. R. Ridgway 1979) and they can now be seen to span between them a couple of centuries between *c.* 700 and *c.* 500, in other words to run parallel to Ha C and D. In round terms their dates are: Ca'Morta A, *c.* 700; Sesto Calende A, *c.* 600; Sesto Calende B, *c.* 575; and Ca'Morta A, *c.* 700; Sesto Calende A, *c.* 600; Sesto Calende B, *c.* 575; and Ca'Morta B, *c.* 500–450 BC.

Ca'Morta A (Bertolone & Kossack 1957; Woytowitsch no. 129) is a cremation grave, not strictly a vehicle grave but containing a pair of bits and a *Kesselwagen* model as well as a number of bronze vessels, including some of Central European Late Urnfield types, and an Italian bronze fluted bowl of the type mentioned at the beginning of this chapter as occurring in Ha C contexts such as Kastenwald in Alsace or Stadtwald, Frankfurt. Another such bowl in the Ca'Morta grave was mounted on four ten-spoked, broad-felloe wheels as a *Kesselwagen* with twisted wire cross-supports like those of the Ha C ritual vehicle of Strettweg already described. Finally, paired draught was represented, as in so many Ha C graves discussed above, by a pair of iron bits with bar cheek-pieces with a curved knobbed upper end, comparable for instance with those from Thalmässing in Franconia (Kossack 1953a, fig. 23A). Ca'Morta A has been dated to the end of the Golasecca Ib phase, which has covering dates of *c.* 800–700 BC (Peroni *et al.* 1975, 346; F. R. Ridgway 1979, 482) and its Ha C affiliations would support a date around 700 or a little later.

The earlier of the two Sesto Calende graves, A, is the well-known Tomba di Guerriero found in 1867 (Randall-MacIver

1927, 69–74; Ghislanzoni 1944; Frey 1969, 47–50; Woytowitsch no. 109). It is a warrior grave of a cremation accompanied by bronze helmet, greaves, daggers, etc. and a situla decorated in pointillé including animals and a man on horseback. One wheel of a presumable original two was represented by the fragments of an iron hoop-tyre *c.* 80 cm diameter, with forty-eight nails up to *c.* 7 cm long and so indicating a likely plank felloe rather than one of single-piece bent wood. An iron naveband *c.* 15 cm diameter was similarly nailed, and the iron linchpin, with flat double-expanded head, is a type known from Etruscan chariot graves (cf. Woytowitsch nos 98, 105, 106). A pair of extraordinary tubular curled iron 'horns', *c.* 44 cm overall, their only counterparts being in the Sesto Calende B grave, remain inexplicable, but some sort of chariot hand-hold seems more likely than yoke terminals. The remains of a pair of iron bits have slightly curved bar cheek-pieces broadly within the Hallstatt range. The grave is dated to the end of Golasecca I c, transitional to IIA, and so *c.* 610 BC (Peroni *et al.* 1975, 306; F. R. Ridgway 1979, 482). The second warrior tomb at Sesto Calende B, again had the full martial equipment of helmet, greaves, sword, spear and phalerae (Ghislanzoni 1944; Woytowitsch no. 110) as well as a *cista a cordoni* containing the cremation (Stjernquist 1967, no. 100) and a bronze model *Kesselwagen* on four six-spoked broad-felloe wheels (Bertolone 1954; Woytowitsch 1978, no. 131). A chariot was represented by two wheels with bronze sheathing evidently on the outer ends of the nave, not dissimilar to those of Hradenin 28 without the outer collars, tubular bronze sheathing at the base of the eight spokes, and anchor-shaped iron linchpins with bossed ornament. The iron tyres *c.* 75 cm diameter, were thought to be each in four segments, but analogy would favour normal hoop-tyres, with *c.* 42 nails up to 11 cm long, certainly indicating a segmental plank felloe. There were tubular

113 Reconstructed wagon, Golasecca culture, from Grave B at Ca' Morta, Como, Italy

bronze mountings including 'horns' as in Grave A, somehow belonging to the chariot box. The two bronze bits have cheek-pieces of long narrow U-shape, already mentioned in connection with those in situla art and with counterparts north of the Alps discussed below. Sesto Calende B is dated to Golasecca IIA, a little later than Tomb A, and so about 590 BC (Peroni *et al.* 1975, 316; F. R. Ridgway 1979, 482).

The latest Golasecca vehicle burial is that of Ca'Morta B (Baserga 1929; Ghislanzoni 1930; Saronio 1969; De Marinis 1969; Barfield 1971, pl. 60; Woytowitsch no. 112), a *113* cremation grave, with the ashes in a bronze stamnos and other bronzes including a shallow bowl and fibulae. Considerable remains survived of an elaborate four-wheeled carriage of chestnut wood with the body, however reconstructed in detail, incorporating two tiers of turned balusters, and decorative sheet bronze with ribbed and ring-and-dot ornament. The naves of the four wheels have multi-piece bronze sheathing and an ogee-moulded profile, recalling as we saw the similar classical mouldings of the Vix nave-sheathing. The eight spokes have bronze sheathing for a length of 11 cm from the nave, recalling those of the Sesto Calende B chariot, but the enormously thick plank felloes of the current reconstruction (cf. Ghislanzoni 1930, fig. 10; Barfield 1971, pl. 60) are quite unjustifiable, and an alternative 10 cm felloe, with the spokes sheathed for only a part their length, is to be preferred. The unusually thick (2 cm) iron hoop-tyres, 95 cm diameter, have no fastening-nails and had been held by shrinking-on alone, so no clue to the felloe thickness is provided here. There were simple bronze axle-caps of 'hat-shaped' type with ring-and-dot ornament on the upper part, and plain anchor-shaped linchpins. Bronze binding of draught-pole and yoke were found, with the terminal knobs of the latter, but its present reconstruction has no validity. The grave is dated to the end of the sixth century by Woytowitsch and to the Golasecca IIB/IIIAi transition, *c.* 500BC, (Peroni *et al.* 1975, 326; F. R. Ridgway 1979, 484), which would equate it with Vix, with which it shares the ogee-moulded naves and the provision of axle-caps. Barfield, on the other hand, would lower the date on the grounds that the grave also contained 'an imported Attic red-figure cup datable to the early fifth century', but there are no good grounds for supposing the cup to have come from the wagon grave itself. (Barfield 1971, 133; Saronio 1969; De Marinis 1969 and F. R. Ridgway *in litt.* 1981).

There remains for additional comment the peculiar type of bit with exaggerated U-shaped cheek-pieces from Sesto Calende B, which we saw was to be dated soon after 600 BC—say 590—or equivalent to Early Ha D north of the Alps. This type may well be represented in certain examples of situla art such as at Vače and Kuffarn, as noted above, but even more interesting are actual examples further afield. In a Ha D hoard (or perhaps a grave-group) of bronzes from Woskowice Małe (formerly Lorzendorf) in southern Poland were a pair of similar bits,

and three *ciste a cordoni* of the type found in the grave of Sesto Calende B itself (Ebert 1924–32, VII, pls 207–8; Stjernquist 1967, no. 67). In a series of chariot graves in south Bohemia (Hořovičky, Mirkovice, Nevězice, Zelkovice and Sedlec-Hůrka), of the very end of the local equivalent of Ha D and the beginnings of Early La Tène, come rather developed forms of such bits, with decorated bronze phalerae and the remains of two-wheeled vehicles: at Hořovičky there are the well-known phalerae decorated with Early Style La Tène masks with 'leaf-crowns', and also an Etruscan iron fire-dog and bronze dishes probably imported from the same region (Dehn 1966; Soudská 1976; Megaw 1970, no. 47). A pair of such fire-dogs, with their roasting-spits, were in the Beilngries vehicle grave of Ha C (Stary 1979b). Dehn further pointed out that a separate but comparable series of bits with U-shaped cheek-pieces came from Early La Téne chariot graves in the Marne (Écury-sur-Coole, Berru, Mairy: Joffroy & Bretz-Mahler 1959, 17, fig. 12; Joffroy 1973) and that from these local derivative cheek-piece types develop in the west (eg. from 'the Marne', Jacobsthal 1944, no. 102; Anloo, De Laet & Glasbergen 1959, pl. 44) and these are further discussed in the next chapter. The chronology suggests that we should look to northern Italy for the origins of this type of horse-bit, in the context of the less spectacular objects of transalpine trade and exchange, side by side with the more precious pieces such as the beaked flagons or other bronzes, in Late Hallstatt and Early La Tène times from the sixth to the fifth century BC.

Vehicle burials in East Europe

Beyond the eastern Hallstatt world, in the sixth and fifth centuries BC, we move into a large area of locally differentiated but interlinked cultural provinces from Slovakia to the Black Sea. As Powell put it (1971b, 187)

'To the north of the Alps was the Hallstatt culture... peopled mainly, but not exclusively, by Celts. Eastwards, on the plains, were Thracian peoples, and perhaps occasionally more oriental intruders. It is hard to find a purely archaeological definition for these that might correspond to Hallstatt... Scythian material elements are certainly in evidence, and the beginnings of the Greek Black Sea contacts'. Over this wide area the burial of horse-gear in graves, in the Ha C manner, is widespread, from Yugoslavia to Romania, and horse burials also occur as at Chotin in Slovakia (ten burials, three with men) or Szentes-Vekerzug on the River Tisza in Hungary, (eight horse graves, three with a pair of animals), but for our purposes only two sites with vehicle burials seem recorded, those of Atenica on the fringe of the east Hallstatt province in western Serbia, and in the Szentes-Vekerzug cemetery just mentioned. A general survey of the Yugoslav material of the period has been made accessible by Alexander (1972), and valuable general discussions are included in the excavation reports of Dušek (1966), Vulpe (1967) and Párducz (1952, 1954, 1955). The horses have been described in detail and the evidence summarized by Bökönyi (1968).

The Atenica vehicles came from 'princely tombs' with cremation graves under large stone and earth barrows lying close together near Čačak 120 km south of Belgrade (Djuknić & Jovanović 1966). The central grave of Barrow I was assumed to be that of a woman, with personal ornaments and no weapons, and imported Greek objects assigned to the end of the sixth or beginning of the fifth century BC. A two-wheeled vehicle (which like all the grave-goods, had been *114* burnt in the funeral pyre) was represented by iron nave-bands and iron tyres *c.* 75 cm diameter, with ten fastening-nails, some clinched over at the ends or with terminal washers, indicating a felloe some 10 cm broad, and so not of bent-wood construc-

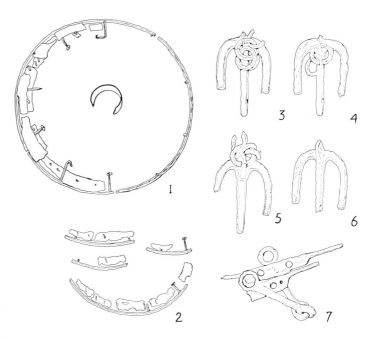

114 Finds from vehicle graves, 6th–5th century BC: *1, 2, iron-sheathed wheels of chariot, Grave I; 3–6, iron linchpins; 7, iron bit from wagon, Grave II, Atenica, Yugoslavia*

tion. The felloes were further sheathed on both sides by curved iron plates, a constructional feature without parallel, except as we shall see, in one context in Spain. Iron rings probably belonged to harness, and there were fragments of two iron bits, of a type better preserved in the second vehicle burial at Atenica, in Barrow II. This was again a cremation grave, but with iron sword and spear fragments as well as bronze socketed arrow-heads of three-winged 'Scythian' type, and again Greek imports of the late sixth century. A four-wheeled vehicle was represented, with simple iron tyres of unrecorded diameter and widely-spaced nails, and no felloe sheathing and four elaborate iron linchpins, trident-shaped with attached jangling rings, very much in the tradition of the Bohemian Ha C examples described earlier. Other iron rings and looped rods belonged to the vehicle or to the vehicle or to harness. A pair of iron bits, of the same type as those with the first burial, are of a type widely dispersed in East Europe and

185

115 Wagon burial with iron tyres and nave-bands and pair of horses, 6th century BC, *at Szentes-Vekerzug, Hungary*

an otherwise lost indigenous tradition, looking back in part, as the elaborate linch-pins suggest, to Hallstatt C traditions, and not only to the Bylany culture but to those of Hallstatt Grave 507 itself. The iron felloe-sheathing, however, if the Spanish example from a cemetery near Granada is really parallel (Arribas 1967), would make us turn, as we shall see, rather to the exotic and ultimately Near Eastern traditions which seem to lie behind Spanish chariotry in the fifth century BC.

The recent publication of a cremation grave under a barrow (Tumulus A) of Middle Phrygian date at Gordion in Asia Minor, with a vehicle burial, almost certainly a chariot, may be relevant here (Kohler 1980). The wheels have iron tyres and nave-bands, and probable felloe-clamps, and there are iron bits. The grave is dated to *c.* 525 BC, and rather earlier graves at Gordion have paired horse burials, as in Tumulus KY (Sandars 1976). In view of the ancient tradition of Phrygian origins in south-east Europe we may have here a hint of reciprocal connections.

The Szentes-Vekerzug cemetery in Hungary had a group of eight horse-graves concentrated in one part of the burial area, three containing a pair of horses, one of which (with two mares of five-six years old, 120–25 cm high at the withers) was accompanied by a four-wheeled vehicle (Grave 13). The *115* cemetery was dated by the excavator to the end of the sixth century BC. The carriage had been burnt and the remains lay on top of the unburnt horses, the wheels represented by iron tyres, *c.* 75 cm diameter, rather wide (3.2 cm) and with probably eight fastening-nails, the naves capped at both ends by iron bushes, 11.5 cm across and with an axle-perforation of 5 cm diameter. The draught-horses had iron bits of Vulpe's Type C2, in common with Atenica as we saw, and small phalerae lying along the cheeks and evidently belonging to reins or head-stall in a manner shown on the well-known Early La

classified by Vulpe as Type C2 in the Fergile cemetery in Romania, and known as far afield as Chotin in Slovakia, Szentes-Vekerzug in Hungary, and nearer to Atenica, in the Brezje cemetery in Slovenia, (Vulpe 1967). The trilobate arrow-heads usually associated with the Scythians have an even wider distribution (and in Western Europe may be due to Scythian mercenaries in Greek service: Piggott 1965, 183 with refs. and map, fig. 106). The Greek imports to Atenica, like the more spectacular finds further to the south such as Novi Pazar and Trebenište may, Boardman suggests, have as well come via the Adriatic cities as from those of Macedonia (1980, 236). But the vehicles must be the sole representatives of

Tène sword-scabbard from Hallstatt on ridden horses (Jacobsthal 1944, no. 96; Megaw 1970, no. 30; Dehn 1966).

The foregoing scanty examples of wheeled transport before or around *c*. 500 BC in Eastern Europe give us a hint of the progenitors of the later efflorescence in the historical Thracian province of vehicle burials culminating in the extraordinary series of Roman date, referred to again in the next chapter (Venedikov 1960): more immediate antecedents seem not to date before the last two or three centuries BC (cf. list in Dušek 1966, 22). The difficulties of cultural nomenclature pointed out by Powell are underlined by the descriptive labels attached by the excavators to the various cemeteries mentioned—'Illyrian' at Atenica, 'Thracian' at Chotin, 'Hallstatt' at Szentes-Vekerzug and Fergile. Discussing the last site Vulpe (1967, fig. 34) made a brave attempt at constructing a chronological table correlating the various East European regions from Bosnia to the Dobrogea, between themselves and with the canonical Central European Hallstatt and La Tène sequence, from *c*. 700 to 300 BC, with the vehicle graves just discussed in phases roughly centred on 500 BC. The degree to which elements of more easterly cultural provinces, broadly 'Scythian', can be detected is difficult to assess, though Vulpe and others have drawn attention to the *akinakes* type of short-sword, cheek-pieces with horse-head terminals, and arrow-heads among other items of material culture as pointing in this direction, quite apart from actual Scythian presence in the west so dramatically demonstrated by the gold stags from Tápioszentmárton and Zöldhalompuszta in Hungary, or the Witaszkowo (Vettersfelde) treasure, as well as by other evidence (cf. Sulimirski 1945; 1961) but not by vehicle burials. Bökönyi, in his study of the horse remains of Iron Age Central and Eastern Europe (1968), showed that east of a line approximately from Vienna to Venice the horses from the sites

we have been discussing belong to the 'Scythian' type, as do those of Pazyryk in the Altai, with an average withers height of *c*. 136–37 cm, as against the somewhat smaller western group of *c*. 127 cm.

The Mediterranean: Greece

In the last chapter brief reference was made to the Mediterranean situation in the ninth and eighth centuries BC, contemporary with the later phases of Urnfield Europe north of the Alps. It was, as we saw, a period marked by oriental contacts extended from the Assyrian world westwards via the Levant by intermediaries concerned with maritime trade, the Phoenicians foremost among them, and the emergence of two new centres of power and patronage, with literacy and figurative art, created by the Greeks and the Etruscans respectively. The study of wheeled transport in these nascent civilizations is naturally of a different order from that of the barbarian Europe of Late Urnfields and Ha C and D, but as we have already seen, essential to an understanding of it. Apart from their inherent qualities, the early civilizations of Greece and Italy have of necessity become the provinces of archaeologists and art historians equipped to deal with their respective problems in a manner proper to their individual and complex source-material, and a prehistorian can do no more than apologize for intrusions which, particularly to Hellenists, might be thought not only impertinent but impious. What follows will then be no more than a superficial survey of aspects of vehicle technology necessary to our fuller understanding of the European scene in this critical period of interplay between an increasingly civilized eastern and central Mediterranean and its still stubbornly barbarian northern hinterland.

In its Dark Age following the collapse of Mycenaean civilization, Greece became a promontory of non-literate prehistoric

187

Europe precocious in its early acquisition of iron technology and geographically placed in proximity to the historical Near Eastern world from the Bosporus to the Nile and with close links established by its Ionian territories and early Levantine trading posts. As we saw in Chapter 4 the evidence for wheeled transport (with the exception of that from a couple of graves in the Kerameikos cemetery) consists of representations on Late Geometric painted pots from about the mid eighth century BC, the interpretation of which has traditionally been bound up with the literary problem of the status and function of the chariot in Homer, and not primarily in terms of comparative technology: the pioneer study of Anderson (1961) on ancient Greek horsemanship has been followed by that of Vigneron (1968) for the wider Graeco-Roman world but we have yet to see a survey of the vehicles on the lines of Littauer and Crouwel's treatment (1979a) of the ancient Near Eastern evidence. Such a corpus is badly needed and would enable the Greek evidence to be assessed in the detail it deserves in the history of technology. Again, we saw earlier that three main positions have been adopted when considering Greek chariotry from its first post-Mycenaean appearance in eighth-century vase painting; that the vehicles are derivative from those of Mycenaean times, and owe nothing to contemporary oriental types (Crouwel 1978); the functional and social, but only incidentally technological distinction between one-man 'racing' (or processional) chariots, indigenous to eighth-century Greece and war chariots with two passengers, unknown there at the time (Greenhalgh 1973); and the absence of a war chariot in the Dark Age, but the reintroduction of the racing or processional vehicle in the late eighth century (Snodgrass 1964; 1971).

All the evidence at our disposal is confined to representations in two-dimensional figural art on pots dated between *c*. 760–700 BC. At the risk of wearisome reiteration we must remind ourselves of the concept of schemata used by artists and the problem of what constitutes 'drawing from the life'. As with prehistoric European rock art or that on the Alpine situlae already discussed, so in Greek vase painting, a conventional schema for a vehicle (or anything else) may be acceptable to artist and patron alike as a representation of reality. A common problem in vehicle representations which like those of early Greece, cannot be checked against archaeological reality, is that of the four-spoked wheel: does this represent constructional detail or is it an acceptable schema for a spoked (as opposed to a disc or cross-bar) wheel in certain circumstances? The question may not be answerable, but it remains as a warning. And a second problem, of course, arises when as in eighth-century Greece a previously aniconic art (pot painting) becomes figural, in a geographical position where a long tradition of figural art exists in closely adjacent areas. We have then to consider whether a schema is of Greek origin, or borrowed and adapted from outside. 'It may be wrong to discount altogether the possibility that the example of eastern works had already stimulated Greek figure styles in the Geometric period... even the impetus to develop any form of figurative art may derive from observation of eastern models', though 'the total translation to the Greek idiom suggests that the influence was generalized rather than a matter of model and copy' (Boardman 1980, 78). This can, in fact, be appositely tested in the context of vehicle conventions. Snodgrass had looked for possible Egyptian prototypes for the Late Geometric processional chariots but, as has been pointed out, the eighth-century chariot throughout the Near East was wholly different (Littauer 1977b). If there were Greek artists in search of a schema for chariots at this time their most likely source would have been the Phoenician Egyptianizing metal bowls which reached both Greece and Etruscan Italy: even if surviving examples are in contexts that cannot be dated before 750 BC (D. Ridgway 1981, 23), the type need not have been wholly new at this time. The chariots on the Etruscan imports are conveniently assembled by Woytowitsch (pls 47–48) and are vehicles with a vaguely Assyrian body with a diagonal bow-case and sometimes a parasol, eight-spoked wheels and a driver and a warrior. They are completely unlike the chariot schema of painted Geometric pottery, with four-spoked wheels and a minimal stripped-down body with high front and looped sides, carrying one man, increasing the likelihood of this being an indigenous convention.

If one takes Greece from the second millennium BC as a part of the European scene of wheeled transport as reviewed in early chapters, one would expect, despite the absence of direct archaeological evidence, the early establishment of ox-drawn vehicles with disc or tripartite disc wheels, certainly as four-wheeled wagons and probably also as two-wheeled carts. The light spoked two-wheeler coming into use in continental Europe by the middle of the second millennium is, of course, directly attested at this time by the specialized Mycenaean chariot for war and parade, and the spoked wheel would be adopted, as elsewhere, for carriages and prestige vehicles in general, and since it appears in Italy at this time, the cross-bar wheel of oriental derivation, to be *116* depicted from the sixth century in Greece in non-aristocratic contexts, (and known from an actual surviving example of this date from Gordion is Asia Minior) could be thus early in Greece (Littauer & Crouwel 1977a) though not fulfilling functions which would lead to its inclusion in court art or documents. At all events, by the eighth century one would expect that in common with the rest of Europe, Greece would possess the full range of vehicles from disc-wheeled ox-draught to spoked-wheeled horse-drawn

116 Vehicles in Attic Greek vase painting: A, *chariots, late 8th century;* B, *four-wheeled carriages, mid 8th century;* C, *chariot, mid 7th century* BC

carriages of prestige with four or two wheels, as well as the cross-bar wheel already described in Chapter 4. This leads to a controversy to which we must briefly turn. The point at issue is whether the Geometric vase painters used two separate schemata, one for a two-wheeled and one a four-wheeled horse-drawn vehicle, and depicted both on the pots, or whether they were prepared to use two lateral profile schemata,

189

one with two (and so in reality four) wheels and the other with one (and so denote a pair of wheels), even combining them in the same scene. Greenhalgh 1973 has recently upheld a view going back to Von Mercklin in 1909 that all vehicles depicted on Late Geometric pots are two-wheelers, whatever schema is used, while Snodgrass accepts 'two-wheel' schemata as denoting four-wheeled carriages and of a Late Geometric oenochoe from the Athenian Agora, with 'two-wheel' and 'one-wheel' schemata in the same frieze, writes 'It is surely inconceivable that the same artist should have used two different conventions for a two-wheeled chariot on the same vase' (Snodgrass 1964, 161, pl. 2; 1971, 432; other 'contentious' vehicles fig. 138; Greenhalgh 1973, figs 7. 26). The latter view is consonant with the conventions employed for vehicles in figural art outside Greece at the time, for instance in Assyria, and to see the Late Geometric funeral scenes which embody such depictions as in parallel with the archaeological evidence for prestige carriages or hearses in contemporary continental Europe burial rites, horse-drawn and four-wheeled, but with two-wheeled chariots present in the same cultural milieu, seems in accord with the general technological and archaeological evidence and avoids inventing an unparalleled circumstance in ancient art. The schematic renderings make detailed interpretation of the vehicle bodies difficult, but the chariots must have been D-shaped in plan, with a high front and sides with prominent rear hand-holds. The four-wheeled carriages would have had a rectangular body with minimal sides (as in Hallstatt vehicles) but the arched 'croquet hoop' at either end is impossible to understand. A good range of illustrations, from Geometric to Red Figure vehicles, is conveniently accessible in Tarr 1969, Chap. 5.

Finally, a technological note on the wheels depicted in early Greek vase painting may be added. The problem of the four-spoked wheel as schema or reality has been mentioned, and it is one recurrent in much ancient art, including the depictions of Mycenaean chariots. On the whole one is prepared to accept it as representing at least one contemporary 'reality' where it occurs without excluding the parallel existence of wheels with more spokes than four. It appears in its simplest form on Late Geometric pottery (Greenhalgh 1973, figs 6, 7, 8, 9, 10 for instance). Marked lateral braces to the spokes at their junction with the felloe are an alternative (figs 4, 12, 14) and have on the one hand Mycenaean prototypes (Wiesner 1968 figs 6a, 9e) and on the other become a standard feature of canonical Black Figure chariot representations (as Greenhalgh 1973, figs 16, 17). All these wheels are shown with narrow or relative narrow felloes, but Kossack drew attention to an alternative wheel type, with a very broad or double felloe, which he compared with the actual wheels as inferred from their metal mountings in Ha C, as we saw in an earlier section of this chapter. (Kossack 1971, fig. 30a; Wiesner 1968, fig. 18b; Greenhalgh 1973, figs 4, 12, 18). A chariot wheel with an extraordinarily heavy felloe and six spokes on a Late Geometric Attic pot (Greenhalgh 1973, fig. 11) seems inapposite for its light stripped-down body and one is uncertain whether one is dealing here with a misapplied borrowed schema or whether the reality behind it could reflect some sort of construction akin to that of the wheels of the Atenica chariot described earlier, with its iron felloe-sheathing.

Etruscan Italy

Owing to a fundamental difference in religious concepts and consequently in burial customs, when we move from Greece to Italy the archaeological evidence for vehicles in the seventh and sixth centuries BC is completely different and, with the burial of prestige chariots in Etruscan tombs, inferentially in connection with a journey to the underworld, actual remains rather than ambiguous representations become available (cf. Blazquez Martinez 1958). In addition, a number of models, going back to the sixth century at least, amplify the picture, and the evidence is conveniently assembled as a corpus in Woytowitsch (1978), and referred to below by his catalogue numbers. It is here treated in outline only, for as with the preceding Greek evidence, Etruscan archaeology and art history form a complex field of study in themselves but, as we have already seen, one which cannot be ignored by the prehistorian of barbarian Europe to the north. The evidence falls into three classes, the first the indigenous tradition of domestic vehicles with disc and cross-bar wheels continuing from the second millennium BC; the second for the adoption from Asiatic sources of the chariot for parade and war; and the third for another foreign borrowing, the racing chariot and the sport it implies, adopted from Greece.

As we saw in earlier chapters, disc-wheeled vehicles go back to the third millennium in north Italy, with tripartite discs in the second, as at Mercurago, a site which also produced the only surviving prehistoric cross-bar wheel in Europe (Littauer & Crouwel 1977a), though a sixth-century example has recently been published from Gordion in Asia Minor (Kohler 1980, fig. 32). Representations show its persistence into the later first millennium BC and beyond, normally on two-wheeled carts. As we have seen carts, in contrast to wagons, carriages and chariots, are remarkably elusive in the archaeological record of prehistoric Europe, and therefore the bronze *117* model 28 cm long from Bolsena (168), attributed to the sixth century BC is of all the greater interest. Its painstaking details, confirmed by a less elaborate model from Civita Castellana (280), show interesting features. The wheels, represented as discs with a pair of crescentic ribs (as on 280), presumably denote the open cross-bar type, and are

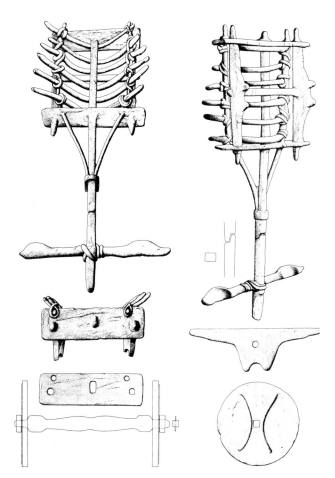

117 Bronze model of cart, 6th century BC, from Bolsena, Italy

fixed to the square ends of an axle rotating with them, while the rectangular undercarriage has a pair of blocks or pads with half-round notches into which the axle fits and turns, so that the body can easily be lifted from the wheels-and-axle unit. This is a constructional feature precisely paralleled in the chariots from the Cypriot tombs of seventh and sixth century BC date (e.g. Karageorgis 1967, 22) and for that matter surviving in recent Sardinian A-frame carts with disc wheels (Tarr 1969, fig. 47; Haudricourt & Delamarre 1955, fig. 47); in modified form in the similar Anatolian carts (Piggott 1968, fig. 9) and again in Portuguese ox-carts (Galhano 1973). The draught-pole of the Bolsena cart continues as a perch to the

191

back of the undercarriage, and a light open-work body, clearly for bulk goods rather than passengers, is provided by six round-sectioned rods bent round this in a shallow U, and lashed to the sides. The draught-pole has a pair of braces at its junction to the body, and a yoke for paired draught (presumably oxen) lightly lashed towards the far end. Unique in its accuracy of detail the model gives us a technical glimpse of what may have been a widely distributed type of farm vehicle in prehistoric Europe.

When we turn to Etruscan chariots, we move into an aristocratic world of a superior social status to that of the Bolsena hay-cart, though the carefully executed model suggests it was a votive offering in circumstances in which the agricultural background that supported the warrior aristocracy was not forgotten. Stary (1979a, 190–91, 1980) has demonstrated how the chariot for parade and warfare was a part of a wholesale adoption of 'Near Eastern elements of arms, armour and warfare', beginning in the late eighth century and continuing until the middle seventh, when Greek practice and equipment began to take over. This orientalizing episode lasted scarcely more than three generations, and spread from Etruria to Latium, Umbria and the Ager Faliscus by the early seventh century, from the middle of the century to the east coast and Picenum, and by 600 to the Abruzzi. North of the Alps, Stary sees Hallstatt warfare affected by Etruscan orientalizing warfare in the replacement of the Ha C long sword by the dagger and spear in Ha D, and we have seen how the development of Hallstatt iron tyre technology looks back to that of Etruscan chariotry.

The surviving remains of chariots in Etruscan or allied contexts in Italy are therefore of frankly Asiatic type, with a D-shaped body, high front, and arched panelled sides with orientalizing repoussé ornament, as in the best known, that from Monteleone di Spoleto (85) in the Metropolitan Museum in New York: others include Capua (6) and Ischia di Castro (36) and S. Mariano (75–84); the Tomba dei Carri at Populonia (66, 67) has openwork mounts and the Vulci model bronze sheathing (34) has zones of stylized men and horses between bands of ultimately Greek cable guilloche pattern, with counterparts, as we saw, in the Hallstatt world so far as the figures go. The range of date of all lies between the late seventh and mid sixth centuries BC and, although versions of such chariots appear in situla art, there is no evidence of such forms of body-work affecting regions north of the Alps, but the wheels, with nailed-on iron hoop-tyres are, as we saw, another matter. Before returning to these a few other features may be noticed in passing, such as pole-ends in the form of a lion's head (9 from Praeneste; 35 from Vulci) or horse and griffin protomes (70, 71 from Chiusi). The Praeneste piece, as Brown originally pointed out (1960, 21), is of bronze with iron inlay in an oriental technique (and one known incidentally in Hallstatt Europe). Bronze axle-caps occasionally occur, and the four plain 'hat-shaped' examples from the Regolini-Galassi tomb at Cerveteri (31) imply an exceptional four-wheeled vehicle in this instance; the lion-headed pair from S. Mariano (72; Brown 1960, 114) have transverse rather than the normal vertical linchpins. The exceptional annular axle-caps with vertical linchpins from Belmonte (91) of the sixth century, have as we shall see counterparts in fifth-century Early La Tène graves in east France.

The wheels of the orientalizing Etruscan chariots show a variety of forms: those from Monteleone (85) have massive bronze sheathings which may have carried a wood felloe on which the iron tyre fitted; the larger vehicle from the Tomba dei Carri at Populonia (66) has unique metal-sheathed wheels with eight spokes and an inner, openwork ring which is not a functional felloe. The nailed-on iron hoop-tyres, more than once referred to, fall into two groups, those with close-set stud nails from six relatively early sites (within the short timespan available), nos 15b, 23, 28, 34, 67 and 100, going back to the seventh century, with another five (15a, 86, 89, 95, 98) almost wholly within the sixth century, and with more widely-spaced nails. The possible bearing of this technological sequence on the comparable development in Ha C and D has already been discussed, as has the ultimately oriental origin of the stud-nailed tyre. A few Etruscan examples of joint-braces or mounts of the spoke-junctions with the felloe have been recovered (eg. 89, 94, 95) suggesting felloes of thicknesses from *c.* 6 to 8 cm.

Of the last group of vehicles, the Greek racing-chariot, little can be added to Bronson's study (1965). Apart from representations in Etruscan art from the mid sixth century, and the situla versions already discussed, the bronze models (e.g. 173, 174, 175) all seem assigned a Roman date, but the well-known 'Tiber' piece (175) does show in detail the leather cross-strapped floor and lashed-on apron to the front of the body framework which one can assume to have been characteristic of the type since its earliest development.

Iberia

The later evidence for wheeled transport in the Iberian Peninsula is somewhat confused and awaits comprehensive publication, but it can be seen that we are dealing with at least two traditions, one indigenous and one resulting from the same process of orientalizing contacts we have discussed in the previous section and to be associated in the main with Phoenician or Punic trade and colonization. It will be helpful briefly to review the background to the seventh and sixth centuries, noted at several points in earlier chapters. We start from an indigenous tradition, probably from at least the second millennium BC, of sledges of square

plan and slide-cars of A or Y plan, depicted in the rock paintings of the Peñalsordo-Badajoz region (Chapter II; Breuil 1933), which similarly include A-frame carts with spoked and in one instance cross-bar wheels, and one square-framed spoked-wheel wagon. A disc or tripartite disc-wheel tradition is not only implicit in its survival in the modern ox-carts of the Peninsula (Galhano 1973), but in some of the votive model vehicles from the well-known Iberian sanctuaries, such as the disc-wheeled square-bodied carts from Despeñaperros and Collado de los Jardines, and one with a Y- or A-framed body from Santa Elena (Serra Rafols 1948; Blazquez 1955; Cuadrado 1955; Pijoan 1953, 424, figs 652, 653), though the sanctuaries in general hardly antedate the fifth century (Arribas n.d. 135). The bronze votive car from Monte de Costa Figuera, Douro (Portugal), dated to *c.* 350–250 BC, has cross-bar wheels (Cardozo 1946) and others, such as the well-known boar-hunt on wheels from Merida, have spoked wheels of normal type (Blazquez 1956; Forrer 1932). An unsatisfactory find of a bronze-bound wheel with tyre and stud-nails, of simple cross-bar form, from Catoira, Pontreveda, should be mentioned here: it is of Cork Oak *(Quercus suber)* and has a radio-carbon date of 3670±45 bc, and clearly demands reappraisal (Galhano 1973, fig. 11; Arribas *in litt.* 1976).

In Chapter 4 we dealt with the late eighth-century stelae with chariot depictions, the 'Tartessian' group which represent Levanto-Cypriot contacts implicit also in the other objects depicted with them, such as shields and fibulae, shortly before 700 BC. From this evidence for the importation of the chariot for parade or war from ultimately Near Eastern sources, we can now turn to the actual surviving remains of slightly later date, beginning with the chariot burials from the La Joya cemetery at Huelva on the coast between Cadiz and the Algarve (Roiz & Garcia 1978). Remains of the bronze mountings of vehicles were found in Tombs 17 and 18, better preserved in the first: the associations included bronze flagons of a type found elsewhere in Spain (Garcia y Bellido 1969) and in Etruria (Rathje 1979), regarded as Phoenician imports and dated from the later seventh century into the sixth. Decorative mountings, and anomalous bits with cheek-pieces were alone found, but there were no metal tyres. In Tomb 17 were a pair of bronze axle-caps modelled as lions' heads and with transverse linchpins, and four stemmed perforated discs interpreted as rein-terrets on a yoke by the excavators. The axle-caps recall those from S. Mariano described in the last section, again with transverse linchpins, but are stylistically dissimilar, the Italian pair, with closed mouths, being probably local Etruscan work (Brown 1960, 113) and those from Huelva, with open mouths, belonging to a different tradition: Professor Brian Shefton (*pers. comm.*) would assign them to Cyprus. If so, it might be remembered that the Cypriot chariot tradition of the eighth-seventh centuries BC did not include the provision of metal tyres, and conceivably the Huelva perforated discs could be simplified versions of the floreate discs assigned to the yokes of the Salamis chariots from Tombs 3 and 79 (Karageorghis 1967, CCXXIII; 1973, fig. 10; criticism of reconstruction, Littauer 1976b). At all events we have vehicles imported, or built by foreign craftsmen, with the Phoenicians once again the obvious intermediaries in the east-west trade.

Claims for considerable contacts, expressed in metal harness-elements, between the Hallstatt culture province and the Iberian Peninsula have been advanced by Schüle (1969), but the material seems too scattered and diffuse to bear the somewhat far-reaching conclusions that have been drawn from it. The cemetery of Alcacer do Sal, in Portuguese Estramadura, has the remains of probably two vehicles, with bronze-sheathed rectangular-sectioned spokes, iron tyres *c.* 90 cm diameter and simple bronze annular nave-caps, and probably dates from the sixth century (Arribas n.d. 181; Schüle 1969a, 128, pls 106, 107). Other finds from cemeteries, all of the sixth and fifth centuries, include remains of wheels or tyres from Tugia (Jaen), Tutugi (Granada), Vera di Badasa, Cabecico de Verdolay (Murcia) and Mirado de Rolando (Granada). The last site appears to have had iron felloe-sheathing in the manner of the Atenica wheels, now misplaced and corroded to the outside of the tyre (Cuadrado 1955; Arribas n.d.; 1967). The remarkable chariot wheels from the monumental Iberian tomb at Toya (Peal de Becerro), with elaborate metal sheathing, seem to mark a local development of the oriental theme; their massive felloes, 10 cm broad, must have been of plank construction, and the very long nails of the Montjuich iron tyres (probably not earlier than the third or second century BC) imply the same thing.

So far as it goes, the Iberian evidence for wheeled transport seems to show, in common with Italy, the adoption of spoked-wheel prestige vehicles of parade-chariot type by the local Iberian aristocracy in a manner reminiscent of Etruscan Italy, and with ultimately Near Eastern origins and perhaps Cypriot connections, transmitted westwards from the late eighth century by Phoenician trade, later with intensified contacts following colonization. A final minor point may be made, which is that among the not uncommon scenes of warriors and warfare on the well-known Iberian painted pottery, chariots are never shown, nor indeed in any other context.

6 The Early Iron Age: La Tène and the Celtic world

Cultures and chronology

At the beginning of the last chapter we saw how, with the conventional archaeological phases of Hallstatt C and D, we moved technologically from the use of bronze for edge-tools and weapons to that of iron, and chronologically spanned about a couple of centuries ending a generation or two after 500 BC. Thenceforward nomenclature shifts from 'Hallstatt' to 'La Tène', the latter taken from the large assemblage of material, probably largely votive deposits, recovered from 'the shallows' at the east end of Lake Neuchâtel in Switzerland in the last century, and subdivided by Reinecke into four phases, A—D, of which the last was terminated by the imposition of Roman rule in the various areas of barbarian Europe where La Tène material culture was dominant. The change of nomenclature, based as it was on novel and distinctive features not present in the preceding Hallstatt cultural tradition (notably the sudden emergence and rapid development of a new and distinctive art style) has perhaps itself contributed to an interpretation in terms of a drastic cultural dislocation in the fifth century BC and an underestimate of essential continuity.

Definition is not rendered easier by the application of the general term La Tène to material from a wide range of varied territories from the British Isles to south Russia: we saw in Chapter 5 a similar misuse of the term Hallstatt in, for instance, south-east Europe. Further confusion rather than illumination has resulted by uncritically conflating archaeological evidence with the historically documented Celtic peoples and their movements, when material culture need not equate with language or tribal entities. English archaeologists in discussing the problem have been at pains to avoid the concept of 'culture' in archaeology as originally propounded by Childe, and have taken refuge in the 'La Tène culture group' (Collis 1975) or 'La Tène complex' (T. Champion 1975), but stressed the element of basic continuity from Hallstatt onwards. It is reasonable to take the well-documented instance of Britain as typical of much of Europe, with its conclusive demonstration that the agrarian pattern embodied in the 'Celtic fields' of immediately pre-Roman times goes back well into the second millennium BC (Fowler 1981); the inferences of ranked societies with 'ostentatious graves' and defended hill-settlements made for Hallstatt can equally be made for much of the La Tène area at one or another point of the five centuries of its duration. Iron-working, important in vehicle manufacture, is developed and Driehaus (1965a) suggested a correlation between the distribution of Early La Tène 'princely graves' and natural iron resources in the Rhineland. The use of standardized 'double pyramid' ingots of a type going back to Ha C is widespread by Late La Tène times and a marked uniformity in utilitarian iron-work such as cauldron chains and firedogs is apparent (Piggott 1965, 247 with refs.; G. Jacobi 1974, 245–53, map fig. 57; Pleiner 1980, map fig. 11:7; Collis 1976). In certain areas such as that of the Hunsrück-Eifel culture of the Rhineland and that of Bylany in Bohemia and the group of graves near Plzen, a clear continuity from Final Hallstatt to Early La

Tène existed, including in both instances burials now with two-wheeled rather than four-wheeled vehicles. Discontinuity is perhaps more apparent at the upper than at the lower end of the social scale, both in terms of material culture such as fine metalwork and the art style it carried with it, and in the geographical distribution of such pieces. As the writer said elsewhere of these phenomena, 'aristocratic art can be discontinuous, peasant crafts more often continuous', pointing to gift-exchange as a plausible mechanism (Piggott 1976). Fighting methods changed in the fifth century, with the return to long swords after the adoption in Ha D of the dagger, probably as a result of Etruscan contacts, but conversely the widespread use of the long oval shield with spindle-shaped boss, an Etruscan form known from the eighth century and already in use in Ha C as on the Strettweg cult-group referred to in Chapter 5 (Stary 1979, a, c). The continuance into La Tène of the tradition of richly furnished vehicle graves from Ha D is of direct concern to us, with the prestige cart or chariot with two wheels replacing the four-wheeled wagon or carriage of the earlier periods. In such graves Italian imports again continue into Earlier La Tène, and afford the starting-point for the chronology of the period (Wells 1980).

The original classification of LT A—D was itself subdivided into a total of eight phases applicable only to the Rhenish heartland of the culture, inapposite to a wider survey and for our present purposes largely otiose. Déchelette for France proposed a simple tripartite division, I–III, and this broader concept of Early, Middle and Late La Tène will, where applicable, be used here. Side by side with these chronological phases run the 'Celtic' art styles of Jacobsthal (1944) which, as he was careful to point out, are not a strict diachronic sequence, but fluid, overlapping and regional styles within the general ambience of 'Early Celtic' art. The beginnings of La Tène (and of the

Early style of art) is marked by the appearance in 'princely graves' of a selected type of Etruscan bronze vessel, the beaked flagon or *Schnabelkanne,* of which more than fifty are known north of the Alps with a concentration in the Rhineland Palatinate and outliers in Bohemia and Austria to the east, and France from Burgundy to the Marne on the west, a shift northwards of the centres of purchasing power of Ha D. Unfortunately these flagons cannot be dated with precision in Italy beyond the late sixth to fifth centuries BC, with Vix as an early example and several others dating vehicle graves to the Early La Tène phase (Jacobsthal & Langsdorf 1929; Frey 1969, 84 with map, fig. 49 and list, 115; Wells 1980). In a restricted area shallow bronze bowls with squared handles are contemporary imports (Schaaf 1969) as are exceptional pieces such as the tripod and stamnos from Bad Dürkheim. Datable Attic pottery imported via northeast Italy confirms these fifth-century dates, as the cup from Klein Aspergle of *c.* 450 and the late fifth-century bowl from the chariot burial of Dürrnberg (Austria) and the red-figure cup of *c.* 430–420 from that of Somme-Bionne, though doubts have recently been cast on the original association of this with the grave (Kruta 1978, 160 n. 32). The vessels from Rodenbach and Motte St Valentin are of the middle fifth century (Dehn & Frey 1979). The beginnings of Early La Tène should then date around 475–450 BC.

Difficulties are then presented by the thinning-out of Italian and Greek imports north of the Alps, including Mediterranean coral, frequent from Ha D (S. Champion, 1976). The latest imported bronze is the probably Tarentine bucket from the Waldalgesheim chariot grave, variously dated from *c.* 380–370 BC to the end of the fourth century. A date in the middle of the century seems best in view of the other relations of the Waldalgesheim art style and its Italian connections (and perhaps origins) with, for instance, characteristic La Tène metalwork

in Picene graves, such as the Waldalge-
sheim-style scabbard at Moscano (Fabriano)
with Greek pottery of *c.* 350–325 BC (Frey
1971; Dehn & Frey 1979; Kruta 1978,
1979). But, from the fourth century BC and
the transition to a Middle La Tène period,
absolute dating becomes at best inferential
and proportionately hazardous. Appeals to
the historical evidence of the movements of
Celtic peoples in classical writers are weak-
ened, as we saw, by the uncertainties inherent in equating these with archaeology, and
with areas beyond their range of reference,
though in our particular field of enquiry,
useful indications of vehicle usage are given,
for instance, by the recorded employment of
the war chariot by Celtic tribes in the engagements of Sentinum in 295 and Clastidium
in 222 BC (D'Arbois de Jubainville 1888;
Frey 1976). Again in historical terms the
Waldalgesheim style and the La Tène metalwork it takes with it must antedate in Italy
the Roman destruction of the Senones in
283/282, or of the Boii defeated at Telamon
in 225 and finally reduced in 191 BC. Burials
of Middle La Tène type as at Ceretolo
should lie before the last date, confirmed by
the depiction of appropriately contemporary shields, helmets and items of chariotry
among the Galatian trophies on the Pergamon reliefs set up in 181 BC (De Navarro
1972, 312–27). The position of Jacobsthal's
Plastic Style and Hungarian Sword Style as
subsequent to Waldalgesheim in the third
century BC (Klindt-Jensen 1949; 1953) is
reinforced by the recent recognition that the
bronze mount of a drinking-horn in the
former style from a grave at Jászberény in
Hungary very closely copies a conventiona-
lized 'sea serpent' motif in third-century
Hellenistic art: it was found with an iron
razor with Sword Style ornament (Krämer
& Schubert 1979). So too the parallel Swiss
sword style might start late in the third century, though in the main later—'a generation after 150' until the Helvetian migration and so *c.* 60 BC (De Navarro 1972,

326). With Later La Tène imported Italian
bronzes, now Roman, such as the flagons of
Kelheim type of *c.* 50 BC, as far afield as Britain, we are taken archaeologically into the
history of the Roman conquests in barbarian
Europe. For the vehicles, iconography
comes to our aid with the chariot representations in Italy, including the Padua stelae,
of the third century BC, and in Gaul on the
coins of the Remi of *c.* 60 BC, as well as on
certain Roman coinage, and this evidence is
discussed later. (Frey 1976). The La Tène
tradition has its longest survival in the British Isles, where the chronological problems
are acute, but so far as the chariot evidence
is concerned, none need date before the
second century BC (Hodson 1964; Stead
1979). Finally, a few dendrochronological
dates have been obtained, not inconsistent
with the foregoing scheme. From the site of
La Tène itself a pile probably from the Pont
Vouga gave a felling date of *c.* 278 BC, one
from the Pont Desor 65, and a wooden
shield fragment 256. From an Early La Tène
grave at Altrier in Luxemburg two felling
dates of 473 and 461 BC were obtained from
wood samples, suspected of being old when
deposited in the grave (De Navarro 1972,
354; Thill 1972). There are two radiocarbon
dates one from a burnt wood plank over the
chariot grave no. 1 at Hamipré in the Belgian Ardennes of 450 ± 55 bc (Hv—4784), *c.*
490 BC, and the other from chariot grave no.
2 at the same site, of 625 ± 75 bc
(Hv—5389). Here old timber may be involved and neither date can be accepted at
face value (Cahen-Delhaye 1974.)

The vehicles: disc and tripartite disc wheels

Our last encounter with disc wheels, representing ox-drawn agrarian vehicles rather
than aristocratic grave-offerings, was in
Chapter 4, with the Urnfield examples of
Buchau and Biskupin. It is no more than the
accident of survival in suitable conditions

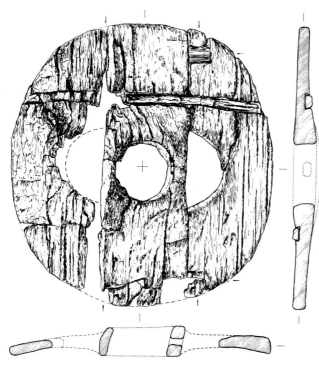

*118 Tripartite disc wheel with inserted nave,
external battens and lunate openings, Late La
Tène, from Mechernich-Antweiler, Germany*

such as peat-bogs, and of the sporadic application of radiocarbon dating to such finds, that deprive us of examples assignable to the Hallstatt period of the seventh and sixth centuries BC. The later material, which we can now deal with, almost all comes from peat-bogs or waterlogged sites on the northwest edge of, or beyond, classic La Tène territory; the Low Countries, the North European plain, Denmark and Ireland, and have been described by Van der Waals (1964), Lucas (1972), Hayen (1973) and Rostholm (1977). Typology is of no help in dating, for, as we have seen, the main forms of single-piece and tripartite disc wheels were established by the fourth millennium BC and the modification of the inserted rather than the integral nave by the second. Radiocarbon dating for unassociated finds affords us the only means of chronological distinction: the simplest forms may be post-Roman (as those from Alt-Bennebek in Schleswig: Struve 1973) or even modern. A further dat-

ing difficulty is that inherent in the manufacture of disc wheels from planks of mature trees up to a couple of centuries or so old, so that a radiocarbon assay from near the heartwood will differ materially from one near the sapwood, as Hayen showed in the instance of the wheels from Glum near Oldenburg (Hayen 1972). The dates quoted below, from samples taken from unknown points on the wheels' diameters, should be treated with this reservation in mind.

Within the available range we may start with a massive single-piece disc, with inserted nave and two small lunate openings, 70 cm diameter, from Tindbaek, Viborg, in Denmark, with a date of 360 ± 80 bc (K-2895) or *c.* 425 BC. The cut-out lunates are too small to lighten the wheel and their presence at all suggests the copying of a tripartite disc, in which such features on a rather larger scale are common from the second millennium BC onwards. An important find of a tripartite disc wheel is that from a settlement site of Early La Tène C, *c.* 200 BC, at Mechernich-Antweiler, Kr. Euskirchen. It is 74 cm diameter, with lunate openings and held by external battens in grooves, one on each face, and had an inserted nave, now missing (Joachim 1979). The remainder of the wheels now under consideration are of similar construction with inserted naves, normally with externally dowelled battens, straight or curved as at Mercurago or Buchau though with small internal dowels towards the rim of the wheel, except for the Irish wheels from Doogarymore and Timahoe (Lucas 1972) and the undated Scottish wheel from Blair Drummond (Piggott 1957), which have internal dowels in tubular mortices. The Doogarymore wheel has dates of 450 ± 35 bc and 365 ± 35 bc (GrN-5991, 5990) or *c.* 490–430 bc and has lunate openings, as has that from Dystrup in Jutland, with a comparable date of 470 ± 110 bc (K-823) or *c.* 530 BC (Tauber 1966; Rostholm 1977). The Ezinge fragments of five tripartite wheels come from a

118

198

stratified settlement site or *terp* in the Netherlands, with a range of archaeologically determined dates from Early La Tène to the second-first century BC (Van der Waals 1964). From a series of remarkable votive deposits originally in a lake, now a peat-bog, at Rappendam in Seeland, Denmark, came fragments of forty wheels, eighteen tripartite with inserted naves, and a radiocarbon date of 70 ± 110 bc (K-1113), *c.* 100 BC, was obtained, but such offerings were made over a long period of time (Kunwald 1970a, b: Glob 1969b, 167, fig. 62; Rostholm 1977). The range of diameter of all the wheels in question is about 1.0 m to 55 cm. The only prehistoric disc wheel from England is an anomalous little thick biconcave disc of oak, 45 cm diameter, from the foundations of Mound XXX in the so-called 'Lake Village' at Glastonbury of the first century BC (Bulleid & Gray, 1911, 321, fig. 84).

The vehicle graves

As with the earlier phases of Ha C and D in the seventh and sixth centuries BC, the evidence for vehicle technology in La Tène A—D, from the fifth century to the eve of the Roman conquest, comes from graves where the practice of burying a prestige vehicle continued among certain sections of the population. What changes is first the type of vehicle sanctioned by rite and custom for this purpose, with the replacement of the four-wheeled carriage by the two-wheeled chariot. A few border-line cases at the Late Ha—Early LT overlap occur, such as Bell and Les Jogasses, described at the end of the last chapter, or the 'Phantomwagengrab' with four wheel-pits at Bassenheim (Driehaus 1966b, 38), and in Late LT, outside the strict northern bounds of the canonical culture-area, an interesting little group of four-wheelers in funerary or ritual contexts appear. But apart from these exceptions, the two hundred or more vehicle burials recorded from the fifth to the first centuries BC are of what we have argued should reasonably be called chariots rather than carts, though not thereby denoting their use in warfare. As Harbison, defending traditional usage in this context says 'the term "cart" is possibly best applied only to vehicles akin to the modern agricultural vehicle of that name' (1969, 34), and the much-debated question of the 'Celtic war-chariot' involving textual rather than archaeological evidence, is taken up in a later section of this chapter. To return to the graves, the second observable change is that, from Early LT times onwards, the richly furnished 'princely graves' of the Hallstatt tradition are replaced by burials in which the chariot itself constitutes the only evidence of inferred social status, without other distinctive or costly grave-offerings; a tendency already perceptible in Ha D vehicle graves such as for instance Offenbach–Rumpenheim.

The overall chronology of the graves presents difficulties in detail after the fifth century which need not concern us in a study of vehicle technology, and we may reasonably divide the material into two phases, Early

119 Remains of disc wheel with inserted nave and external battens, 1st century BC, from votive find in bog at Rappendam, Denmark

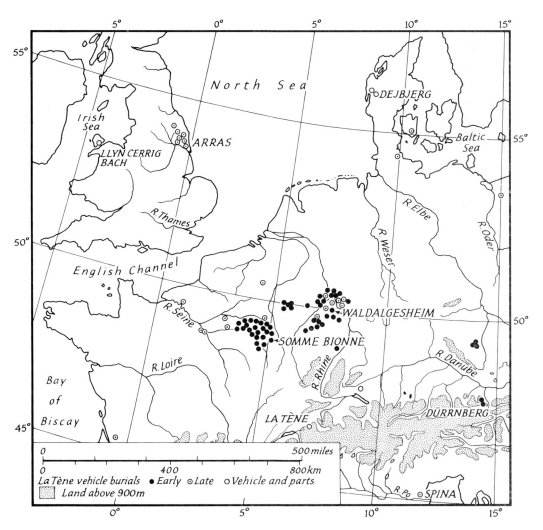

120 *Map of La Tène vehicle burials, 5th to 1st century* BC

and Late, with an approximate dividing line around 300 BC, marked by the cessation of imported Etruscan or allied bronzes in the graves, and the transition from the Waldalgesheim art-style to the Plastic and Sword-Styles and their eventual successors. In the original scheme of Reinecke the boundary would come approximately between his LT B2 and C1. In the first, fifth century, part of the Early phase, come the 'princely graves', many with Etruscan imports, as well as others, as in the Belgian Ardennes, with no rich objects; the fourth century is represented by hardly more than the Waldalgesh-

eim grave itself and a handful of linchpins decorated in the eponymous style. The Late phase is marked by the chariot being now increasingly represented in the graves in part only, essentially its wheels, and it may mark the only social differentiation among the graves of a cemetery, as at Hoppstädten or Gransdorf (Haffner 1969; Schindler 1970), as indeed was the case in the Early phase with such poorly furnished chariot graves at Hamipré and other Belgian sites (Cahen–Delhaye 1974). General surveys, with catalogues and distribution maps, of La Tène chariot burials, are those of Joffroy

126

& Bretz-Mahler 1959; Stead 1965a, 1979; Harbison 1969; Joachim 1969, and a specific study of the early transitional Hunsrück–Eifel material in Haffner 1976.

The miscellaneous imported Etruscan bronze-ware of Ha C and D at the end of the latter phase in the late sixth century becomes largely replaced by a single preferred type (at least for deposition in graves), the trefoil-mouthed or beaked flagon *(Schnabelkanne)* of which an early example was in the Vix grave. Of the fifty or so known examples of the fifth century surviving and recorded north of the Alps, at least a dozen come from chariot graves (Jacobsthal & Langsdorf 1929; Frey 1969, 115) and with them may be taken a type of shallow bronze bowl with or without handles, also of Etruscan or Italian origin, from about half that number (Schaaf 1969), the exceptional bronze tripod and stamnos from Bad Dürkheim and a couple of imported Greek cups. Less spectacular Etruscan imports from such contexts are the iron firedog from Hořovičky and the roasting-spits from Somme Bionne, repeating the earlier Ha C occurrence of firedog and spits in the wagon grave of Beilngries (Stary 1979b). Another link between the earliest La Tène chariot graves are the local tall conical bronze helmets of Berru type, with possibly ultimately oriental antecedents (Schaaf 1973), and bronze situlae of a type probably originating in the Golasecca culture-area (Kimmig 1964b; Penninger 1972). These occurrences, summarized in the table overleaf give the impression of a closely linked group of centres of purchasing power and prestige, sharing common sources of luxury objects over probably not more than a few generations, and all sharing also in the use of the chariot for parade and probably for war.

The Late phase of chariot burials should start by 300 BC but the evidence is extremely thin, consisting of little more than a handful of linchpins decorated in Jacobsthal's Plastic style: the Waldalgesheim grave of the later fourth century was dug out in pouring rain on a November day in 1870 and we know hardly more than that the elaborate bronze and iron mountings of a chariot and fragments of iron tyres were found. The linchpins referred to have a wide distribution, from what appears to have been a chariot burial in Paris to one in the dromos of the Hellenistic chamber-tomb of Mal-Tepe, at Mezek in Bulgaria, with in both instances other bronze chariot-fittings, probably reinterrets, discussed later (Krämer & Schubert 1979). Thereafter dating becomes hazardous except in terms of local relative chronologies established by pottery or fibula types, the latter, however, showing the continuance of chariot burial up to the time of use of the Nauheim fibula, from the mid first century BC to Caesarian-Augustan times (e.g. Hoppstädten 14: Haffner 1969). An important technological development was the abandonment, some time probably in the third century BC, of fastening-nails from the iron hoop-tyres, which henceforward hold to the felloe solely by reason of their shrinkage when cooling from red heat. For the last two centuries BC the vehicle evidence is usefully augmented by the metalwork from settlements such as the *oppidum* of Manching, Bavaria, occupied from LT C, some time in the second century, to the mid or late first (Collis 1975, 104; G. Jacobi 1974) or by probably votive finds such as La Tène itself (Vouga 1923; de Navarro 1972), Kappel in Saulgau (F. Fischer 1959) or Llyn Cerrig Bach in Wales (Fox 1946), or quite exceptional finds such as the bronze-smith's debris from casting harness parts at Gussage All Saints in southern England (Wainwright 1979; Foster 1980). In the continental graves of the last phase, the adoption of the cremation rite led to the cramming of the folded-up metal parts of the vehicle, rescued from the funeral pyre, into a small grave.

In the Early phase, the chariot graves are mainly concentrated in the Rhineland Palatinate, and in east France in the Marne and

	Flagon	Bowls	Greek cup	Helmet	Situla	Miscellaneous
Austria						
Dürrnberg XVI	X					
Dürrnberg 44		X	X	X	X	
Czechoslovakia						
Hořovičky		X				Firedog
France						
Berru				X		
Bussy-le-Château		X				
Châlons-sur-Marne				X		
Cuperly				X		
Écury-sur-Coole				X		
Pernant		X				
Prunay				X		
Sept-Seaulx	X					
Somme-Bionne	X		X			Spits
Somme-Tourbe	X					
Germany						
Armsheim	X					
Bad Dürkheim	X					Stamnos & tripod
Bell					X	
Besseringen	X					
Dörth	X					
Hillesheim	X					
Horshausen	X					
Hundheim	X				X	
Kärlich 1906	X				X	
Kärlich 1932	X					
Theley	X					

Dürrnberg 44 has a locally made bronze-mounted wooden flagon, one of a group with prototypes in Etruscan spouted bronze types (Dehn 1969).

Champagne regions, but with important outliers in Austria and Bohemia to the east, and Belgium to the west. The Late phase burials (including those of an intermediate middle period if that can be distinguished), extend the distribution beyond the Rhineland and France to Belgium north-west of the Marne and as far as the Seine, as well as to Yorkshire. The Early grave-type continues the Ha D tradition in Germany, with roofed pits, some with timber lining as at Kärlich 1932, under a barrow within a circular ditch (Günther 1934; Rest & Röder 1941), or Hillesheim, and with a wooden chamber and stonework at Hundheim (Haffner 1976, 182; Kimmig 1938). Dürrnberg 44 was under a barrow but on the old ground surface (Penninger 1972). Wheelpits are normal (as too in the graves near Plzen in Bohemia: Soudská 1976), but while whole vehicles were buried as in Kärlich 1932, that at Hillesheim was certainly not complete, leading us to a wider problem to be discussed later. In France, we encounter the well-known massive pit-graves, of *121* which Somme-Bionne and Châlons-sur-Marne and at least two others had a separate *123* T-shaped trench to accommodate the yoke and pole-tip, and there is no question of anything but complete chariots (Stead 1965 a, figs 5, 6, for comparative plans). The recent excavations in the Belgian Ardennes have produced comparable arrangements *124* though with less rectilinear accuracy and neatness than the published French plans of *125* around a century ago, which may reflect higher standards of field survey and less schematic afterthoughts. The placing of bits

and other elements of harness in vehicle graves was, as we saw in the last chapter, becoming rare in Ha D, and in the Early LT graves under discussion the practice is virtually abandoned except in the East French graves and those of the Plzen region in Bohemia such as Hořovičky.

The large grave-pits of the Marne, with rich grave-goods and a complete chariot, must have carried a horizontal timber roof in the manner of the Czechoslovakian graves of Ha C and their contemporary counterparts, and indeed the formal plan of the Gorge-Meillet grave, trapezoidal and *122* with a 'stepped' floor, comes very close to that of Grosseibstadt 1 of Ha C (Fourdrin- *86* gier 1878; Kossack 1970). Roofing is also to be expected in less rich chariot graves, as

121 Original sketch-plan of Early La Tène chariot grave at Somme-Bionne, France, 1876

122 Original sketch-plan of Early La Tène chariot grave at La Gorge-Meillet, France, 1878

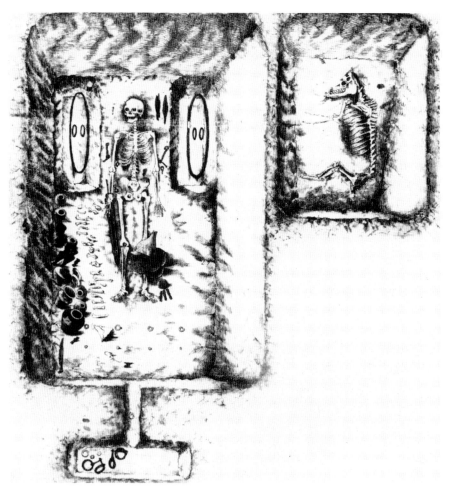

123 Original sketch-plan of Early La Tène chariot grave and adjacent boar burial at Châlons-sur-Marne, France, 1906

those in Belgium. The early excavators of the French graves claimed that a large proportion had been disturbed and robbed in antiquity, and the very recent (1980) excavation of such a grave at Quilly in Champagne confirms this generalization, which would also support the supposition of roofed chambers from which the contents could be extracted (Stead, *pers. comm.*). At Berru the grave was within a circular ditch (Déchelette 1927, fig. 426; Joffroy 1973) and at Quilly within a square one and originally under a barrow. Ploughed-out or otherwise destroyed mounds, and enclosing square or circular ditches, may well have been features

unappreciated or undetected by the early excavators. They are known to have dug at least 130 chariot graves in the region, usually with the minimum of record and leaving only three plans, Somme-Bionne, *121* Gorge-Meillet and Châlons-sur-Marne, with *122* any pretensions to completeness or approxi- *123* mate accuracy: none is to scale (Joffroy & Bretz-Mahler 1959; Morel 1876; Fourdringier 1878; Lemoine 1906). Barrows and enclosing ditches are commonly reported in Germany: at Hillesheim the grave measured only 2.20 × 2.15 m and was under a barrow, and again probably roofed, and although it had the iron tyres and nave-bands of a pair

204

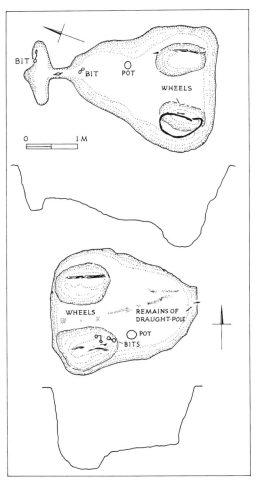

124 *Plans and sections of Early La Tène chariot graves 2 and 3, Hamipré, Ardennes, Belgium*

the iron tyres and nave-bands of the wheels together with a pair of fitments of a type later discussed, frequently found by the axle or fore-edge of the chariot body (Schindler 1970; Haffner 1976, 30). Another comparable example, though perhaps of earlier date, is the similarly proportioned chamber under one of the Kärlich barrows (1939), again with tyres and nave-bands but no wheel-pits (Rest & Röder 1941). Two explanations have usually been put forward to explain these and comparable instances; the burial of the chariot without its draught-pole, or the deposition of two detached and dismantled wheels. Now, while the removal of the draught-pole articulated to a four-wheeled vehicle is feasible, and as we saw appears to have been a not uncommon prac-

of wheels, could never have contained a whole vehicle: the grave-goods included a bronze beaked flagon and a gold armlet and finger ring (Steiner 1929; Haffner 1976, 182). This introduces us to the problem of graves containing evidence for no more than the wheels of a chariot, which becomes more frequent in the inhumation graves of the Late phase.

An excellent and well-recorded example is the burial in Grave 17 of the Late LT cemetery of Gransdorf, with a grave-pit containing the bedding-bench for a square wooden chamber about 2.0 m across, with a pair of wheel-slots in the floor near one wall, and

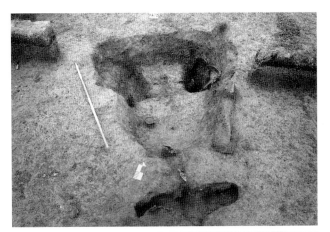

125 *Excavated Early La Tène chariot burials 2 and 3, Hamipré, Ardennes, Belgium*

126

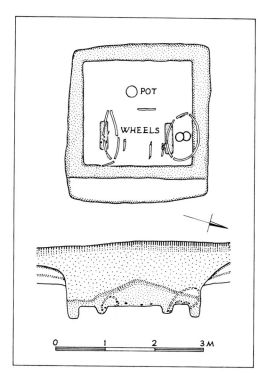

126 Plan and section of Early La Tène chariot grave, Barrow 17, Gransdorf, Germany

tice in Ha D vehicle burials, the pole of a chariot is an integral part of its structure, as a component part of the floor of the box. It is suggested that a third explanation is more probable, and that in the graves under discussion what was buried was a wheels-and-axle unit from which the body and draught-pole in one could have been removed with relative ease. This does not preclude the dismantling of wheels in some instances—the evidence of the Yorkshire graves, for instance, is clear here—but could be applicable to paired wheel-pits or of wheels themselves found at what we shall see had become the standard gauge of 1.30 m.

The practical advisability of relieving the weight borne by lightly constructed wheels, to avoid their distortion when not in use, was pointed out by Powell (1953b, 165) and by Gordon (1978, 145), citing Homeric instances (e.g. Iliad 8·441, 435; Odyssey 4·42) and one way in which this could be done

was to have a body which could be lifted from the axle-bed, complete with draught-pole, as is implied by the separate storage of chariot bodies from that of wheels in pairs in the Mycenaean inventories (Ventris & Chadwick 1959, 361). Archaeological confirmation is provided by the details of the seventh-sixth-century chariots from the tombs of Cypriot Salamis, where the body had a pair of U-shaped blocks to fit over the axles, from which it could be lifted (Karageorghis 1967, 22, 51; pls XVIII, CXV), and the bronze cart-model from Bolsena, *117* described in Chapter 5, shows in miniature the same arrangement (Woytowitsch no. 168). If such a construction was present in at least some La Tène chariots it would be possible to provide a *pars pro toto* representation of a vehicle in a grave by the simple expedient of taking off the body with its integral draught-pole from the wheels-and-axle unit, and burying the latter alone. As we shall see, there is no good reason for assuming a uniform and standard 'Celtic chariot' over the whole area of La Tène culture and beyond in the last five centuries BC, and variations in burial rite may in part at least reflect differences in chariot-builders' practice.

In the latest series of cremation graves, with chariots as at Hoppstädten (Haffner 1969) or four-wheeled wagons or carriages as at Husby (Raddatz 1967), the iron tyres and other fittings of the vehicle were evidently recovered from the funeral pyre, and bent and folded so that they could be crammed into the relatively small grave: at Husby the draught-pole was included as its terminal iron sheathing was present, but at Hoppstädten and many other graves only the tyres of a pair of wheels survived.

A final mention must be made of grave-goods in chariot burials so far as they may indicate the status of the deceased and so by inference the social function and use in life of the vehicles. Especially in the Early series of graves we are clearly, as with the Hallstatt vehicle graves, dealing with the élite of a stratified society in a position to acquire luxury goods by import or patronage of local craftsmen. The incidence of objects implying warrior rank is curiously unequal, and was one of the factors influencing Stead in his adoption of 'cart-graves' rather than 'chariot-graves' as an appropriate name for the class. Helmets have already been mentioned, and swords were present at Dürrnberg 44 and in twenty or so Marne graves, including Châlons-sur-Marne, Somme Bionne and Gorge-Meillet (but not at Berru). The heads of what are probably throwing spears or javelins often accompany swords, but may be the only weapons, as in the Belgian Ardennes or in Germany at, for instance, Kärlich 1906 and 1939 and Hilleshem. In the Yorkshire chariot graves no weapons were found. A number of the Marne graves excavated in the last century were claimed as those of women on the grounds of the absence of weapons and the presence of ornaments, particularly torcs, not found in demonstrable warrior graves, and Stead is prepared to admit at least a proportion of them. A clear example in Yorkshire is that of Arras A 3, a chariot burial with an iron mirror, recalling the undoubted woman's grave (not a vehicle burial) with a bronze mirror and other rich grave-goods of Early La Tène at Reinheim near Saarbrücken (Stead 1979, 25, 98; Keller 1965). We may accept the fact then that chariot burial was not an exclusively male prerogative. There are two or three ill-recorded references to horses being buried though it is not clear that they were actually in the chariot grave (Stead 1965a, 17). The differential distribution of the custom of placing bits and harness ornaments such as phalerae with the burial are discussed below when dealing with harness as a whole.

The food-offerings recorded in a few instances in the graves are of interest in themselves and for the reminiscences they convey of the practice of Ha C vehicle burials, notably the joint of pork with a chopping knife

at Dürrnberg 44 and in the Marne. Some of the French graves offered an elaborate menu—pork, poultry and eggs at Gorge-Meillet, and at Châlons-sur-Marne beef, pork (or boar), hares, pigeons, ducks and smaller birds and a cock; one pot containing cooked frogs (the skeletons of 110 *succulents batrachiens* were counted) and others peas, lentils and the residue of mead (Lemoine 1906). Domestic poultry and probably fighting-cocks must rank as minor oriental imports originating in the wild Indian jungle fowl (*Gallus* sp.) (Zeuner 1963, 443) and coming into the Greek world in the orientalizing period, and so thence to continental Europe in the general context of the Adriatic trade route. A few bones are as early as Late Hallstatt (Heuneburg; Praha–Michle; Peške 1976) and in the Reinheim grave just mentioned was a fibula in the form of a cock. Pigs and wild boars play an important part in later Celtic iconography and mythology: the entire carcase of a boar was buried in a separate pit beside the Châlons-sur-Marne grave. *123*

The chariots

As occasionally in Ha C and D, and indeed beyond this into Urnfield times. the evidence of the La Tène graves, from the fifth century BC onwards, is that of a prestige vehicle used in funeral ceremonies of a two-wheeled type, going back to the second millennium BC as in the graves of the Sintashta cemetery in the Urals. By the first millennium similar burials are known outside Europe in, for instance, Phrygian Asia Minor in the eighth to sixth centuries BC, with chamber tombs under barrows, paired horses, or actual chariot burials (Sandars 1976; Kohler 1980) and in Cyprus (Karageorghis 1967, 1969, 1973), or again, as we saw in the last chapter, Etruscan Italy. The European graves could be, but were not necessarily always, those of warriors, and include women (as too at Atenica, described in the last chapter) and could have grave-goods of varying degrees of luxury, or no more than the common run of offerings in graves of cemeteries where the vehicle alone differentiated the burial. On analogy, we may suppose the chariot was a symbolic vehicle to bear the dead to the other world, and we need not look outside temperate Europe for the chariot or its place in the graves, where it represented a long-established tradition.

The question remains, however, as to what we are really seeing in what remains for archaeological recognition in the graves, and the possibilities seem three. The first is an actual status-symbol two-wheeler which could on occasion serve as a war chariot, or for civil purposes—'for travelling and in battle they use two-horse chariots', observed Posidonius of Gaulish Celts in the second century BC (Diodorus Siculus v. 29). A second could be a hearse, or vehicle wholly intended for funerary purposes, and a third, an everyday vehicle modified on occasion for burial. Stead, discussing this problem (1965b), is persuasive on this last option. 'It does not necessarily follow' he writes 'that the two-wheeled carts in Marnian and Rhenish graves had not originally been chariots. The surviving fittings usually show signs of wear, and they may have been used for other purposes before the funeral. But even if this had been so, they would probably have been modified for a funeral ... Extended skeletons indicate that the funerary carts had neither front nor back.' The placing of the deceased on the vehicle was not, in fact, universal: it was well attested in Dürrnberg 44 and inferred in many Marne graves, but the burials seem to be under the chariot in some instances, such as Kärlich 1932 and Gransdorf, and perhaps at Somme-Bionne. Two years after Stead wrote Karageorghis published the seventh-sixth-century vehicle graves in Cypriot Salamis, recognizing only one 'real' chariot (chariot β from Tomb 3; later, chariot β

from Tomb 69 was found: Karageorghis 1967, 1973; Littauer 1976b), and interpreting the others, with simple plank platforms, as specifically for burial purposes. Reviewing the first Salamis publication the writer asked the question, in respect of the La Tène graves, 'Have we ... the Celtic war-chariot at all, or rather a simplified funeral vehicle which will allow the occupant to lie full length in the tomb?' (Piggott 1969b). If we consider as plausible the idea put forward earlier and partly based on the Salamis evidence, that some chariots could have bodies easily removable from the wheels-and-axle unit, the modification suggested by Stead would be a simple matter.

The chariot bodies

With the foregoing reservations, that the consistent platform about a metre square recorded or to be inferred in many graves may be a simplified funerary modification, there remains other evidence to consider for chariot bodies or boxes. The adaption may not have been a universal practice over the whole area of recorded chariot burials in Europe over five centuries, and again Stead made the point (1965b) that there is no evidence to support or reason to assume a standardized 'Celtic chariot' when iconography suggests an eclectic use of local forms in, for instance, north Italy or Galatian Asia Minor, and a platform type, or one open at back and front, is not excluded as one variant among many. The direct archaeological evidence from the graves is exclusively that of a metre-square platform with the axle set centrally and made without any use of metal nails or clamps, but with two main types of looped metal attachments discussed below. The chariot floor, particularly in the simpler forms, is likely to have been of leather strap-work, as in the Italian racing-chariot model (Woytowitsch no. 175) and in ancient Egypt. But remembering the shallow bodies of the Hallstatt carriages, with a low sur-round often decorated with appliqué metal, and corner-knobs, a similar square chariot-platform might be visualized and would explain, for instance, the knobbed and baluster-moulded iron and bronze pins such as the four at Berru (Joffroy 1973) or the two at Somme-Bionne, recalling the Hüns-ruck-Eifel sets as at Bell or Hennweiler (Driehaus 1966a). Shallow boxes with low sides might again, as in Ha D wagons, provide a place for the fragments of decorative, often openwork, mountings from several Early La Tène graves such as Besseringen, Dörth, Horhausen, Kärlich 1932 and Waldalgesheim (Günther 1934; Jacobsthal 1944, nos 153–56), or in the Late phase, the Neuwied grave of *c.* 30–10 BC, compared to the decorative bronze-work on the Dejbjerg carriage (Joachim 1973). A group of looped and linked attachments associated with the floor or box of the chariot will be discussed shortly, but in the meantime the iconographic evidence for the side-panels present on certain chariot types must be considered.

Discussion of the original nature of prehistoric European chariots goes back over a century and the early excavations in the Marne, but the first attempt to consider the matter in modern archaeological terms was that of Cyril Fox (1946) in publishing the Llyn Cerrig Bach votive find from Anglesey, combining the evidence from the find itself with that from earlier excavations, and interpretations including reference to a frequently cited Celtic coin type of the Gaulish *127* tribe of the Remi discussed below. Fox's well-known drawing, and its subsequent realization in model form, was a pioneering piece of work, at once practical and perceptive, but naturally subject to criticism as knowledge of the subject advanced. Our present realization of the lightness of ancient chariot construction has been emphasized in earlier chapters, and Fox's vehicle is certainly too heavy a piece of carpentry; Stead (1965b) made further cogent criticisms, particularly with regard to the

127 Late La Tène chariots on native and Roman coins: A, coin of the Remi, c. 60 BC; B, coin of L. Hostilius Saserna, c. 50 BC; C, coin of Julius Caesar, c. 50 BC; D, coin of Scaurus, 92/118 BC

pair of semi-circular wickerwork side-screens inferred from the Celtic coin evidence, to which the writer had added two comparable depictions on Roman coins *127* (Piggott 1952). The earliest Celtic gold coinage is based on the stater of Philip II of Macedon, with a chariot drawn by two horses on the reverse, and 'the likelihood is that the philippus was chosen simply because the Gauls preferred a gold type with horse and chariot', because it represented to them a potent status symbol in their own social order (Allen & Nash 1980, 69). With such a classical prototype, distinctively La Tène features in the vehicle depicted on so many Celtic coins are absent except in the one Remic type referred to.

It was unfortunate too that Fox's reconstruction (which he was at pains to point out was a tentative essay based on the then available evidence) became widely accepted as the canonical 'Celtic chariot', the more so when it was realized by 1968 that we had

128 Stele with arched-side chariot, 3rd century BC, *from Padua, Italy*

though, like the whole Celtic coin series, it owes its charioteering scene to classical models, it is alone in showing the double-hoop side-screen or any other distinctively non-classical features. A Celtic coin type, closely related to the 'first generation' copies of the gold stater of Philip II of Macedon, shows an unaccompanied and seated charioteer with a whip or goad. Unlocated examples are widely distributed in collections, but its authenticity is suspect and it cannot be used in evidence (Allen 1978, no. 49). The Roman coins (Piggott 1952) are two, the first a denarius of Hostilius Saserna of *c.* *127* 50 BC showing a double-hoop sided chariot with a naked warrior standing facing backwards with shield and spear, and a driver with a whip crouched well forward. The second coin, of Julius Caesar and again *c.* 50 BC, shows a display of Celtic trophies on the conventional tree-trunk and propped up against it a double-hooped chariot rather schematically rendered. Other chariot scenes on coins with warriors carrying identifiable Celtic weapons include one of Scaurus of 92 or 118 BC and thought to be associated with the foundation of Narbo and the defeat of the chieftain Bituitos: it shows again a standing naked warrior throwing a spear and carrying a long shield and a *carnyx* or war-trumpet, with no features of the side or front of the chariot, the body being a mere platform behind the waving tails of the horses. On both the native and the Roman coins showing chariots with double-hooped sides, the wheel is consistently shown as set rear of centre.

all—Fox, Stead and the writer—misinterpreted the numismatic evidence. The 'chariot horns' or hand-holds of Fox's reconstruction are open to doubt and capable of alternative explanations. They are discussed with yoke-fittings in a later section.

Clarification came with the publication by Frey (1968) of two carved stone grave stelae from Padua, the earlier probably of the third *128* century BC, and showing a man and woman sitting in a two-horse vehicle, with the man holding a spear and with a long oval shield of La Tène type; the second stele, of rougher workmanship, had a similar vehicle with a standing warrior with a similar shield brandishing a sword. (For shields of this type in Italy cf. Stary 1979a, 200). The side of both vehicles is shown not as a single, but a double inverted 'U' hoop, as we can now see this was what the coins depict, which had previously been interpreted as being attempts at a perspective rendering of a single hoop on each side, or (Stead 1965b) of one side and the front of the vehicle. The coins in question are as follows. The Remic *127* coin is of a bronze type of *c.* 60 BC (Allen 1978, no. 75; Allen & Nash 1980, 100) and

The sum of the iconographic evidence is then that one of the chariot types current in Later La Tène Europe, and attested both in northern Italy and in the territory around Reims, had double-hoop side screens and *129* was almost certainly open front and back. The first Padua stele may represent the journey to the otherworld in Etruscan fashion (cf. Blazquez Martinez 1958) and is certainly not a scene of chariot warfare, underlining the peaceful use of the vehicle as described by Posidonius, and indeed reminiscent of those who 'rode into the country with their wives in two-horse chariots' in early Cumae, as we saw in Chapter IV. Though widespread the type was neither unique nor universal among Celtic-speaking peoples, as the iconographic evidence from Civita Alba, Chiusi and Pergamon shows.

Before turning to the remaining elements of the chariot such as its wheels, draught-pole and yoke there remain certain metal attachments of distinctive type and already mentioned in passing. The first are simple eyed bolts of iron, found in pairs in Early La Tène graves at the rear of the body (Somme-Bionne): the front (Gorge-Meillet) or at the axle at Hamipré 1 (Cahen-Delhaye 1974) as well as in unassigned contexts as in the Late phase cremation grave at Urmitz (Joachim 1969, fig. 6, 14–20) and elsewhere. The loops imply some perishable passed through them but the function is unknown. More important and puzzling are more elaborate looped link attachments, found in pairs at the fore edge of the body or body floor: their rear end shows evidence of original fastening to wood and is sometimes half-round, and a jointed single or two-piece bar with looped end is attached. They have been discussed and listed by Haffner (1976, 31; add Hamipré 2; Cahen-Delhaye 1974) from fifteen or more Early graves in the Rhineland and the Marne, who points out that as several are elaborately decorated, they were meant to be seen, whatever their function. This function has been a matter of dispute since the nineteenth-century discoveries, and Déchelette (1927, 694) thought they held straps across the open front end of the chariot body. Jacobsthal, illustrating an example from the Dörth grave (for a second, Joachim 1978), described it as the 'roll of swingle-tree' (1944, no. 153). This was an unfortunate and misleading anachronism: the swingle-tree is a pivoted free-swinging bar taking the traces of horses har-

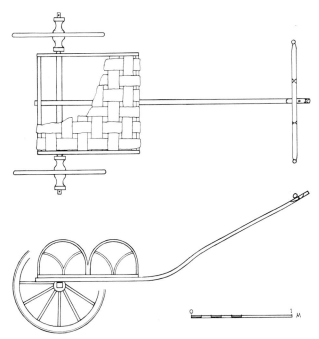

129 *Constructional diagram of a type of Late La Tène chariot*

nessed between shafts with a rigid collar, not known in Europe until late Roman times and essentially a medieval and modern arrangement (Jope 1956). An alternative is the splinter-bar, rigid and forward of the body on the pole for paired draught in vehicles of the type of the nineteenth-century 'curricle' or 'Cape cart' (Adams 1837; Philipson 1882, 91). Fox in his reconstruction introduced a splinter bar, but following Jacobsthal called it a swingle-tree. Archaeologically there is no evidence for such a bar, and trace attachments are difficult to account for in the normal chariot-harness of the ancient world (cf. Spruytte 1980). Their purpose, therefore, still remains unknown. A terret from an Early La Tène grave at Laumersheim, described below, has a mounting similar to Dörth.

The wheels: sizes and spokes

The approach to standardization of measurements which we saw beginning in Ha

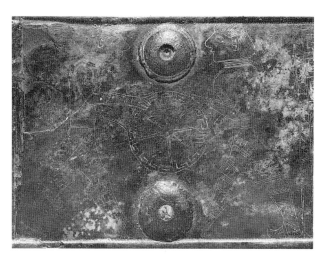

130 *Figures holding symbolic wheel, engraving on bronze sword-scabbard, Early La Tène, Grave 994, Hallstatt cemetery, Austria*

D becomes marked in La Tène, as if a norm of common workshop practice among wheelwrights had become established within the limits of non-industrial craftsmanship and was maintained over some four centuries in time, and in space over an area which not only included such peripheral regions within the general La Tène tradition as Britain, but also areas outside its ambit, as in the four-wheeled 'Cimbric Wagons' later described from Schleswig and Denmark. The general reviews of Haffner (1976, 29–34) and Stead (1979, 42–44) render detailed enumeration unnecessary. The range of thirty or so records of wheel diameters, from the Hunsrück-Eifel and Early LT series to the Roman period, extend from exceptionally large wheels (early at Dürrnberg 44 and Bad Dürkheim, *c.* 1.10 m; late at Attichy and Nanterre, 1.20 and 1.30 m) to those with a diameter of 80 cm (Kärlich 1932; Dürrnberg XVI, Châlons-sur-Marne) or even 70 cm (Kärlich 1928). The great majority, however, approach a standard of 90 cm. Similarly in the gauge, so often ascertainable from the pits or slots for the reception of wheels in the floors of graves, there is no significant

departure, within 10 cm or so, from a standard of 1.30 m.

The number of spokes is infrequently known, owing to the absence of metal nave or spoke sheathing in the Ha D manner, and the only Early LT record, possible owing to the decorative bronze rings at the base and half-way along the spokes, is that of eight in Kärlich 1932. In Late LT contexts actual wooden wheels survive, as at La Tène itself (ten spokes) and the 'Cimbric' four-wheelers of Dejbjerg in Denmark, one with twelve-spoked wheels and the other with fourteen. The British material, reviewed in more detail in a later section, is all late, with probably nothing before the second century BC and shows a range of from twelve spokes (the Garton Slack chariot burial) to nine in the (undated) wheel from Ryton.

Two wheel-representations, one from the Early and one from the Later phase should be mentioned: in both a detached wheel is being displayed as a cult object. The first such scene is on the engraved bronze sword-scabbard from Grave 994 at Hallstatt, of La Tène A (fifth-century) date (Megaw 1970, no. 30; Dehn 1970) and is duplicated, a pair *130* of men at the two ends of the pictorial frieze confronting one another, grasping a small wheel with nave, eight elegantly tapered spokes, and felloe decorated with the Greek key-pattern, also used on shield-rims and decorated dividing-bands elsewhere in the composition. Another symbolic wheel is that held by a figure with a horned helmet among the complex scenes of elaborate religious ceremony on the silver cauldron from Gundestrup in Jutland, half of which is shown, with nave, moulded and reeded spokes (an original sixteen seems intended) and felloe and tyre differentiated. Two decorative wheel motifs rather than accurately delineated examples appear on another panel. The cauldron probably dates from *c.* 100 BC and its stylistic affinities are with contemporary Dacian and Thracian silver and gold-work and a place of manufacture in south Slovakia seems best to fit its complex artistic background (Klindt-Jensen 1949; Powell 1971b; Olmsted 1979).

Naves, axle-caps and linchpins

As we saw in Chapter 5, the end of Ha D is marked by a reduction in the bronze or iron sheathing of the wheel naves, becoming 'less demonstrative and more utilitarian' with those classified as Type D. This simplification is maintained in La Tène, with a return to pairs of nave-bands harking back to Urnfield types such as those at Heathery Burn in Britain. This lack of metal sheathing deprives us of a knowledge of the wooden core of the naves in the Early phase, but when in the Late period we encounter actual surviving wheels as at La Tène, Dejbjerg, Glastonbury or in the Roman period, we find the cyma mouldings of the Ha D Type D as at Vix maintained, and it is reasonable to assume its continuity from that time, with simple or moulded metal bands at the outer and inner ends alone. Sizes are relatively standardized and consistent, between 14 and 18 cm in diameter from Hunsrück-Eifel to those of the Marne or the Belgian Ardennes in Early La Tène (Haffner 1976, 29–30; Joffroy & Bretz-Mahler 1959, 13: Cahen-Delhaye 1974), and again in such late examples as the Yorkshire nave-bands or those of Llyn Cerrig Bach (Stead 1979; Fox 1946). Lengths are less certain but seem to lie between *c.* 30–35 cm. Exceptional examples are the naves in Kärlich 1932, with bronze rings alternating with bronze and iron chequerwork inlay (recalling the Frankfurt Ha C yoke and its congeners), and in Hamipré 1, with iron bands and sheathing in a not dissimilar manner. Both seem to look back to the Ha D cylindrical ribbed naves of Type B (Günther 1934; Cahen-Delhaye 1974). In the two Dejbjerg carriages the naves have a feature without parallel in the technology of the ancient world. 'The axle bearing, about 7 cm in diameter, is round,

but it has a number of longitudinal grooves close together, rather worn ... but clearly round. When discovered there were cylindrical sticks in these grooves. The excavator considered they had served to reduce the wear, but to me it is more probable that they rotated with the wheel and thus acted as a roller bearing like the modern device' (Klindt-Jensen 1949, 89; cf. Jope 1956, 551). Such a technological innovation as roller bearings in north-west Europe around the first century BC is quite extraordinary, but the evidence admits of no other interpretation, and it stands as an unique phenomenon. Lathe-turning of naves can now be assumed as standard practice, and that from Kärlich 1939 is composite, in four sections held by iron clamps (Rest & Röder 1941; Kossack 1971). The Roman period (Antonine) wheel in the native manner from Bar Hill in Scotland (Macdonald & Park 1906) has at either end of the nave a circular crimped fillet of iron driven into the wood, a feature exactly paralleled in more delicate bronze-work on the turned wood base of the bronze-sheathed tankard of the late pre-

131 Nave with iron nave-bands as excavated, Early La Tène, chariot-grave 1, Hamipré, Ardennes, Belgium

213

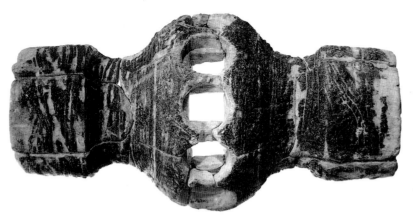

132 *Maple-wood nave of ten-spoked wheel with pitch painting, Late La Tène, Bad Nauheim, Germany*

Roman Iron Age from Trawsfynydd in Wales (Fox 1958, pl. 65), and so perhaps to be accounted a native technique, though also present in a Roman period wheel from Thrace, noted again below. A nave of maple wood *(Acer)* from a Late La Tène settle-
132 ment of about 100 BC at Bad Nauheim had been painted with pitch as preservative or colouring (Dr H. J. Hundt *in litt.*). The use of axle-caps is sporadic and appears to be more or less confined to the Early Phase in the Rhineland, with simple cylindrical 'canister' or flanged 'hat-shaped' types in the long-standing Urnfield and Hallstatt C tradition, as at Besseringen, Bad Dürkheim, Horhausen, Dörth and Theley, and at Hořovičky in Bohemia (Haffner 1976; Jacobsthal 1944, nos 153–55). These do not appear in the Marne or the Belgian Ardennes, but at Somme Bionne and Sillery are two quite exceptional and highly decorated pairs in bronze and iron, of annular form and likely to be the products of one craftsman (Jacobsthal 1944, nos 157, 158; Joffroy & Bretz-Mahler 1959, fig. 8). Plain annular caps of the same basic form come from the sixth-century chariot grave, the Tomba del Duce at Belmonte (Woytowitsch no. 91).

Linchpins in metal are exceptional in Early La Tène graves, and in the Rhineland are small and peg-shaped, sometimes with a decorated head as at Dörth (Jacobsthal 1944, 153b) or simple curved iron types in

the Marne as at Prunay and Condé-sur-Marne (Stead 1979, 45) or Cuperly (Joffroy & Bretz-Mahler 1959, fig. 7). A sudden efflorescence of linchpins with decorated heads appears, in terms of art styles, at the end of the Waldalgesheim and into the Plastic style of ornament, perhaps *c.* 300–250 BC, in a group in bronze and iron with a curved and knobbed pin and box-shaped head decorated with curves and tendrils often incorporating a human mask, from Waldalgesheim itself, Grossdraxdorf, the Erms at Urach, Niederweis, La Courte, 'Paris' and

133 *Linchpins with owl-heads, mid La Tène, from Manching, Germany*

214

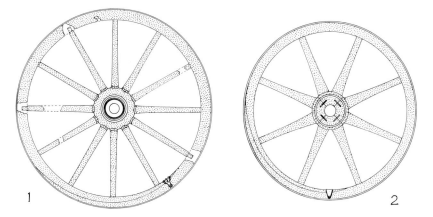

134 Constructional details of wheels with single-piece felloes; 1, Dejbjerg, Denmark; 2, Kärlich, Germany

1

2

133 Mezek in Bulgaria; there are more realistic human heads from Donnersberg and again Urach, and a splendid pair with owl-heads at Manching (Jacobsthal nos. 159–64; Mariën 1961, 40–44; Krämer & Schubert 1979). Ultimately derivative forms occur in Late contexts as at Nanterre, but the fashion appears otherwise to have been a short-lived one, reminiscent of the flamboyant linchpin episode in Ha C Bohemia noted in Chapter 5. In general, Late period linchpins are small and inconspicuous, in iron with double-looped heads in the 'Cimbric Wagons' such as Husby and Dejbjerg described in a later section. The British series is specialized, rod-shaped or curved and in general derivative from Late period continental types; these also are discussed later in connection with British chariotry as a whole (Ward Perkins 1940, 1941; Stead 1979, 45–46).

Felloes and tyres

Single-piece bent-wood felloes of the type familiar from previous chapters were early recognized in the later European Iron Age following the publication of those from La Tène itself in 1882 and Dejbjerg in 1888, and commented on by Déchelette (1927, 692). The first modern study devoted to this form of construction was that of Kossack (1971), in connection with the wider issue of Hallstatt wheels already summarized in Chapter 5, making two main points, the critical proportions within which it is practical to bend a felloe, and (following the Ha C iron fittings), the use of iron or bronze joint-braces or felloe-clamps over the long *134* overlapping scarf joint of the bent circle of wood, though the use of these had been earlier appreciated (Fox 1946; Stead 1965a, 30–31). These are U-shaped and by no means universal, but appear from Ha D at Vix, and in Hunsrück-Eifel vehicles such as Bell, and sporadically in Early La Tène (Somme-Bionne, Gorge-Meillet and other Marne graves; Hamipré 1; Kärlich '1927' and '1939') and again in Late contexts (Husby, Dejbjerg). Perishable bindings (and glue) were presumably used elsewhere: there must have been something of the sort in the similar joints of the very large two-piece felloe wheels at Pazyryk in the Altai. Romano-British versions of single-piece felloe wheels as at Newstead or Bar Hill have *136* butt joints held by an iron cleat or cramp (Piggott 1965, fig. 137; Stead 1979, fig. 12). Together with such single-piece felloes we must assume the continued use of those with segmental felloes cut from the plank, already as we saw in Chapter 4, attested in north Germany early in the first millennium BC. From Holme Pierrepont in England a complete twelve-spoked wheel of this type, with shrunk-on iron hoop tyre, is probably *135* to be dated to the second century BC (Musty

215

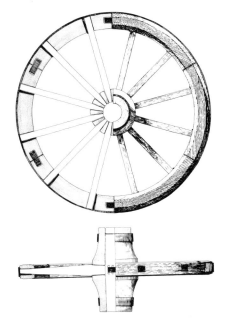

135 Wheel with segmental felloe, iron tyre and nave-bands, 2nd century BC, from Holme Pierrepont, Nottinghamshire, England

136 Wheel with single-piece felloe, iron tyre and nave-bands, 2nd century AD from Antonine Roman fort at Bar Hill, Dumbartonshire, Scotland

216

& MacCormick 1973; Stead 1979, 43); the presence of segmental felloes in Roman contexts may not be relevant to ultimate La Tène wheel technology. In Denmark and Sweden a number of spoked wheels with very massive segmental felloes occur as bog-finds but there is no sure dating before the Roman Iron Age of Scandinavia, and some are certainly later (Von Post *et al.* 1939; Witt 1970).

The discussion of Hallstatt tyres in the previous chapter examined the question of the iron hoop-tyre made in one piece and 'shrunk on' to the felloe of the otherwise completed wheel by contraction when cooling from its expanded circumference when at a red heat. It was noted how stud-nails were replaced by smaller fastening nails set at wider intervals, so that only six for instance, were used in the wheels of the wagon in the Hohmichele 6 grave. The La Tène evidence does not present a simple progression by further lessening the number of nails in the Early Period chariots: Dürrnberg XVI has *c.* 20 nails and 44 some 25–30; Bell and the two Hundheim vehicles in the Hunsrück-Eifel culture *c.* 12–15. But in the Marne, Somme-Bionne has six nails, Gorge-Meillet only four, and in the Belgian graves Hamipré 1 has few or none, Grave 2 only two 'nail-rivets' (Cahen-Delhaye 1974). Records are often unsatisfactory, and in Kärlich 1932 no nails were recorded in the original report (Günther 1934) but re-excavation produced a fragment of a nailed tyre (Rest & Röder 1941. By Late La Tène, nails had been wholly dispensed with (as had been precociously the case in the north Italian Ca'Morta B grave of *c.* 500), but it is difficult to fix the point of technological transition. If the tyre fragments, fairly certainly without nails, at La Courte in Belgium could be shown to be associated with the Plastic Style linchpins from the same site, a third-century date would be suggested, but the finds did not come from controlled excavations (Mariën 1961). The tyres from La

Tène have no nails, but dating objects from a long-used votive site is hazardous. It is reasonable to assume that the technology of the shrunk-on iron hoop-tyre without fastening nails was achieved, probably in the western La Tène world, by the second century BC, and certainly by whatever date the British series, uniformly without nails, was established.

Draught-poles and yokes

In all ancient chariots the draught-pole is structurally a part of the floor of the body or box of the vehicle and therefore at its rear end is only slightly higher above ground than half the diameter of the wheel. In order to keep the floor of the chariot horizontal the pole had to be curved 'to compensate for the difference between the height of the pole where it was set directly over the axle and under the floor, and its height at the yoke, chariot floors being kept as low as possible in order to keep the centre of gravity low and to facilitate mounting' (Littauer & Crouwel 1979a, 55). In none of the La Tène vehicle graves can more than the approximate length from axle to yoke be established, ranging from *c.* 3.0 m at Châlons-sur-Marne, and *c.* 2.7 at Gorge-Meillet and Somme Bionne in France, and 3.0 at Al Vaux, 2.30 m at Hamipré 2 and 2.0 m at Hamipré 3 in the Belgian Ardennes (Joffroy & Bretz-Mahler 1959; Stead 1965a; Cahen-Delhaye 1974, 1979). In the Rhineland the pole-length in the Kärlich 1932 grave was estimated at *c.* 2.0 m and at Cawthorn Camps in Yorkshire about the same (Günther 1934; Stead 1979). Fox in his reconstruction assumed a pole of about 2.6 m long, allowing as in all the foregoing instances about 50 cm for half the length of the chariot floor and *c.* 75 cm 'dashboard to rump'. He, in consultation with zoologists, used 116 cm for the likely withers height, but the average of the small 'western' Iron Age horse can now be seen to be higher, at

126–27 cm (Bökönyi 1968). Cahen-Delhaye in reconstructing the chariot in the Hamipré 2 grave (1974, fig. 18) failed to consider this factor and, giving the vehicle a straight pole, made allowance for a draught-animal only 60 cm high at the yoke. The same mistake was made in the original reconstruction of the second-millennium chariot from Lchashen described in Chapter 3.

Little is known of pole-mountings. In the Early La Tène grave of La Bouvandeau in the Marne decorative bronze mountings were found in the pole-trench in front of the grave (Stead 1965a, 36) and by the original excavator combined as an elaborate pole-tip sheathing (cf. Déchelette 1927, 697, fig. 505). Mariën (1961) and the writer (Piggott 1969a) regarded one of these (Jacobsthal no. 168) as a yoke-mount of a type shortly to be discussed—it is a curved tubular mounting with incised ornament and a globular knob-terminal—but it now seems preferable to take it, with Stead (1979, 53), as, in fact, a pole-tip, while interpreting the accompanying pair of openwork mounts (Jacobsthal 1944, no. 171) as yoke-decorations as argued below. In England, a simple bronze mount, almost certainly from a pole-tip, comes from one of the Arras chariot burials in Yorkshire, and some sort of an iron pole-cap from Garton Slack (Stead 1979, 53; Brewster 1971). The Llyn Cerrig Bach find included an utilitarian type of iron sheathing for the tip of a pole of hawthorn *(Crataegus)* or cherry-wood *(Prunus* sp.) with parallels at La Tène and in the 'Cimbric wagon' of the Husby find described later (Fox 1946, 23; Vouga 1923. pl. XXXIX, 20–21; Raddatz 1967, fig. 7). The nature of the pole-tip in Kärlich 1932 is uncertain (Günther 1934).

The nature and details of the yokes to which the chariot horses were harnessed present a problem and some uncertainties. None survive from the graves, though as we saw several of the Marne and Ardennes graves have explicit provision for pole-tip and yoke cut in the subsoil beyond the main

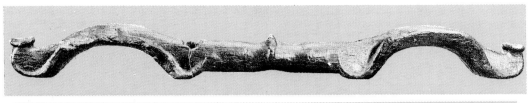

137 Wooden yoke, 2nd century BC., *from site of La Tène, Lake Neuchâtel, Switzerland*

burial pit. A certain number of actual wooden yokes of Later La Tène date survive—two from La Tène and one from the Dutch settlement-site of Ezinge—and unassociated finds from Denmark, Britain and Ireland which might be contemporary (Piggott 1949b; Fenton 1972). All of these Fenton takes as head-yokes for oxen; of the two from La Tène the larger (of oak) was originally thought to be an ox-yoke, though Ferdinand Keller thought it for horses, the smaller (of ash) was found lying on a horse's skull and near to a second skull (Vouga

137

1923, 95–96; pl. XXV, 1, 2; Jacobsthal no. 172). Jacobsthal and the zoologist Max Hilzheimer regarded the larger as an ox-yoke (and the slightly asymmetric carving of a yoke among the Galatian trophies at Pergamon as for an ox and a cow); Mariën (1961) by implication classed the larger La Tène specimen as a horse-yoke. In fact, no surviving La Tène or broadly contemporary yoke can be assigned to horse traction with any confidence.

We are left with metal mountings which may have been parts of lost wooden yokes. A pair of curved iron strips from the tomb of La Bouvandeau, already mentioned, have been seen as yoke reinforcements (Joffroy & Bretz-Mahler 1959, fig. 16). The two fine decorative bronze mounts were thought by Jacobsthal (no. 171) to be 'mountings of leather-covered wooden *hames* of a type still seen today on draught-horses', but as the writer pointed out this involved an anachronism, as hames form part of the rigid horse-collar and a type of harnessing between shafts not known in early antiquity (Piggott 1969a; Keegan 1978, 59). Fox (1958, 126) similarly misinterpreted a fragmentary bronze mount from East Anglia. A preferable explanation would be yoke-mounts, as Mariën (1961) had already suggested. An attempt to combine the openwork bronze mounts from Malomčřice, Brno, into yoke ornaments rather than the metal elements of a wooden spouted flagon (Radnóti 1958) has

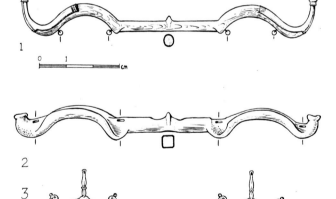

138 1, reconstructed yoke, wood with bronze terminals, Early La Tène, Waldalgesheim chariot grave, Germany; 2, wooden yoke, La Tène; 3, reconstructed yoke, wood with bronze terrets, Late La Tène, La Courte, Belgium

218

not been found convincing. There remain for consideration the bronze mountings which in the first instance gave rise to the 'chariot horns' or hand-holds of Fox's reconstructed chariot of 1946. These were subsequently discussed by Mariën (1961) and the writer (1969a) and the fine double-curved pair of tubular bronze mounts with cupped terminals from Waldalgesheim (Jacobsthal, no. 156a) seem most satisfactor-
138 ily accommodated in Mariën's reconstruction as yoke-terminals, the cupped ends recalling those of the Ha C yokes described in Chapter V on the one hand, and those of the La Tène ox-yoke on the other. The writer, following this interpretation, applied it to the British bronze terminals which Fox had classified as 'chariot-horn caps' and suggested these were more likely yoke-knobs. This remains an assumption of function as unsubstantiated as any other suggestion and as has recently been observed, 'there is no evidence at all that they were ever attached to chariots' (Spratling in Wainwright 1979, 134). But a pair of bronze mounts from Lough Gur in Ireland (Fox 1950) seem more convincingly interpreted as yoke-terminals than a pair of pole-tips (cf. Stead 1979, 53).

Bits and harness ornaments

The infrequent occurrence of horse-bits in Ha D graves was commented on in Chapter 5, and in Early La Tène the burial-custom varies, with a predominance of bits and harness equipment in the chariot graves of the Marne but a comparative paucity in, for instance, contemporary graves in the Rhineland. These western bits are of simple forms, usually of iron, and normally with two canons, exceptionally joined by a central link (as at Gorge-Meillet, in bronze) or ring (Ciry-Salsogne, Sogny, La Croix-en-Champagne). A few have a single bar canon (Wargemoulin, Hamipré 1) and all a simple pair of lateral rein-rings. These main types are illustrated in Joffroy and Bretz-Mahler,

1959 fig. 19; Stead 1965, 38–41; Cahen-Delhaye 1975. A unique bit from Sept-Seaulx has a large decorative openwork disc cheekpiece (Jacobsthal pl. 251, f.). The 'three-link' type has an archaeological, rather than technological, importance in the assessment of the continental affinities of the British and Irish Iron Age horse-bits and will be discussed in a later section. For our present purposes it is profitable to return to the point made by Balkwill (1973) in connection with Urnfield bits and discussed in Chapter IV, the variability of mouth size in the horses as implied by the bits themselves. The Urnfield bits imply a mouth width of *c.* 10 cm, with some as small as 8 cm, and some variability is shown among the La Tène series. For the Early Marne bits Joffroy and Bretz-Mahler (1959, 22) give an average of *c.* 14 cm, but published scale drawings suggest that many are around 12 cm; for the Belgian Ardennes two-canon bits range from 12 to 9 cm, and a single-canon example is 10 cm (Cahen-Delhaye 1974, 1979). The Later La Tène iron bits, now of simplified and indeed almost standardized two-canon form, average *c.* 10 cm at La Tène itself and *c.* 9–10 cm in the large Manching series (G. Jacobi 1974, 175). The British and Irish bits as we shall see tend to average around 12 cm in width, with an exceptional narrow-mouthed group of *c.* 8.5 cm in north Britain, evidently functional, as several show signs of wear. Mouth width need not be correlated with withers height, as both 'broad-faced' and 'narrow-faced' breeds may evolve and co-exist (Ewart 1911; Stevenson 1966, 42; Macgregor 1976, 25).

A curious and distinctive form of bit with exaggeratedly large U-shaped cheek-pieces or lateral pendants was mentioned in the last chapter in connection with an early sixth-century example from Sesto Calende wagon-grave B in the north Italian Golasecca culture. Versions of such bits appear in a group of chariot graves of Late Ha D and Early La Tène in Bohemia, south of

Plzen (Dehn 1966; Soudska 1976) but present difficulties of functional interpretation if the one or two moulded links joining the U-shaped bars are to be regarded as mouth-pieces, as while at Hořovičky the width is *c.* 9 cm, it is only 6 cm at Zelkovice and 5 cm at Sedlec-Hůrka, and the same would appear to apply to an elaborate example from Donauworth in Bavaria (Zürn 1972; Pauli 1980, no. 32). Their counterparts in the Marne (Berru, Écury-sur-Coole, Mairy) seem to have been pendants, the last with the bars joined by two links *c.* 12 cm across (Joffroy & Bretz-Mahler 1959, fig. 12; Stead 1965a, 39); and the same is probably the case at Anloo in the Netherlands (De Laet & Glasbergen 1959, pl. 44) and another elaborate example from the Marne (Jacobsthal no. 102). Decorative side-pendants of quite different form come from La Tène (Vouga 1923, pl. XXXVIII; Stead 1965, fig. 21; Drack *et al.* 1974, 127). A restricted class of Late La Tène bits with omega-shaped side-pieces or pendants has been recognized and identified as probably for cavalry rather than traction-horses (Krämer 1964; G. Jacobi 1974, 175 with map fig. 49; Frey 1976).

The problematical function of the varied group of looped discs known as phalerae has been discussed in Chapters 4 and 5 in connection with continued occurrence of such objects in vehicle graves from Urnfield times onwards. Of the Plzen graves just mentioned, six contain phalerae, simple or decorated, in all but one instances associated with the U-shaped bit attachments, and Dehn (1966) drew attention to the discs shown decorating the reins of the riders on the Early La Tène sword-scabbard from Grave 994 at Hallstatt as suggesting their position on the harness. In the Marne graves *121* however, where records of the position of similar phalerae exist, only at Somme-*122* Bionne were they with the bits in the 'yoke-*123* trench' forward of the grave, while at Gorge-Meillet and Châlons-sur-Marne they were at the feet of the skeleton by the side of

the helmet, a position which suggests they were part of the warrior's armour, recalling the comparison made in Chapter 4 with the discs on armour of the *kardiophylakes* type (Stead 1965, 37; cf. Joffroy & Bretz-Mahler 1959, 17). Some of the Marne phalerae are particularly fine pieces of Early La Tène art in openwork based on intricate compass-drawn circles: there is a good range illustrated in Jacobsthal nos 179–89; De Laet & Glasbergen 1959, pl. 44; A sub-variety (nos 190–95; Cahen-Delhaye 1975) has a layout based on a D with three circles along the bar and it is possible to recognize an ultimate reminiscence of this in a harness-mount from Polden Hill in Somerset, hardly before the middle of the first century BC (Brailsford 1975, pl. XXII, lower).

There remain for mention terrets or fairleads for reins. There is no evidence for specialized harness-mounts of this type in Ha C or D, as we saw in the last chapter, though pairs of simple rings with the Ha C yokes may have performed this function. In Early La Tène we have two or three quite exceptional finds, notably the highly decorated openwork disc on a mount semi-circular in section from the Waldalgesheim grave (Jacobsthal no. 156b) and an analogous mount with simple ring from Laumersheim (Kimmig 1950). With those and at a slightly later (third-century) date would go the elaborate pair in the Plastic Style, probably from the Seine at Paris (Jacobsthal no. 173; Megaw 1970, no. 167) and presumably the massive ring-mounting from Mezek (Jacobsthal no. 176). Other Early La Tène graves, as in the Marne, have produced no recognizable terrets: at Gorge-Meillet and *122* Châlons-sur-Marne, a line of four plain *123* bronze rings was laid across the grave floor, not with the bits but forward of the feet of the skeletons. If these are to be rein-rings on harness in the approximate position they would have occupied in life, they could only be on a pair of straps equivalent to loin- or hip-straps in recent trace-harness (Philipson

220

1882, pl. IX; Keegan 1978, figs. 4, 5).

In Late La Tène contexts terrets appear unequivocally and commonly, and presumably indicate some change in harnessing or driving techniques (cf. Déchelette 1927, 701, fig. 510; Mariën 1961, 45, 173; Joachim 1969, fig. 6; G. Jacobi 1974, 175). The small moulded rings, sometimes with a central pillar, and the method of yoke attachment, in no way resemble the main British D-shaped series, as we shall see, but the type continues into Roman times, as was well shown for Pannonia by Alföldi & Radnóti (1940), and these Roman provincial forms seem to lie behind the north British 'massive' type (Macgregor 1976, 39–41).

Britain and Ireland

Intermittent reference has been made throughout this chapter to the British evidence in relation to that of the Continent, but a short survey of the material in its own terms is desirable: Ireland presents a separate problem. The information the insular finds contribute to the knowledge of vehicle technology, our primary concern, is small, even though they highlight some of the classic archaeological problems of the later prehistory of the British Isles, in terms of the European relationships of their first iron-using economies and the adoption of chariotry. It is here that we enter a world where the polemics associated with the loaded terminology of 'invasions' and 'immigrations' can engender more heat than light, and where misunderstandings may well have arisen from the projection of the irrelevant concepts of modern nation-states into prehistoric antiquity and of thinking of narrow seas as inevitably barriers rather than as providing routes of contact alternative to those on land. Britain can best be regarded as an outlying (and necessarily insular and individual) province of adjacent European culture areas, linked to them by short Channel crossings and coast-wise traffic, as Cham-

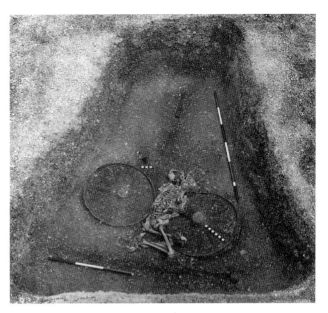

139 Chariot burial with dismantled wheels, Late La Tène, Garton Slack, Yorkshire, England

pion has suggested in the present context of the Iron Age (1975), and movements within such a loose cultural continuum, transmitting technological innovations of the types with which we are concerned, are no more than those, for instance, between Bohemia, the Rhineland and the Marne, and less than those across the Alps. In our specific enquiry we have one conclusive contact to consider, the establishment of chariot-burying communities of the Arras culture in Yorkshire, unambiguously expressed by Stead. 'The arrival in Yorkshire' he writes 'of artefacts from west-central Europe could be explained away by trade, but the arrival of ideas—complex burial-rites—must surely mean the arrival and settlement of people ...As to the burial rites, the differences are perhaps no more marked than those between the various groups on the continent ...Arras stands quite happily alongside the others as one member of a widespread family of La Tène cultures' (Stead *139* 1979, 92–93).

The information on vehicle technology from the Yorkshire graves has been admira-

making such objects in a settlement of the Iron Age at Gussage All Saints in Dorset discussed below. General surveys, with lists and maps, of the main types of metalwork are contained in Macgregor 1976: bits 51, map 1; terrets 60–72, maps 5–11; linchpins 73–76, map 12.

The chronology of the British chariotry and horse-harness shares the unfortunate imprecision of that of the whole period of which they are a product; the dating of the types of bit, for instance, 'epitomises the problem of British Iron Age chronology' as Stead remarked (1979, 50). But if the material is taken in connection with the sequence of vehicle and harness technology on the continent, discussed earlier in this chapter, some limiting factors of comparative chronology can be suggested which may slightly clarify the position. We may conveniently start with the iron *tyres* from the Yorkshire graves, Llyn Cerrig and elsewhere. With the exception of one dubious early nineteenth-century record all the British examples are shrunk-on iron hoop-tyres with no fastening-nails. Their advent in Britain must therefore be later than the achievement of this piece of technology in northwest Europe, and we saw that although difficult to date with precision, fastening-nails were given up on the Continent probably in the third and fairly certainly by the second century BC. Single-piece bent felloes can be assumed, as these appear in north British wheels of the Roman date, and the U-shaped joint-braces or felloe-clamps are sporadic in Early La Tène contexts and absent in Late wheels except in the 'Cimbric' wagons such as Husby or Dejbjerg. The *ogee-moulded naves*, with an ancestry going back to the end of the sixth century at Vix, of course continued in Europe, and the wheels from La Tène (whatever their precise date), with nail-less hoop tyres and no joint-braces to their single-piece felloes, would offer the best likely comparisons with the British series. Ogee-moulded naves were being lathe-turned at

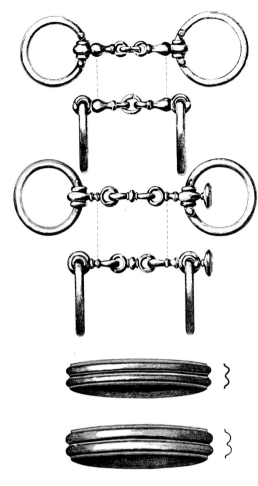

140 Bronze bits and nave-bands, 1st century BC, *from votive deposit at Llyn Cerrig Bach, Anglesey, Wales*

bly presented and assessed by Stead (1965, 1979), and is supplemented by other finds in hoards or in isolation, and Fox's study of

140 the Llyn Cerrig Bach votive find in Anglesey (1946) still remains a classic survey and discussion. Other hoards are usually of harness-trappings, bits and terrets, such as Stanwick (Macgregor 1962) or Polden Hill (Brailsford 1975) or the votive finds from the pre-Roman temple on Hayling Island (Downey *et al.* 1979), and of exceptional importance both from the technological and socio-economic points of view has been the discovery of the debris of a bronze-founder

222

Glastonbury in the first century BC, and afford an interesting glimpse of the circumstances in which a small settlement of five to seven households, as we now recognize the site to have been at any one time, supported a wheelwright among its craftsmen (Bulleid & Gray 1911, 321; Clarke 1972). The British iron and bronze *nave-bands* are of varied profiles which could be matched anywhere in the continental La Tène series, and the *linchpins,* sporadic in Britain as in Europe, include in their range of types curved forms reminiscent of more than one phase of La Tène, but are in the main insular forms. So far as a time-range can be estimated for these elements of vehicles, it is difficult to date any before the second century BC, and they continue into and beyond the period of Roman occupation.

The harness, which in the form of *bits* and *terrets,* constitutes our main source of knowledge for equine management and paired-horse draught in Iron Age Britain, presents some peculiarly intractable problems. The *bits* with the mouthpiece of three linked canons (the 'three-link' bit) normally of bronze and often elaborately decorated, have for long shared, with so much of British (and Irish) 'Early Celtic' art, the difficulties of explaining, in any practical terms, the preservation of an eclectic range of archaisms in style into what is a demonstrably late phase of prehistory, and the combining of them in wholly insular and individual modes. The Early La Tène bits of three-link form from Gorge-Meillet have been turned to again and again as a prototype in the fifth century BC for British bits demonstrably manufactured in the first century BC as in the founders' workshop at Gussage All Saints, together with distinctive insular terret and other types quite unknown on the Continent. Here, as we have seen, by Later La Tène times a standardized simple, utilitarian iron two-link bit, already present in earlier contexts, has become the dominant type, and any prototypes for the British series must be sought before the widespread adoption of this, in 'lost' three-link bronze bits with a date later than the fifth but before about the end of the third century BC. The Irish bits pose a special problem to be discussed shortly. The British terrets are of completely individual insular types, and whatever their typological origins, these cannot be found in the different but equally individual terret series which characterize the Late La Tène vehicle burials which occur as north-western outliers in Belgium and France (Joachim 1969; map. fig. 7, 2, 4, 24, 28, 31; Duval 1975) and so in relative proximity to south-east England. Any British relations in chariot equipment with these continental regions should therefore be before the appearance and development of the local terret forms which on Joachim's chronology would first appear in graves of La Tène C/D1, a point notoriously difficult to date in absolute terms, beyond putting it around 100 BC (cf. Collis 1975, 5). The British terrets, variations on large D-shaped knobbed or moulded loops, might be insular enlargements of small loops such as one from La Courte (Mariën 1961, fig. 20, no. 61), or, as was hinted at in an earlier chapter, a wholly indigenous development from Late Bronze Age prototypes such as those in the Parc-y-Meirch (Abergele) hoard (Savory 1976, fig. 9). The long-standing tradition of 'phalerae' as presumptive harness-ornaments continues in distinctive insular forms, such as decorative openwork discs as at Stanwick, and the Polden Hill mount already referred to, both with archaic reminiscences though of later date, and other 'horse-brooches' and mounts of unknown function (Fox 1958, pls 66–74). A possible fragment of an openwork yoke-mounting in the manner of the La Bouvandeau pieces has been mentioned above (Fox 1958, 126; pl. 69b). The unique bronze 'horse-cap' from Torrs in southern Scotland could have adorned a pony for riding or driving (Atkinson & Piggott 1955).

140

140

The technological unity of the British Iron Age harness industry has been dramatically brought home by the find of debris from a bronze-founder's apparently short-lived operations on a settlement site of the first century BC at Gussage All Saints in Dorset in southern England (Wainwright & Spratling 1973; Wainwright 1979; Foster 1980). The material recovered, about three cubic metres of well over 7,000 fragments of fired clay investment moulds for *cire perdue* casting and about 600 fragments of broken crucibles, had been consigned to a rubbish-pit of the middle phase of an unpretentious settlement within an enclosing ditch, typical of many in Wessex and estimated (from the capacity of its grain-storage pits) to have had a population of somewhere between thirty-sixty persons, or six to twelve households. The settlement can in no way be classed as a *Fürstensitz*, still less as an *oppidum* in Late La Tène terms, any more than the contemporary and comparable Glastonbury hamlet with its wheelwright, just described. The Gussage All Saints debris was virtually entirely from the casting of horse-bits, terrets, linchpin heads and strap-junctions; the 'estimate that the moulds represent production of fifty sets of chariot equipment is certainly not improbable', and the circumstances have been interpreted as indicating 'a temporary workshop set up by itinerant metallurgists to accomplish one particular job' (Foster 1980, 37). The social status of the customers for whom the harness-trappings were produced, the position of the ancillary craftsmen such as wheelwrights and leather-workers, and the mechanics of the organization of standardized production are all unknown—as indeed are the circumstances of the provision of broken-in and trained horses. The Gussage All Saints find serves to remind us of how very little we know of the realities of chariotry in late prehistoric Britain.

Finally, Ireland presents a quite different picture from Britain, with archaeological evidence of Iron Age horse-management in the form of numerous versions of 'three-link' bronze bits, but none of chariotry except a pair of possible yoke-terminals from Lough Gur, Co. Limerick (Fox 1950) and an undated crescent-head bronze linchpin from Dunmore, Co. Galway (Pryor 1980, no. 167). As Fox originally pointed out (1946, 33) the Irish bits are typologically distinct from the main British series, where the canons or links have their outer perforations at right angles to those of the mouthpiece, whereas in the Irish bits all perforations are in the same plane: two or three British bits share this feature. Despite this, and the other insular characteristics whereby the Irish can be distinguished from the British pieces, the only reasonable relationship to assume is that the Irish bits derive from the southern British three-link type just as in turn the north-west British 'small mouthpiece' type already mentioned is again derivative but displaying its own local features this side of the Irish Sea. Twenty-five years ago Jope found it difficult to establish more than paired draught in Iron Age Ireland, insisting that the evidence 'does not of itself show that the vehicles were light two-wheeled chariots such as were used for fighting by the continental and British Celts' (1955, 37), and in the most recent and complete review Haworth wrote 'the facts seem to me to require us to take seriously the possibility that the bronze bits were used for riding on horseback' (1971, 42). He also discusses the bronze 'Y-shaped pieces' or 'leading-pieces' on occasion found with bits, which can hardly have had the function of yoke-saddles (one suggested purpose) and seem to be a local elaboration in metal of some sort of leading-rein of leather, thought appropriate to processional use rather than chariotry. If the Irish bits are considered as derivatives of British, the question arises as to whether horses were also imported in view of the absence of indigenous Irish equids discussed in Chapter 3, as also the chronology of any such events—presumably at the beginning of the British bit series, while the differentiation of link perforation was still fluid, perhaps at the beginning of the second century BC. Certainly some of the Irish bits, such as those of Haworth's Type E, have, as Jope (1955) pointed out, ornament derived from Romano-British motifs of the second or third century AD, and the Irish Iron Age fibulae are similarly best explained as derivatives of southern British types (Jope 1962), but other elements of chariotry, such as terrets, let alone remains of wheels, nave-bands or tyres, are conspicuously absent from the Irish archaeological record. What remains a problem is the interpretation of the references to chariotry in the earliest stratum of the Irish vernacular literary tradition, and this is discussed in the final section of this chapter. Kenneth Jackson subtitled his original stimulating essay on the subject (1964) 'A window on the Iron Age', and we shall then have to ask on to which Iron Age do the magic casements open.

The 'Cimbric Wagons'

A small group of North European vehicle finds, beyond the geographical limits of Late La Tène culture may appropriately, from their concentration in the Danish peninsula be termed 'Cimbric' from the tribe which we know historically to have occupied that area at the end of the second century BC. General aspects of the north European archaeology and history of the period are discussed in Rowlett 1968 and Todd 1975, and the status and date of the vehicles by Klindt-Jensen (1949; 1953); Raddatz (1967) and Joachim (1969). The historical migrations of the Cimbri and Teutoni between 113 and 101 BC are the subject of a recent study by Champion (1980). South Scandinavia, as we saw in Chapter 4, had intermittent contact from as early as the fourth millennium BC with the Danu-

bian-Carpathian area, and from the third century BC onwards this is expressed in fine metalwork such as the Brå cauldron, with its Czechoslovak affinities in the Plastic Style of Early Celtic art, and later by the Gundestrup cauldron and its relation to Dacian and Thracian fine repoussé silver. While in the main the contacts seem to have been along, and east of the River Elbe, contacts with more westerly regions are not excluded, but it is perhaps significant that the initial thrust of the Cimbric migration took the tribe from Jutland to the territory of the Boii in Bohemia, and in the later stages of their extraordinary wanderings they are recorded by classical writers as being accompanied by their baggage-wagons and chariots.

Of the vehicle-finds (cf. map in Todd 1975, fig. 4) four come from cremation *120* graves, and the fifth is the well-known votive find of two wagons or carriages from a peat-bog at Dejbjerg. The earliest grave is the most easterly and the least well recorded, from Rosenfelde in West Pomerania, where burnt remains of a vehicle and a fibula belong to Joachim's earlier group, at the La Tène C/D1 transition (Raddatz 1967, 43; Joachim 1969, 110). The Husby find in Schleswig is admirably published (Raddatz 1967) and was an exceptional burial in a cremation-cemetery, a massive stone cist only 70 × 80 cm internally into which has been crammed the iron-work fittings of a four-wheeled vehicle which had been burnt. It had hoop-tyres with no fastening-nails, forcibly bent to reduce their size but originally 93 cm in diameter, and four joint-braces showing the felloes to have been single-piece, *c.* 4 cm thick; the nave-bands were 12.5 cm in diameter, and the linchpins had decorative double-S heads. There was a draught-pole terminal mounting in the form of two plates bolted together, comparable with those from La Tène and Llyn Cerrig Bach, and a pair of simple iron two-canon bits with a mouthpiece of *c.* 10 cm. The human cremation was contained in a bronze

141 Carriage from votive deposit, 1st century BC, *from Dejbjerg, Denmark*

cauldron with the burnt claws of a bearskin in which presumably the body had been wrapped on the pyre. This distinctive rite, characteristic of the northern Late Iron Age, as Raddatz points out, occurs again in a Rhineland chariot grave at Neuwied with bronze-work comparable to that at Dejbjerg and dated to latest La Tène, *c.* 30–10 BC (Joachim 1973) and again, though not with a vehicle, in a grave at Welwyn Garden City in southern England of *c.* 50–10 BC (Stead 1967). The Husby vehicle falls into Joachim's second and later group of vehicle graves, of Final La Tène. The two Danish vehicle burials both offer no more than the burnt fragments of the metal mounts of four-wheeled wagons. That at Langå in Fyn, found in 1877, was like Husby an exceptional grave in a cremation-cemetery, the cremation contained in a bronze iron-rimmed cauldron, and with a single-edged iron sword, three daggers, two shield-bosses, two gold finger-rings and the remains of a fourth-century BC Vulcentine bronze stamnos which must have been an antique when buried. Of the wagon there were iron tyre fragments (one with a nail), seven of an original eight nave-bands, and bronze joint-braces of bent felloes as well as bronze and iron mountings of what appear to have been a perch and draught-pole as at

Dejbjerg, the whole being assigned to the last century BC (Sehested 1878, 172, 307, pls XXXVII–XXXIX; Klindt-Jensen 1949, 100–2; Albrechtsen 1954, 103, 125). The Kraghede, Hjørring burial was in a cremation-grave and an adjacent pit in a cemetery, with an iron fibula, lances, knife and a primitive type of scissors, found in other graves and causing Klindt-Jensen to look to 'eastern impulses ... from Celtic sources'. In the pit were the bones of 'at least two horses, two pigs and a sheep' which had been cremated as whole carcases. The wagon remains were badly burnt, and in the 2.5 kg of fused bronze there were recognizable nave-bands, nails and sheet metal (Klindt-Jensen 1949, 102, 203).

Of all the Danish finds the most famous and informative (but still technologically puzzling) is that of the parts of two ceremonial carriages from Dejbjerg in western Jutland. The vehicles were found as dis- *141* membered fragments in two groups in a peat-bog, within a stake enclosure, in 1881 and 1883, and were published with admirable scale drawings by Henry Petersen (1888). They were again reviewed by Klindt-Jensen, but as a part only of a much wider study of the outside influences of the Danish Iron Age (1949, 87–100). In fact, there is no modern detailed technological study of this century-old discovery, and it is to be hoped that this will be undertaken (Kossack 1971, fig. 28 is a redrawing of Petersen). The wheels are 95 cm in diameter, four with fourteen and four with twelve spokes, and the turned oak naves with moulded bronze nave-bands are of the ogee-moulded form which goes back to Ha D at Vix and continues into the Roman period wheels in Scotland, as does the circular mortice with rectangular tenon slot: the spokes are of hornbeam and we have earlier noted the unique provision of roller-bearings. The single-piece bent felloes are of ash, with long overlapping scarf-joints and bronze joint-braces, and the iron hoop-tyres have no fastening nails. Two unpretentious iron linchpins with double-looped heads, removed when the wheels were dismantled, survived.

The axles, however, present a problem. In the 1881 group of finds one axle-tree was found, and two more in the 1883 group, all rather slender in proportions, with long arms and *c.* 1.5 m overall. In attempting a reconstruction of the elaborate pole-and-perch undercarriage which survived virtually intact in the second find, Klindt-Jensen dismisses these—'Nothing of the axles was found: two simple cart axles in the find cannot belong to this ornamental cart' (1949, 91), though surely the coincidence of the remains of two wagons and three unrelated axles being found together strains credulity. The pole and perch, both Y-shaped and with elaborate openwork bronze decorative sheathing, were convincingly combined as a 'double-Y' undercarriage by Klindt-Jensen, allowing for the necessary vertical movement of the draught-pole, *c.* 3.6 m overall when joined, the pole being 2.0 m long and the overall length of the body with its wheels, 2.20 m. But he then has to assume two (missing) bowed iron axles and a king-pin for a limited pivoting at least of the front axle, a point discussed in the previous chapter in the instance of the

Hallstatt wagons, which the Dejbjerg carriages so closely resemble. Like them, it could have been a quarter-lock wagon with a resonable turning circle. Great uncertainty remains as to the exact arrangement and disposition of the wooden members of the body or box, elaborately ornamented, like the pole and perch, with openwork bronze panels, but two points are clear. The first is that the sides and ends of the body were nowhere more than 30 cm high, and secondly, 'the curious thing about these sides is that they were not joined or nailed to any other part of the cart. There are no traces of joining anywhere' (Klindt-Jensen 1949, 93). We are, in fact, seeing a shallow tray-like box simply laid on top of the Y-perch undercarriage in precisely the manner assumed for the Ha D ceremonial carriages from such graves as Bad Cannstatt, Hochdorf or Ohnenheim, a similarity reinforced by the T-shaped projecting mounts on the sides of the Dejbjerg carriage and the fact that it carried an alder-wood chair with turned legs, both T-shaped mounts and chair being closely paralleled at Ohnenheim. The elaborate decoration of pole and perch suggests that both were intended to be at visible, at least on occasion. Connection has often been made between the Dejbjerg find and the description in Tacitus (*Germania* 40.2) of the sacred vehicle of the goddess Nerthus, kept in a holy wood on an island in the west Baltic, drawn by heifers in procession, and ceremonially washed in a lake by slaves who were then drowned. The disc-wheeled vehicles of the Rappendam votive find already referred to may again be related to such cults (Todd 1975, 185).

Whatever the details of their reconstruction, what we are surely seeing in the Djebjerg carriages is, in fact, a direct continuation of the sixth-century type of four-wheeled ceremonial vehicle into the first century BC, and inferentially this would apply to the other 'Cimbric Wagons' just described. Owing to the replacement of the

227

four-wheeled carriage by the two-wheeled chariot as the favoured vehicle for ceremonial burial, the archaeological record between Ohnenheim and Djebjerg fails us, but in the North European area itself the representations of wagons with 'double-Y' pole-and-perch construction on the Pommeranian 'Face-Urns', described in the last chapter, show us the type was in use there in the fifth century BC, and perhaps the long oval shield of La Tène type shown on the Grabau urn adjacent to the wagon scene could be taken with the actual wooden shields in the Hjortspring votive find in Denmark, and serve as another link between the canonical Hallstatt and La Tène culture-areas and those on its northern fringe. Reciprocal contacts maintained by trade or gift-exchange could have existed for some centuries until the migrations of the Cimbri at the end of the second century BC brought them successively into contact with the Boii, the Scordiscae and the Tauriscae and settlement in north-east Gaul where the Atuatici, in Caesar's time, claimed descent from Cimbri and Teutoni (*BG*. II. 29).

Thracians and Scythians

We saw in the last chapter the confusion of nomenclature that has arisen among East European archaeologists in classifying the regional variants of the Iron Age in the area now roughly comprised by Bulgaria and Romania, and by the mixture of traditions they embody. From the early sixth century BC, in addition to indigenous cultures, we have to reckon with the Greek colonies of the Black Sea coast (Boardman 1980, 238) and the Persian presence, 'almost continuous in European Thrace and Macedonia from about 513 till the end of the Greco-Persian wars in 479, with an aftermath that lasted down to 450' (Sandars 1971, 103; 1976). Eastwards, in south Russia, from the seventh century BC, the Scythians had established themselves as another barbarian peo-

ple to be recognized by the Greeks, and their wagons became famous from the accounts of Herodotus. Both Thracians and Scythians sacrificed and buried horses, and on occasion vehicles as well, but archaeological information is sadly deficient, all too often the result of incompetent excavation and inadequate publication, old and new. The Thracian material is summarized by Hoddinott (1975; 1981) and that from the Scythian tombs by Minns (1913) and Rolle (1979). Thracian vehicle burial continued to have a remarkable efflorescence in Roman times, when Venedikov (1960) listed over forty finds, eighteen recognizably of wagons and fifteen of chariots, and similar burials extend as far as the Dobrogea (Harţuche 1967).

So far as can be gathered from the scrappy published sources, most of the Bulgarian pre-Roman vehicle-burials hardly antedate the last three centuries BC, but two of the better recorded finds are of the fourth. Outside the entrance to the monumental chamber tomb at Strelcha, east of Panagyurishte, was a burial of a four-wheeled wagon or carriage, with two draught horses and a third, presumably for riding, sacrificed by it. The richly furnished tomb is dated to the fifth-fourth century BC and Hoddinnot has tentatively suggested it may be that of the historical Thracian ruler Amadokos I (*c.* 410–385 BC) (Kitov 1979; Hoddinott 1981, 121). No technological details of the vehicle have been published, but it had iron tyres and nave-bands. A second princely tomb is that of Vratsa, between the Stara Planina and the Danube, of the mid-fourth century BC, where again there was a four-wheeled vehicle, with a pair of horses with iron bits, as well as the riding horse with a silver bit and harness ornaments. The gold and silver-work from the grave included greaves closely similar to those from another rich tomb at Agighiol in Romania, with red figure Greek pottery of the early fourth century, where three horses with their bits were buried in the dromos or antechamber of the tomb, but no vehicle (Berciu 1969, 33–6; Hoddinott 1981, 113). The vehicle burial with bronze fittings in the Plastic Style of La Tène art, from the dromos of a chamber tomb at Mezek, has already been mentioned. Venedikov in his study of the Thracian horse-bits (1957) pointed out that while indigenous types related to those from for instance the Atenica vehicle graves described in Chapter 5, form a background, the great majority derive from Greek bits (cf. Anderson 1961, Chaps 3–5). The possibility of some influence from oriental types via Persia must also be reckoned with. Roman types of bit naturally take over in the vehicle graves of that date.

The Scythian vehicles (Dušek 1966; Rolle 1979; 1980, 25) are mainly known from old excavations and technological details are hardly recorded. Both two-wheeled and four-wheeled vehicles are represented, and their presence together in some burials gave rise in the past to claims of six-wheeled wagons, as at the fifth-fourth century BC barrow at Elisavetinskaya Stanitsa on the Don: another half-dozen or so occurrences of vehicles have been listed by Rolle (1979, 113–21). At Alexandropolsk on the Dnieper, six wheels must represent the same state of affairs; these had diameters ranging from 93 cm to 1.23 m, with iron tyres fastened by dome-headed nails, and the burial dates from the third century BC (Dušek 1966, 22). There are also the well-known models of covered wagons or house-wagons from Kerch (Minns 1913; 50; Rolle 1980. 122).

It remains finally to comment briefly on the Roman period chariot and wagon burials of Thrace, though Venedikov's elaborate publication (1960) leaves many technological questions unanswered. (cf. also Botusharova 1930 for a well-recorded burial). It appears that both nailed and unnailed hoop tyres were in use, on wheels averaging a little under a metre in diameter, but how the number of spokes (eight) in the restorations were decided is not clear. One nave (pl. 88) had, between the iron nave-band and internal bush, a wavy inset iron band, very reminiscent of that on the Bar Hill wheel from Scotland described earlier, *136* and raising questions as to the origin of the technique. There appears to be no evidence whether the Thraco-Roman wheels had segmental or bent felloes, though the former is more likely. Venedikov 'presented a remarkably plausible reconstruction' of a four-wheeled vehicle 'as having a body suspended by straps depending from metal hooks or rings protruding from each side of four vertical bronze sockets', and this has influenced almost all subsequent students of later carriage design, and has been accepted as the origin of the sprung or suspended coach body. More recently Lynn White (1970) has shown that this reconstruction cannot be accepted, and that suspended coach-bodies must rank among the technological innovations of the western Middle Ages.

The lexical and literary evidence of the classical writers

The references to Celtic (almost wholly Gaulish or British) vehicles in the classical writers, and the rich vocabulary of Celtic loan-words for them in Latin, are topics strictly outside the scope of archaeology, and the province of scholars in the appropriate literary and linguistic fields. But some notice must be taken of them here, however superficially and inexpertly, while it is to be hoped that authoritative studies will be made by others, on the lines initiated by Rivet (1979) on the vexed question of the 'scythed' chariot. The question of 'borrowed names for borrowed things' has been discussed in another Romano-Celtic context by Wild (1970) and the Celtic loan-words in Latin by Schmidt (1967) and others. The main literary references for vehicles in the Roman world (but with omissions) can be found in De Azevedo 1938 and the normal dictionary sources, and a general treatment is in Vigneron 1968, 288–96. In any attempt to connect vehicle terminology with vehicle technology we must remember that we are dealing with writers who are conscious literary stylists, and when poets, working within the constraints of scansion and metre. In such circumstances they are, in Fowler's words, 'open to the allurements of elegant variation' and their choice of vocabulary is not necessarily dictated by ethnographic accuracy. The usually imprecise use of 'cart' and 'wagon' in modern English is a case in point.

Before turning to the complex Latin usages, we must remember that the earliest references to Celtic vehicles are in authors writing in Greek, many of them based on or quoting with acknowledgement the otherwise lost works of the Stoic philosopher Posidonius, writing at the beginning of the first century BC of the events of the second (Tierney 1960; Nash 1976). His philosophical standpoint led him to look for qualities of ancient virtue among barbarians and this seems to have included the recognition of a warrior-aristocracy that could be compared with the heroic age of Homer and the Iliad, and Diodorus Siculus makes explicit comparisons between Celts and Homeric Greeks: the 'champion's portion' at the feast recalls Ajax (V. 28.4) and the Britons 'preserve in their ways of living the ancient manner of life' including the use of the chariot as in Homeric warfare (V. 21.5). For such writers there was then a ready-made terminology to hand, that of the Iliad with the war-chariot (ἅρμα) as opposed to the four-wheeled carriage (ἅμαξα, ἀπήνη): Delebecque 1951, 169–82). Diodorus (V. 29. 1), when saying of the Gauls 'for their journeys and in battle they use chariots, the chariot carrying both charioteer and chieftain' uses ἅρμα as does Dio Cassius (LX. 19.1) for Boudica, and for Caractacus and Togidubnus (LX. 20.2), though Strabo writing

of British chariots (IV. 5.2) uses ἀπήνη presumably as a 'vehicle' word rather than a four-wheeled carriage. Strabo also uses ἀπήνη in his shortened version of the Lovernios story, whereas Athenaeus in his fuller account has ἅρμα (IV. 37). The earliest author to describe Celtic vehicles in the battlefield, Polybius (II. 28.5) writing *c.* 150 BC of Telamon (225) and unconcerned with Homeric reminiscences, writes of ἅμαξαι καὶ σμνωρίδες, wagons and chariots, the latter word denoting a pair of horses yoked to a two-wheeler and so exactly cognate with the Latin *biga* (used only once for Celtic chariots, in Pomponius Mela III. 52). The latest reference to the use of chariotry in the prehistoric world of Western Europe is that to the chariots (using ἅρματα) with small fast horses combined with infantry by the north British tribes of the Caledonii and Meatae against the Roman army in the Severan campaigns of the first decade of the third century AD (Dio Cassius, LXXVII, 12). Jackson (1964, 35) regards these tribes as Pictish. None of these references conveys any technological or practical information on the vehicles concerned.

The Celtic loan-words in Latin, and the comments on Gaulish and British vehicles made directly or indirectly by writers from the first century BC onwards, are more informative but often fraught with ambiguity, obscurity or contradiction. To the dozen or so words for vehicles can be added others associated with horses, including *caballus* itself and *badius*, a chestnut bay colour: Schmidt (1967) compares the situation to that resulting in the adoption by Akkadian and other Near Eastern languages of the second millennium BC of extraneous Indo-European horse and chariot words (cf. Mayrhofer 1966). The Celtic vehicle words in Latin seem roughly divisible into two groups, the majority being those absorbed in the language at a relatively early date and then used for a variety of Roman wagons, carriages and two-wheelers; a plausible origin would be among the Cisalpine Gauls. As no modern comprehensive study of Roman vehicles has been made, little more can be said, except to stress that in no instance can a vehicle name be confidently applied to any of the numerous representations of a wide range of vehicle types depicted in Roman art. Sufficient literary evidence exists to imply certain details, as in *plaustrum* or its cognate *ploxenum* (Quintillian (1. 5. 8) quoting Catullus (97. 6) incidentally noted that *ploxenum circa Padum invenit,* indicating Cisalpine Gaul), which Cato, Varro and Virgil make clear was a heavy-duty wagon drawn by oxen, mules or asses, and which a scholium on Virgil, *Georg.* I. 163 says was disc-wheeled, with iron tyres, and with the axle turning with the wheels. Two other four-wheeled vehicles in Celtic Gaul were noted by Latin writers, *raeda* and *petorritum*—*plurima Gallica valuerunt ut raeda et petorritum* (Quintillian I. 5. 57). The first is from a Gaulish root **redo-, *reda,* cognate with English 'ride', and the second, glossed by Festus *a numero quatuor rotarum,* from **petorroto-, *petruroto-,* a four-wheeler (Evans 1967, 242–51). The word *carpentum,* however, presents a pretty set of problems. It is undoubtedly from the Celtic root **carbanto-* (present in place-names such as *Carbantorate, Carpentorate* (Carpentras) in Gaul or *Carbantoritum* in North Britain: Rivet & Smith 1979, 300) and was in general use in Latin, at least from Propertius to Juvenal and Suetonius, as a civil vehicle especially for women. But it could also be used for Gaulish chariots (Livy XLI. 21. 17, *carpentis gallicis;* X. 20, *mille carpentorum* at Sentinum) and Florus, writing in the second century AD, employs it specifically for the silver-mounted vehicle in which the Gaulish chieftain Bituitos was paraded in his brightly-coloured battle array after his defeat in 121 BC—*nihil tam conspicuum in triumpho quam rex ipse Bituitus discoloribus in armis argenteoque carpento, qualis pugnaverat* (Florus I. 37. 6). Florus also uses *carpentum* for British and

Cimbric vehicles (III. 3. 16, III. 10. 17), and it can hardly be more than a personally favoured word with a generalized 'chariot' connotation, rather than denoting a specific type of barbarian vehicle. Indeed by the fourth century the usage had become so loose that the agronomist Palladius could apply it to a dung-cart (*carpentum stercoris,* X. 1. 2)—though it does not appear in the passage in Columella II. 15 which he is paraphrasing, and to a part of the Gallo-Roman reaping-machine, the *vallus* (VII. 2—though again it does not appear in the earlier account of the *vallus* given by Pliny (*N. H.* XVIII. 296; K. D. White 1967, 157).

In addition to this larger group of assimilated Celtic loan-words for vehicles in the Latin language, two came to be used primarily of Celtic vehicles themselves, in Celtic territory. The earlier 'Posidonian' sources for Celtic ethnography related broadly to the second century BC and La Tène C so far as the Reinecke subdivisions are applicable; the new sources to the first century BC and, in Britain, later, the equivalent of La Tène D on the continent (cf. Nash 1976). These are the result of first-hand encounters in the process of conquest first in Gaul by Julius Caesar from 58 BC, with the raids on Britain in 55–54, and then with the Roman subjugation of Britain from Claudius in AD 43 to the northern campaigns of Agricola from 80. The primary sources are two, Caesar's account of his own wars and the account by Tacitus of those of his father-in-law Agricola in his biography. The two new Celtic words to be adopted in Latin were originally used specifically of Celtic chariotry, and were *essedum* (or *esseda*) and *covinnus;* both then came to be used of civil Roman vehicles as a result of a fashionable whim for things Celtic or barbarian noted in other contexts. 'When Caesar opened up the north, Rome for the first time came face to face with this world, and northern styles and artifacts became fashionable... this fashion for northern styles was an ephemeral fad, soon supplanted by the rage for Egyptian antiquities' (Bonfante 1977, 113).

Essedum presupposes a Celtic root **ensedo-* (Greene 1972), with the sense of something for sitting in, and was used by Caesar of the chariots encountered in his famous engagement with the British tribes of south-east England (most recent review, Hawkes 1977). Before proceeding to discuss further the use of this word we must pause a moment to consider the well-known fact that, however much allowance we make for Caesar's disingenuous handling of facts, so often commented on, it seems reasonable to accept the negative evidence of his narrative for a complete absence of Celtic chariotry in the native forces mustered against him in Gaul. Here perhaps some of us, including the writer, have been misled into too wide an interpretation: what Caesar is recording, quite objectively and as a matter of fact, is no more than that the Gaulish tribes used infantry and cavalry but not war chariots in their engagements with the Roman army, leaving it quite open as to whether particular tribes had them or not for civil use or their own form of internecine warfare. The dual purpose of the Gaulish chariot, 'for journeys and in battle', was stressed by Diodorus Siculus, and even in the small area of Britain Tacitus, presumably on the good authority of Agricola, noted that only some tribes used the chariot for warfare—*quaedam nationes et curru proeliantur* (*Agric.* 12). We should therefore think of a Gaul where possibly some tribes did not use chariots at all; others used them in peace and war among themselves, but none thought them a worthwhile weapon with which to confront the disciplined Roman army. But another usage may take us a little further. Writing his *Georgics* between 37 and 30 BC, Virgil refers to a spirited colt that might be broken in as a racehorse, or for *belgica esseda* (plural because the metre demands it: *Georg.* III. 204). His commentators later underlined this attribution, Servius saying these are Gaulish vehi-

cles, because the Belgae are a Gaulish tribe and the *esseda* were used there; Philargyrius more shortly stating that the *essedum* was a kind of *currus* in which the Gauls used to fight. It was then a current Roman belief shared by Virgil that Gallia Belgica was particularly the home of the *essedum,* and one has to go no further than to compare the maps of Hawkes in his study of the Belgae and their territory (1968) with that of the latest La Tène chariot burials (Joachim 1969) to see the plausibility of equating the Virgilian tradition with the archaeological evidence. That the chariot should also be called *essedum* by Caesar in Kent, a part of 'Belgic' Britain, comes as no surprise. The Romano-British place-name *Manduessedum* (Mancetter) contains not only *essedum* but the *mandu-* prefix is also Celtic, probably related to Latin *mannus,* a pony (Rivet & Smith 1979, 411, Evans 1967, 222). Cicero wrote of *essedum* as British (*Ad fam.* VII. 7. 1), as did Propertius about 25 BC, drawing attention to its carved yoke—*Esseda caelatis siste britanna jugis* (II. 1. 76), but by now he may have been writing of what became a Roman pleasure-carriage of that name: at least 'by the time of Seneca (*Epp.* 56. 4) they were all too common in Rome' (Rivet 1979). When Livy uses the same word for Celtic chariots at the battle of Sentinum (295) he is simply projecting into the past the current popular term. But when Martial, about AD 84–85, writes a poem enumerating the respective merits of carriages for his personal use, and compares the *essedum* and the *carruca* adversely to the privacy of a *covinnus* (XII. 24), we are in a world where foreign names are in use for wholly Roman vehicles, like nineteenth-century London when gentlemen might discuss the relative comfort of the berline or landau as against brougham and tilbury. And Martial introduces us to the second and newest Celtic loan-word, *covinnus.*

Here we are fortunate in having a detailed study of the literary evidence on the lines one would like to see applied to the whole range of the Latin vehicle vocabulary (Rivet 1979). *Covinnus* has a Celtic etymology from **vignos,* the Indo-European root which gives equally Latin *veho, vehiculum,* and English 'wain' and 'wagon'. The initial element may imply two passengers, as with another Celtic vehicle loan-word in Latin where *combennones* are those who share transport in a *benna* (for Gaulish *co-*, Evans 1967, 183). But *covinnus,* in its usage for Celtic vehicles and not the smart Roman adaptations of Martial's poem, is in two of its four recorded occurrences said to be 'scythed', or with *falces* fitted to its axles, a remarkable and sensational description which Rivet set out to elucidate. Putting the particular feature aside for the moment, Lucan, writing about AD 62/63, assigns the *covinnus,* as did Virgil the *essedum,* to the Belgae—*docilis rector monstrati belga covinni* (*Phars.* I. 426) which may involve Gaul or Britain or both, and Tacitus writing in AD 98 of Agricola's campaign in North Britain, uses *covinnarius* for 'charioteer' (*Agric.* 35, 36): in neither instance can we tell whether *essedum* and *covinnus* denote two typologically and structurally different forms of chariot, or whether the words merely record regional variants in Celtic linguistic usage. If *essedum* could be regarded as a Belgic usage in north-west Gaul and southeast England one might quote another likely 'Belgic linguistic peculiarity' common to the same two areas, that of place-names in *Duro-* (Rivet 1980). That fashionable two-wheelers in Rome were differentiated may be irrelevant to the original barbarian situation.

We come finally to the literary crux, discussed fully in Rivet 1979, the 'scythed chariot' of the Britons; literary and not archaeological because up to the present no piece of La Tène metalwork has been found which could be interpreted as such a piece of vehicle armament. The use of the word 'scythe' in English to translate the Greek δρεπανη and Latin *falx* in this context is long-established. One of the earliest discussions of this

question is that of Aylett Sammes in Chapter VIII of his *Britannia Antiqua Illustrata* (1676) who decided that 'the *Covinnus* was the Chariot with the Scyths and Hooks, as Pomponius Mela witnesseth, and their *Esseda* were not Armed Chariots'. But, as Sammes hints, *falx* (and its Greek counterpart) can mean a wide range of curved tools—scythes, sickles and bill-hooks (cf. K. D. White 1967, 69–103)—and a secondary sense of *falx* is of a sort of grappling-hook used in siege warfare. Some type of curved blade, but not necessarily precisely that of a modern scythe, is presumably to be envisaged, comparable with something within the range of such tools from La Tène (Vouga 1923, pls XXIV–XXV). To return to the classical texts, four-horsed 'scythed' chariots are referred to repeatedly as having been used in the Near East from the time of Cyrus in the sixth century BC onwards, and by Galatians in the third, and though unconfirmed by archaeology or iconography are clearly to be accepted as fact. Frontinus (who incidentally was governor of Britain *c.* 74–78) wrote that Caesar had encountered the *falcatas quadrigas* of the Gauls, but as we have seen in this book the nearest example of a *quadriga* is in Armenia of the ninth century BC and it was unknown in western European prehistory. An unclear tradition must be suspected here. Silius Italicus about AD 90 had heard of blue-painted savages in the remote north fighting in *falcigero covinno,* but the earliest and most specific statement is that of Pomponius Mela, writing about the time of the Claudian conquest of Britain, and of the Britons themselves, that they had two types of vehicle, the *biga* and the *currus,* the latter armed with *falces* on its axles and which they called a *covinnus.* That the tradition became confused is shown by the sixth-century paraphrase of the passage by Jordanes, who uses *essedum* instead of *covinnus.* Rivet concludes that in Mela 'the *bigae* are *esseda* ... while the *currus* are (or include) scythed *covinni*'. But if Mela's *et bigis et cur-*

ribus was taken as a phrase comparable to Livy's *essedis carrisque* at Sentinum (X. 28) or that of Polybius of Telamon, ἅμαξαι καὶ σμνωρίδες (II. 28), both descriptive of Celtic warfare, we might have a wider choice of vehicles on which to fix our scythes.

Our concern being with vehicles and their technology we do not have to discuss in any detail Celtic military tactics and the precise manner in which the vehicles were employed, and it should be borne in mind that the classical writers describe only the exceptional engagements of Celts against disciplined Roman armies, not Celt against Celt in their own form of internal warfare. So far as the peaceful use of vehicles is concerned, a couple of references show heavy and presumptively ox-drawn wagons in use for commercial purposes, with information probably from Posidonius. Strabo, writing of the Rhône, says that as it is 'swift and hard to sail up, part of the goods ... rather go by land on wagons', using here the word ἁρμάμαξα, a covered wagon appropriate to such a use (IV. 1. 14), and Diodorus Siculus uses the normal 'wagon' word, ἅμαξα, for those transporting tin in south-west Britain (V. 22). As we saw he also recorded that chariots were used for civil travel as well as in battle, though otherwise we only have references to vehicles in war, where clearly two types were likely to appear on the battlefield or its fringes, the war chariot and the supply-wagon, two-wheelers and four-wheelers respectively, the latter to bring up supplies, including women, and to carry away booty. These two types must be denoted by the ἅμαξαι καὶ σμνωρίδες at Telamon (Polybius (II. 28. 5) and the *essedis carrisque* at Sentinum (Livy X. 8) and (*pace* Rivet) the *et bigis et curribus* of Mela (III. 52). In the famous scene of AD 61, Tacitus puts Boudica, with her two daughters in front of her (and presumably with a driver as well), in a *currus,* which must have been more commodious than a light two-man war chariot, while on the edge of the battle area were the womenfolk in *plaustra* (*Ann.* XIV. 33). Among the Cimbri, Strabo (VII. 2. 3) records that the priestesses sought omens by examing the entrails of their sacrificed prisoners of war 'and during the battles they would beat on the hides that were stretched over the wicker bodies of the wagons and in this way produce an unearthly noise', using here ἁρμάμαξα, as for the covered wagons in the Rhône valley.

The tactics of Celtic chariotry have been much discussed, frequently in the debated context of Homeric practice (e.g. Anderson 1965; Greenhalgh 1973). For our purposes, we need only note that the chariot carried a warrior and a driver, and that while Diodorus Siculus (V. 29) writing of Gaul (and presumably quoting Posidonius) states that charioteers were 'freemen chosen from the poorer classes', Tacitus (*Agric.* XII. 1) writing of Britain (and so probably informed by his father-in-law) says that 'the higher in rank is the charioteer *(auriga),* the dependants *(clientes)* fight' for him. With two different countries and about two centuries separating the statements, they need not be inconsistent. Diodorus is specific on the function of the Gaulish chariot in war in the second century BC. 'When travelling and in war they [the Gauls] use chariots which seat a driver and a warrior (ἅρματος ἡνίοχον καὶ παραβάτην). When they encounter horsemen in war they hurl the javelin against the foe and then dismounting engage the enemy with the sword' (V. 29). Caesar's description of the British chariotry in Kent some two centuries later is explicit and detailed (*B.G.* IV. 33). 'They first drive about in all directions and hurl missiles; by the terror of the horses and noise of the wheels they throw the ranks of the enemy into confusion'. This, as Powell put it of earlier chariot warfare 'was the phase for display and intimidation' (1963b, 166). Then, Caesar continues, they penetrate the ranks of the cavalry and the warriors jump down to fight on foot, the charioteers with-

234

drawing gradually, ready to rescue the warrior if things go badly, thus combining 'the mobility of cavalry and the stability of infantry'. The Britons were here using chariotry and cavalry against Roman infantry and cavalry, and one cannot tell whether such tactics were also employed in inter-tribal warfare. The Britons were employing the chariot component of their forces on a large scale in these engagements. In the following year, after the crossing of the Thames, Caesar remarked that Cassivellaunus 'disbanded the greater part of his force, retaining only about four thousand charioteers', with therefore twice that number of highly trained horses (*B. G.* V. 19). One is reminded of the scale of production of harness implied archeologically by the evidence from Gussage All Saints quoted earlier. The other recorded British chariot engagements confirm Caesar. In the battle between Paulinus and Boudica in AD 61 the intimidation phase of confusing display is recorded, and at Mons Graupius twenty-three years later the noisy careering about of the chariots of Calgacus and their subsequent entanglement in the Roman infantry before the final debacle is vividly described by Tacitus (*Agric.* XXXV–XXXVI). The circumstances at Sentinum in 295 BC, when the Celts standing in their chariots and wagons charged headlong against the opposing forces, seem the exceptional tactics of despair.

One last feature of Caesar's description has long been recognized as an indication of the construction of the chariot-body, the circus-like feat of one of the occupants leaping forward, running out along the draught-pole on to the yoke, and rapidly returning to his place while at high speed: though not stated, it was presumably the warrior who indulged in these spectacular acrobatics while the driver was managing the horses with the training and practice Caesar could not but admire. To do this argues a body with a very low front or open front and

back, which we saw the archaeological evidence from the chariot graves could also imply. Mrs Littauer (1968b) drew attention to a curious convention in nineteenth and twentieth-dynasty Egyptian art in which the heroic Pharaoh-Warrior in his chariot 'stands astraddle the front breastwork of the vehicle, one foot on the pole just in front of the breastwork', and which is repeated on Greek sixth-century black-figure vases, usually by Herakles in a gigantomachy, and something similar may be shown on a Celtic coin (Allen & Nash 1980, 140, No. 510). She suggests that the Greek artists had seen Egyptian prototypes and ends by quoting Caesar with a hint that an ultimate further connection, with artistic convention turned to actual showmanship, might even be possible. With this startling, if not wholly convincing suggestion, it is appropriate to end this section.

The chariot in the Old Irish hero-tales

It has long been recognized that in early Irish traditional vernacular literature the most ancient stratum is that embodying the so-called 'Ulster Cycle' of prose hero-tales, centred on the epic *Táin Bó Cuailnge,* the Cattle-Raid of Cooley. It is a sort of primitive Celtic *Iliad,* with the Men of Ulster as Achaeans and those of Connaught as Trojans, and the champion Bull of Cooley (near Carlingford Lough) however incongruously takes the place, in a simple pastoralist society, of Helen. As with all heroic literature, it reflects a past age with 'a taste for warfare and adventure, a powerful nobility, and a simple but temporarily adequate material culture ... In such conditions the heroic virtues of honour and martial courage dominate all others, ultimately with depressive effects on the stability and prosperity of the society. It is usually during the subsequent period of decline that the poetical elaboration of glorious deeds, deeds that now lie in the past, reaches its climax' (Kirk 1965, 2). In a study which has affected all subsequent thinking, Jackson (1964) conclusively demonstrated that the Ulster group of hero-tales 'might perfectly well have been put together more or less as we have them in, say, the fourth century AD and not earlier'; they reflect a pre-Christian Celtic past with details that again and again recall those of the 'Posidonian' sources for Gaul and Caesar for Britain in the second and first centuries BC. Warfare is the dominant theme and 'the really notable feature is the use of the war-chariot' from which the heroes dismount to fight, with the relationship of warrior to driver as in Diodorus rather than in Tacitus: 'the charioteer his favoured servant and familiar counsellor'. There is a complete absence of references to cavalry or to riding at all, so that chariots are not being used as an adjunct to cavalry as in the early accounts.

At first sight a welcome confirmation of the classical texts from a vernacular Celtic source, the chariotry of the hero-tales proves on examination to be no more straightforward than that in Homer, a subject of contention for generations. In Ireland, vehicles with tripartite disc wheels, as we saw earlier, go back to the fifth century BC, and were presumably drawn by oxen, and the undated yokes which survive all seem to be ox-yokes. The status of the horse in Ireland has also been discussed, in Chapter III, where it was seen to be almost certainly introduced as a domesticate in Beaker contexts of the late third millennium BC—'there is little doubt that domestic horses were imported into Ireland, because for the alternative of local domestication, a large wild stock would be needed which was presumably never present' (Wijngaarden-Bakker 1975, 346). By the early centuries AD a large stock could have been available but there is little existing archaeological evidence for the intervening centuries. A few horse bones are recorded from the Rath of Feerwore, probably first century BC (Raftery 1944) but no reports on the animal bones from the 'royal' ceremonial sites of Navan Fort (Emain Macha, the capital of Ulster in the epics) Tara or Dun Ailinne, contemporary with the presumed period of the events of the hero-tales, have appeared (Wailes 1976; Piggott 1978c). The archaeological evidence for the horse in Early Iron Age Ireland is basically that of the bronze bits, already discussed in this chapter, the undecorated types are not likely to be earlier than the second century BC and the later with ornament of Romano–British derivation in the second or third centuries AD. 'Only the most tenuous evidence' wrote Jope 'hints at paired-draught in Ireland during the earlier part of the period . . . this evidence does not of itself show that the vehicles were light two-wheeled chariots such as were used for fighting by the continental and British Celts' (1955, 37). There is certainly no evidence for British contacts in other elements of chariotry like terrets or linchpins such as were being cast at Gussage All Saints together with bits: of 115 finds of terrets, singly or in sets, mapped for the British Isles by Macgregor (1976) the one Irish example from Antrim is of north British type and obviously a stray, and the forty-one linchpin finds are all British. Terrets of continental types are equally lacking, as are nave-bands or iron tyres. If the Lough Gur 'pole-tips' are better to be considered yoke-mounts, they could equally belong to an ox-yoke or horse-yoke.

Recent examination of the lexical and textual evidence for the vehicle of the hero-tales provides additional anomalies (Harbison 1969, 1971; Greene 1972). The word used in the Old Irish texts is invariably *carpat,* from the Celtic **carbanto-* root which as we saw also lies behind the Latin *carpentum,* with its varied connotations. Four other Irish vehicle words exist but are not used in the hero-tales and when *carpat* is used elsewhere it is always a transport or travelling vehicle, glossed as *currus,* and no cognate of *essedum* or *covinnus* exists. This may not be of great significance if we think of these as regional variants of preferred words for a chariot—*essedum* as Belgic in Gaul and Britain, *covinnus* elsewhere in Britain and *carpentum (Carbantorate).* British *(Carbantoritum)* and Irish. A passage referring to a 'scythed' *carpat* had long been detected as a late interpolation, probably about the eleventh century, from medieval romances of the orient. As to *carpat* 'we may translate it "chariot" if we wish, but to translate it as "war chariot" is to impose a restriction not justified by the evidence'. The descriptions in the texts show 'a simple two-wheeled cart, containing two single seats in tandem in a light wooden frame and drawn by two horses harnessed by bridles to a yoke attached to the chariot pole; the wheels were shod with iron tyres' and apparently spoked (Greene 1972, 61, 70). Persistent references to twin rear 'shafts' can perhaps be best explained by assuming some form of A-frame cart, and Harbison suggests we should visualize 'a type of chariot (possibly a slide-car to which wheels were added) which rests upon a tradition of wheeled vehicles which has nothing to do with and which is possibly older than the chariot used by the continental and Yorkshire La Tène Celts . . . the descriptions of chariots in Old Irish literature should be applied only to the Old Irish 'cart-chariots' but should not be projected back to the continental and Yorkshire chariots known from the chariot-burials' (Harbison 1969, 55). If the modern Irish use of *carpat* and the cognate Welsh *carfan* as 'frame' and 'jaw', or 'gum (of the jaw)' are borne in mind (Greene 1972, 61) one might remark that the lower mandible of a horse closely recalls the plan of an A-frame cart, and that small children's sledges made of such horse-jaws were in use from the fourteenth century in the Low Countries and elsewhere (Ijzereef 1974). Jope, in his study of the archaeology of

paired horse draught in the Irish Iron Age, had gone further, writing 'it would seem at least possible that the people who elaborated this hero-literature in Ireland knew well that great warriors used chariots in the lands from which their forefathers sprang— such may have been indeed an element in an oral tradition brought with immigrants to Ireland in the Early Iron Age—but they did not necessarily use them to any extent in Ireland' (1955, 39). We return to Jackson's arresting sub-title, 'A Window on the Iron Age', asking whether, so far as chariot warfare is concerned, we are seeing from it reality on Irish soil at all?

It may not be irrelevant to go back to the problems of the chariot in Homer, and particularly to views recently expressed by several writers (Finley 1956; Kirk 1964, 1965; Greenhalgh 1973). Very briefly, the question arises from the divergences and incompatibilities between Homer's presentation of chariot warfare in the *Iliad* and the normal use of chariotry to be expected in the Late Mycenaean world in which the siege of Troy is set. As Anderson (1965) pointed out the Homeric version, with the heroes dismounting and fighting on foot, recalls Caesar and Diodorus Siculus rather than the Aegean of the late second millenium BC. It has therefore been suggested that at the time of the composition of the poem as we have it today, in the eighth century BC, 'the true use of the many chariots of Knossos and Pylos had not quite disappeared from the poetic tradition, but only a faint memory survived . . . a fairy-tale, the result of a progressive and radical misunderstanding of a vague and undetailed tradition about the organized warfare of the past . . . the singer of the Dark Age . . . knew by tradition that chariots were used in the great days of his Achaean ancestors—that ownership of a chariot, and its use in war, was one of the marks of the nobleman' (Kirk 1964, 23; 1965, 82). Greenhalgh (1973, 3) has gone so far as to suggest 'that the Dark Age bards have heroized and archaized warfare of their own experience simply by transferring to the more heroic chariot the military role of the contemporary mounted horse', and deliberately avoided mention of cavalry (cf. Delebecque 1951, 76). If it is allowed that chariotry in the Irish hero-tales may present a problem in some degree analogous to its status in the *Iliad* one should obviously not think in terms of simplistic parallels in the transmission of two such disparate texts.

But it cannot be excluded as a possibility that the war-chariot as used in Gaul and Britain never existed in Ireland, and that the *Táin Bó Cuailnge* as we have it today is a composition in terms of what should have happened in an ancient Celtic world elsewhere. The Cattle-Raid itself could be an Irish event as 'historical' as the Siege of Troy is 'historical', but the chariotry may look back to earlier oral traditions on the Continent, or more likely in Britain.

7 Epilogue: 'wheel-going Europe'

When A. W. Kinglake came to write up his Near Eastern travels of 1835 in his enchanting *Eothen* of 1844, he recalled in his opening paragraph how he had stood looking across the frontier to Belgrade and the Ottoman Empire with the words 'I had come, as it were, to the end of this wheel-going Europe, and now my eyes would see the Splendour and the Havoc of the East'. He was to make his journey on horse and camel, and the circumstances in which the two worlds of transport technology came into existence form the subject of Richard Bulliet's penetrating study (1975), which shows how, between the third and seventh century AD, the civilizations of the Near East and North Africa, soon to become part of the Islamic world, abandoned wheeled transport in favour of pack draught by camels. With this profound technological change, the long established primacy of the ancient Near East in the development of wheeled transport came to an end, and was replaced by that of the inventive and innovative societies of medieval Europe, heirs to a tradition which, as this book has attempted to show, had been firmly established since the early third millennium BC. In the early nineteenth century this dichotomy still existed, to evoke the comment of travellers such as the perceptive Kinglake, even though strictly speaking Ottoman Turkey was a transitional zone, retaining ox-carts and the four-wheeled *araba* perhaps from a prehistoric substrate looking back to second-millennium Armenia.

The ultimate origins of wheeled transport can only be guessed at but seem to lie in economic necessities arising in agricultural societies with domestic cattle and probably a knowledge of castration, leading to the use of oxen as draught-animals for a traction plough. The long-standing theoretical sequence from sledge through sledge-on-rollers to sledge-on-wheels, while it cannot be directly substantiated, remains inherently likely as a means of reducing friction to a degree beyond that of the sledge-runners themselves, and such evidence as there is suggests that the horizontal sledge set on four wheels, rather than the tilted travois or slide-car with two, was the earliest form: ox-wagon before cart, as Scheffer hazarded in the seventeenth century. The need such a contrivance satisfied may be assumed to be that of bulk transport over some distance and, in an agrarian context, larger harvests and increased distances between the raw product growing and reaped in the fields and the processing facilities of threshing, winnowing and corn-grinding in the village would be appropriate stimuli to more efficient modes of transport. Constructionally, adequate timber supplies and the skills of carpentry are demanded, both well demonstrated in prehistoric Europe before the invention or adoption of any wheeled vehicles.

The central problem of the earliest wheeled vehicles in Europe from about 3000 BC is that of assessing the respective merits of two hypotheses, that assuming a restricted place and time for an invention subsequently rapidly and widely adopted, and that permitting independent invention of the basic principle of wheeled transport in more than one locality, with subsequent parallel regional development. In specific terms it raises the classic issue of 'diffusion' from an area with a higher degree of technological performance to others with less inventive expertise: the Near East and Neolithic Europe around 3000 BC. The problem is not rendered easier by the fact that we are dealing with wooden structures with a low survival value as archaeological artifacts, helped only by fired clay models among those societies which had a tradition of producing such miniature versions of everyday objects, itself a restricted cultural trait. In the instance of the earliest agricultural communities of south-east Europe from the seventh millennium BC, which did so model humans, animals, houses and even furniture, the absence of vehicle models is at least a suggestive piece of negative evidence for a failure to make this break-through in vehicle technology, despite an efficient agrarian economy and a precocious non-ferrous metallurgy before the beginning of the third millennium. When in that millennium the first European wheels, and depictions and models of wheeled vehicles, appear, radiocarbon dates show us how close in time these are to the comparable evidence for the first appearance in Sumer and Elam of the same invention, and the likelihood of independent discovery in east and west, virtually simultaneously, is sensibly diminished. The thesis of the rapid adoption of a novel piece of transport technology originating in the ancient Near East, as proposed by Childe thirty years ago, still remains the preferable alternative. One of the most recent finds in Western Europe, the wagon from Zürich with disc wheels of the tripartite construction, and a calibrated radiocarbon date of 3030 BC, greatly strengthens this supposition, for the relatively complex technology is precisely that of the early third millennium wheels of Kish, Ur and Susa. Independent invention of simple one-piece wooden plank wheels remains a theoretical possibility, but a closely comparable technique of construction in Sumer and Switzerland can

only mean an original invention adopted by two disparate cultures but originating in one, and the direction of transmission can hardly be in doubt.

The radiocarbon time-scale for this process of invention, transmission and adoption is, however interpreted, a very short one: we are dealing with a 'successful' technological innovation quickly accepted among many European Neolithic societies within a few centuries over a wide geographical area. A comparable phenomenon, again involving the application of rotary motion, is the rapid diffusion of the watermill in north-west Europe between the eighth and eleventh centuries AD, or again, rather later in the Middle Ages, of the windmill. The sparse and unequally distributed evidence for the first European wheeled transport prevents us from making any assessment of the degree to which the new invention was rejected, as well as accepted, among societies of different cultural traditions and in different geographical settings, but there is no reason to assume wholesale adoption, and even in later prehistory and beyond, wheeled vehicles need not have been universally employed over the whole of Europe. As Margaret Hodgen (1939) wrote in the instance of the medieval watermill already referred to, 'we have no way of knowing yet why an element of advanced culture was immediately incorporated into the life of one agricultural community and resisted by an adjoining one'; it is part of a wider problem that has engaged historians, archaeologists and anthropologists alike. On the evidence as it stands, transmission and acceptance from east to west might be inferred from Transcaucasia and the Pontic steppe to Central and north-western Europe on the lines suggested by Van der Waals (1964), with the Corded Ware cultures, from the Netherlands and Denmark to Switzerland, forming one context of acceptance, those of the Funnel Beakers and the Late Hungarian Copper Age another. Once established, the technology of the tripartite disc wheel (and its less complicated version made from a single massive plank) continued unchanged throughout prehistory and into the present, when wheels made to specifications proper to the third millennium BC were being manufactured in Ireland in the 1950s.

The primary European adoption of ox-drawn disc-wheeled wagons and carts was among Neolithic communities whose social structure can reasonably be assumed to be at the level of small villages with their headmen. But almost as early as their use as practical adjuncts to the agrarian economy we find them taking a place as symbols of status and prestige deposited in the graves of a section of the community anxious to display these qualities as a part of funerary ritual—certainly in the south Russian Pit Graves and possibly in Central Europe as well. Once established, vehicle burial in one form or another becomes a recurrent feature in European prehistory up to the time of the Roman conquest, and indeed provides us archaeologically with the bulk of our evidence for vehicle technology. What is interesting is not only the persistence of the practice, but the rapidity of the initial appearance in the third millennium; an almost concurrent modification of agricultural techniques and of religious ritual arising from the acceptance of a new invention which itself conferred prestige on its owners. It is possible, and not inappropriate, to quote a modern and well-documented parallel. The last recorded European folk-movement, that of the gypsies from north-west India, beginning probably in the eleventh century AD, reached Britain in the fifteenth, thereby establishing a small alien ethnic and linguistic minority in the population. Across Europe the gypsies had sporadically and intermittently used wagons or tilt carts, but normally travelled on foot or horseback. In nineteenth-century England and Wales the gypsies adopted the use of the living-wagon

or 'caravan' from travelling showmen (who had only recently developed it themselves) not before about 1850—George Borrow writing in 1873 noted that living-wagons 'have only been used of late years'. By the end of the century the gypsy and his caravan had become, in popular tradition, an indissoluble entity (Ward-Jackson & Harvey 1972). But more than this: the caravan had in a few generations so established itself in gypsy tradition itself that it was forming a part of their funeral traditions and as late as the 1950s was ritually burnt, with its harness, the horses killed and the burnt remains of the vehicle buried, with the broken crockery of the deceased, in a pit (Piggott 1965, 196, quoting *The Times,* 13 January 1953). Here we have an example of a close-knit community following precisely the process of technological adoption and social integration of a wheeled vehicle, at once functional and symbolic, as we have inferred for the third millennium BC, within less than a century.

The use of domesticated equids, in Europe the horse, originally for food and secondarily for draught purposes, brought into wheeled transport a new set of concepts and social connotations. Ox-draught, and the vehicles appropriate to it, were the products of a need for the slow transport of a load which in weight and bulk was beyond the capacity of a man or pack-beast; transport of persons at a walking pace can hardly have been a primary objective. The small swift-footed horse, however, offered potentialities of a different order, if combined with a suitably modified vehicle which allowed these qualities to be exploited to the full, the effective transport of a driver, or a driver and passenger, at a speed up to ten times greater than the plodding foot-pace of an ox-cart. A new social function is fulfilled, one not likely to be integral to peasant agricultural economies but proper to the emergence of a class with a prestige economy demonstrated in display, hunting and war. There were

241

therefore three preconditions in Europe of the second millennium BC which lay behind the new departure in vehicle technology, the development of the fast, light, spoked-wheel prestige vehicle for human transport. The first can only be indirectly inferred, a social structure within which such a vehicle could play a significant part. The second and third can be archaeologically demonstrated, the domestication of equids (horses, hemiones and asses in western Asia; horses in Europe) and the appearance of the light vehicle with two spoked wheels, largely of bent-wood construction and drawn by a pair of horses, which reached its apogee of sophistication in the war and parade chariot of the ancient Near East. The area of origin, satisfying all three of the requirements listed above, has commonly been seen in the oriental civilizations of Western Asia from the beginning of the second millennium BC, with a subsequent adoption of the chariot by peripheral cultures, including Shang China and Vedic India to the east, and barbarian Europe to the north-west. An alternative put forward by the writer would see a wider area of experimentation in light two-wheeled vehicles for fast draught over much of the area of the occurrence of wild equids and their early domestication, including the northern edges of ancient Near Eastern civilization, with the specialized development of the war-chariot taking place in these centres of higher culture, engaged in highly organized internal and international warfare at a professional level. In such a context the lightly-built, spoked-wheeled chariot for parade, and sport, including hunting and raiding, could be adopted by European societies wherever horses were available, either by importing them as a package deal with the vehicle they drew or, more likely, extended domestication among local wild stocks. At all events, by the second half of the second millennium BC such vehicles are known from the Urals (here as vehicle burials in the now traditional manner) to Slo-

vakia, where their presence need owe nothing to those of contemporary Mycenaean Greece, independently adopted in a more sophisticated context from the Levant. In Western Europe the chariot continued to hold its position of prestige into the hierarchically structured heroic societies of the Celtic peoples and their neighbours up to Roman times, while the application of spoked-wheel technology to the wagon led to the development of the more prestigious carriage probably from early in the first millennium.

By now a combination of skills in carpentry and metalworking had reached, in barbarian Europe, a degree of sophistication capable of producing prestige vehicles with two or four wheels in no way inferior to those of the Middle Ages and indeed almost up to the Industrial Revolution. Certain pieces of advanced technology—the widespread adoption of the shrunk-on hoop tyre and the exceptional roller-bearings of Dejbjerg for instance—were, in fact, to be lost and forgotten until they were reinvented in modern times. The 'modern' form of horse-harness with shafts, rigid collars, traces and whipple-tree was being experimented with in the later Roman world as well as in China, and its adoption from at least the eighth century AD in medieval Europe may be the result of the convergence of ideas along more than one route of transmission, including North Africa and Sicily as well as the Central Asian steppe (Lynn White 1962; Bulliet 1975, 197–215). Behind medieval Europe north of the Alps lies the heritage of millennia of prehistoric technological achievement as well as the Roman innovations adopted or imposed with the Empire. The distinguished historian of technology, Lynn White, once toyed with the idea that 'the vigor of medieval technology may have been simply an amplification of a cultural condition pre-existent in Gaul', and not of Roman inspiration (1978, 228). If one extended the geographical area to prehistoric north-western Europe and limited the technological scope to wheeled transport, this book has set out the evidence for the support of such a thesis and thereby perhaps strengthened the case for its acceptance in wider terms.

A final theme links antiquity to the present; the status in human thought and emotion of the two original draught and transport animals, the ox and the horse. Both Eurasiatic domesticates, in the first place for meat and secondarily for other uses such as draught, from their first employment in transport they fulfilled different roles and satisfied different psychological needs. The initial use of the ox was, and continued to be throughout its history, that of a docile working beast of burden, acquiring inevitably the legendary anthropomorphic qualities of humility and patience, with no aspi-

rations above its lowly station in life. In complete contrast, the horse was from the beginning used as a transport animal, whether for draught or for riding, in a totally different spirit, to satisfy vanity and support prestige by its speed and the panache it could confer on its master; proud arrogant and aristocratic from the beginning of the second millenium BC and the charioteering oriental kings, through Celtic chieftains and medieval knights to Royal Ascot today. Equally, in China or the Central Asian steppe or in the Arab world, it served to satisfy this same deep-seated need, and eventually in the New World both Indians and cow-boys alike shared a status-symbol, the latter preserving the same social *142* grading between their steeds and their herds that had been established and maintained over four thousand years.

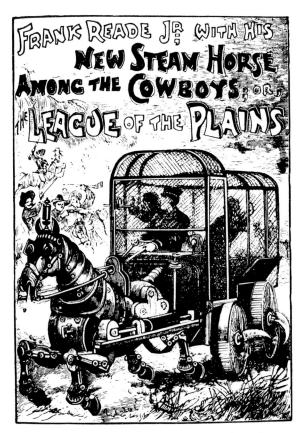

142 The scythed chariot reappears among the cowboys

Bibliography

Titles in Cyrillic script in Roman transliteration

ADAMS, W. B. 1837 *English Pleasure Carriages,* Bath (reprint 1971)

ALBRECHTSEN, E. 1954 *Fynske Jernaldergrave,* Copenhagen

ALCOCK, L. 1961 'The winged objects in the Llyn Fawr hoard', *Antiq.* 35, 149–51

ALEXANDER, J. 1972 *Jugoslavia before the Roman Conquest,* London & New York

ALFÖLDI, A. & RADNOTI, A. 1940 'Zügelringe und Zierbeschläge von römischen Jochen und Kummeten aus Pannonien', in *Serta Hoffileriana,* 309–19, Zagreb

ALLEN, D. 1978 *An introduction to Celtic coins,* London (Brit. Mus.)

—— & NASH, D. 1980 *The coins of the ancient Celts,* Edinburgh

ALMAGRO, M. 1966 *Las Estelas Decoradas del Suroeste Peninsular,* Madrid

ALTHIN, C. A. 1945 *Studien zu den bronzezeitlichen Felszeichnungen von Skåne,* Lund/Copenhagen

ANATI, E. 1960 'Bronze Age chariots from Europe', *Proc. Prehist. Soc.* 26, 50–63

—— 1964 *Camonica Valley,* London

—— 1975 *Evoluzione e stile nell'arte rupestre Camuna,* Capo di Ponte (Archivi, Vol. 6)

ANDERSON, J. K. 1961 *Ancient Greek horsemanship,* Berkeley & Los Angeles

—— 1965 'Homeric, British and Cyrenaic chariots', *Amer. Journ. Arch.* 69, 349–52

ANDREWS, P. A. 1973 'The White House of Khurasan: the felt tents of the Iranian Yomut and Göklen', *Iran,* 11, 93–110

—— 1978 'The tents of Timur: an examination of reports on the Quriltay at Samarqand, 1404', in P. Denwood (ed.) *Arts of the Eurasian Steppelands* (Colloquies on Art & Arch. in Asia 7), 143–97, London

ANGELI, W., *et al.* 1970 *Krieger und Salzherren* (exhibition catalogue) Mainz

—— 1980 *Die Hallstatt-kultur* (exhibition catalogue) Steyr

ANON 1977 'Raboty kakhetskoy ekspeditsii', *Arkh. Otkrit. 1976,* 1977, 478–79

D'ARBOIS DE JUBAINVILLE, H. 1888 'Le char de guerre des Celtes dans quelques textes historiques', *Rev. Celtique* 9, 387–93

ARNOLD, B. 1978 'Gallo-Roman boat finds in Switzerland', in J. du P. Taylor & H. Cleere (edd), *Roman shipping and trade: Britain and the Rhine Provinces* (CBA Research Report 24), 31–35, London

ARNOLD, J. 1974 *The farm waggons of England and Wales,* London

ARRIBAS, A. n.d. *The Iberians,* London & New York

—— 1967 'La necropolis bastitana del Mirador de Rolando (Granada)' *Pyrenae* 3, 67–105

ARTS COUNCIL 1964 *Hittite art and the antiquities of Anatolia* (exhibition catalogue), London

ASHBEE, P. 1970 *The earthen long barrow in Britain,* London

—— *et al.* 1979 'Excavation of three long barrows near Avebury, Wiltshire', *Proc. Prehist. Soc.* 45, 207–300

ATKINSON, F. & WARD, A. 1964 'A pair of "Clog" wheels from Northern England', *Trans. Yorks. Dialect Soc.* 11, 64, 33–40

ATKINSON, R. J. C. 1960 *Stonehenge,* Harmondsworth

—— & PIGGOTT, S. 1955 'The Torrs chamfrein', *Archaeologia,* 96, 187–235

DE AZEVEDO, M. C. 1938 *I transporti e il traffico,* (Mostra Augustea della Romanità) Rome

BAHN, P. G. 1978 'The "unacceptable face" of the western European Upper Palaeolithic', *Antiq.* 52, 183–92

BAKKER, J. A. 1976 'On the possibility of reconstructing roads from the TRB period', *Bericht. Rijksdienst Oudheid. Bodemond.* 26, 63–91

—— *et al.* 1969 'TRB and other C14 dates from Poland', *Helinium* 9, 3–27; 209–38

—— & VAN DER WAALS, J. D. 1973 'Denekamp-

Angelslo', in G. Daniel & P. Kjaerum, (edd), *Megalithic Graves and Ritual,* Moesgård 17–50

BALKWILL, C. J. 1973 'The earliest horse-bits of western Europe', *Proc. Prehist. Soc.* 39, 425–52

BALTZER, L. 1881 *Glyphes des rochers de Bohuslän,* Stockholm

BÁNDI, G. 1963 'Die Frage der Riementeiler des mittelbronzezeitlichen Pferdegeschirrs im Karpatenbecken', *Arch. Értesítő* 90, 48–60

BANNER, J. 1956 *Die Péceler Kultur,* (*Arch. Hung.* 35), Budapest

BARBER, J. & MEGAW, J. V. S. 1963 'A decorated Iron Age bridle-bit in the London Museum', *Proc. Prehist. Soc.* 29, 206–13

BARFIELD, L. 1971 *Northern Italy,* London & New York

BARTH, F. E. 1973 'Zur Identifizierung einiger Gegenstände aus dem Gräberfeld in der Sammlung Johan Georg Ramsauer', *Mitt. Anthrop. Gess. Wien* 103, 48–54

BASERGA, G. 1929 'Tomba con carro ed altere scoperte alla Ca'Morta', *Rivista Arch. Como* 96–98, 25–44

LA BAUME, W. 1928 'Bildliche Darstellungen auf Ostgermanischen Tongefässen der frühen Eisenzeit', *IPEK* 1928, 25–56

—— 1963 *Die Pommerellischen Gesichtsurnen,* Mainz

BEHRENS, C. 1927 *Bodenurkunden aus Rheinhessen I,* Mainz

BEHRENS, H. 1963 'Die Rindskelletfunde der Péceler Kultur', *Acta Arch. Hung.* 15, 33–36

—— 1964 *Die neolithisch-frühmetallzeitlichen Tierskelettfunde der Alten Welt,* Berlin

—— 1965 'Berichtigtes und ergänztes Schema mitteldeutscher neolithischer C14–Daten', *Ausgrab. u. Funde* 10, 1–2

BENEŠOVÁ, A. 1956 'Nález Měděných Předmětú na Starých Zámcích v Brně-Líšni', *Pamatky Arch.* 47, 236–44

BENIRSCHKE, K. 1969 'Cytogenetics in the zoo', *New Scientist* 16 Jan. 1969, 132–33

BERCIU, D. 1961 *Contributii la probleme neoliticului in Romînia,* Bucharest

—— 1967 *Romania before Burebista,* London & New York

—— 1969 *Arta Traco-Getica,* Bucharest

BERG, G. 1935 *Sledges and Wheeled Vehicles,* Stockholm

VAN BERG-OSTERRIETH, M. 1972 *Les chars préhistoriques du Val Camonica,* (Centro Camuno di studi preistoriche, Archivi 3), Capo di Ponte

BERGER, R. & LIBBY, W. F. 1968 'UCLA radiocarbon dates VIII', *Radiocarbon* 10, 402–16

—— & PROTSCH, R. 1973 'The domestication of plants and animals in Europe and the Near East', *Orientalia* 42, 214–27

BERTOLONE, M. 1954 'Ancora sulla seconda tomba de guerriero scopeta a Sesto Calende', *Sibrium* 1, 67–72

—— & KOSSACK, G. 1957 'Tomba della prima età del ferro . . . scoperta alla Ca' Morta (Como) . . . zu den Metallbeigaben des Wagengrabes von Ca' Morta', *Sibrium* 3, 37–54

BICHIR, G. 1964 'Autour du problème des plus anciens modèles de chariots découverts en Roumanie', *Dacia* NS 8, 67–86

BICKNELL, C. 1913 *A guide to the prehistoric rock engravings in the Italian Maritime Alps,* Bordighera

BIEL, J. 1981 'The late Hallstatt chieftain's grave at Hochdorf', *Antiq.* 55, 16–18

BIRMINGHAM, J. 1963 'The development of the fibula in Cyprus and the Levant', *Palestine Exp. Quarterly* 1963, 80–112

BITTEL, K. (ed) 1981 *Die Kelten in Baden–Württemberg,* Stuttgart

BLAZQUEZ, J. M. 1955 'Los carros votivos de Mérida y Almorchón', *Zephyrus* 6, 41–60

BLAZQUEZ MARTINEZ, J. M. 1958 'Caballos en el infierno etrusco', *Ampurias* 19/20, 31–68

BOARDMAN, J. 1971 'A southern view of Situla Art', in Boardman, J. *et al.* (edd) *The European Community in later prehistory,* 121–40, London

—— 1980 *The Greeks Overseas,* London & New York

BOCKSBERGER, O. J. 1971 'Nouvelles recherches au Petit-chasseur, à Sion (Valais, Suisse)', *Jahrb. Schweiz, Gesellsch. f. Ur. u. Frühgesch,* 56, 77–99

BOESSNECK, J. 1973 'Was weiss man von den Alluvial Vorgeschichtlichen Equiden der Iberischen Halbinsel?', in J. Matolcsi (ed) *Domestikationsforschung und Geschichte der Haustiere,* 277–84, Budapest

BOGNÁR–KUTZIÁN, I. 1972 *The Early Copper Age Tiszapolgár culture in the Carpathian Basin,* Budapest

—— 1973a 'The beginning and position of the Copper Age in the Carpatho-Pannonian region', *Actes VIII Cong. Int. Sciences Préhist. & Protohist.* 2, Belgrade

—— 1973b 'The relationship between the Bodrogkeresztúr and the Baden cultures', in Kalousek, F. & Budinský–Krička, V., *Symposium über ... der Badener Kultur,* 31–50, Bratislava

BÖKÖNYI, S. 1952 'Die Wirbeltierfauna der Ausgrabungen in Tószeg vom Jahre 1948', *Acta Arch. Hung.* 2, 71–112

—— 1968 *Data on Iron Age horses of central and eastern Europe (Mecklenburg Collection Part I),* Cambridge, Mass.

—— 1974 *History of Domestic Mammals in Central and Eastern Europe,* Budapest

—— 1980 'The importance of horse domestication in economy and transport', in P. Sörbom (ed), *Transport Technology and Social Change,* 15–21 Stockholm

—— *et al.* 1973 'Earliest animal domestication dated ?', *Science* 182, 1161

BÓNA, I. 1960 'Clay Models of Bronze Age Wagons and Wheels in the Middle Danube Basin', *Acta Arch. Hung.* 12, 83–111

—— 1975 *Die mittlere Bronzezeit Ungarns und ihre südöstlichen Bezieungen,* Budapest

BONFANTE, L. 1977 'The Corsini throne', *Journ. Walters Art Gallery* 36, 111–22

—— 1979 'I popoli delle situle: una civiltà protourbana', *Dialoghi di Arch.* 2, 73–94

BORHEGYI, S. F. de 1970 'Wheels and man', *Archaeology* 23, 18–25

BOTUSHAROVA, L. 1950 'Trakiysko mogilno porebenie s kolesitsa', *Ann. Nat. Mus. Arch. Plovdiv* 2, 101–35

BOUZEK, J. 1966 'The Aegean and central Europe: an introduction to the study of cultural interrelations 1600–1300 BC', *Pamatky Arch.* 62, 242–76

BOYER, M. N. 1960 'Medieval pivoted axles', *Technology and Culture* 1, 128–38

BRAILSFORD, J. W. 1975 'The Polden Hill Hoard, Somerset', *Proc. Prehist. Soc.* 41, 222–34

BREUIL, H. 1933 *Les peintures rupestres schématiques de la Peninsule Ibérïque II,* Lagny

BREWSTER, T. C. M. 1971 'The Garton Slack chariot burial, East Yorkshire', *Antiq.* 45, 289–92

BRIARD, J. 1965 *Les dépôts Bretons et l'Age du Bronze atlantique,* Rennes

BRITNELL, W. J. 1976 'Antler cheekpieces of the British Late Bronze Age', *Antiq. Journ.* 56, 24–34

BRITTON D. & LONGWORTH, I. F. 1968 *Late Bronze Age finds in the Heathery Burn Cave Co. Durham,* (Invent. Arch. G. B. 55), London

BRONSON, R. C. 1965 'Chariot racing in Etruria', *Studi in onore di Luisa Banti,* 89–106, Rome

BRØRNSTED, J. 1938 *Danmarks Oldtid* 1, Copenhagen

—— 1939 *Danmarks Oldtid II: Bronzealderen,* Copenhagen

BROWN, W. LL. 1960 *The Etruscan Lion,* Oxford

BUCHNER, G. 1979 'Early orientalizing: aspects of the Euboean connection', in D. & F. R. Ridgway, *Italy before the Romans,* 129–44, London/New York/San Francisco

BULLEID, A. & GRAY, H. St. G. 1917 *The Glastonbury Lake-Village,* Glastonbury

BULLIET, R. W. 1975 *The Camel and the Wheel,* Cambridge, Mass.

BURCHULADZE, A. A. *et al.* 1976 'Tbilisi Radiocarbon Dates III', *Radiocarbon* 18, No. 3, 355–61

BURGESS, C. 1969 'The later Bronze Age in the British Isles and north-western France', *Arch. Journ.* 125, 2–45

BURNEY, C. & LANG, D. M. 1971 *The Peoples of the Hills: Ancient Ararat and Caucasus,* London

BURTON BROWN, T. 1951 *Excavations in Azarbaijan 1948,* London

CAHEN-DELHAYE, A. 1974 'Nécropole de la Tène I à Hamipré, Offaing. I Trois tombes à char', *Arch. Belgica* 162, 5–48

—— 1975 'Tombes à char et buchers sous tombelles de La Tène I à l'Église-Gohimont', *Arch. Belgica* 177, 17–21

—— 1979 'Nécropole et site d'habitat de La Tène à Longlier-Massul', *Arch. Belgica* 218, 6–38

CARDOZO, M. 1946 'Carrito votivo de bronce, del Museo de Guimarães (Portugal)', *Archivo Esp. de Arqueologia* 19, 1–28

CASSAU, A. 1938 'Ein frühbronzezeitliche oder endsteinzeitlicher Wagenradfund, in Beckdorf. Kr. Stade', *Nachrichten aus Niedersachs. Urgesch.* 12, 63–71

CATLING, H. W. 1964 *Cypriot bronzework in the Mycenaean world,* Oxford

CHAMPION, S. 1976 'Coral in Europe: commerce and Celtic ornament', in P. M. Duval and C. Hawkes (edd), *Celtic art in Ancient Europe: five protohistoric centuries,* 29–40, London &

New York

CHAMPION, T. 1975 'Britain in the European Iron Age', *Arch. Atlantica,* 1, 2, 127–45

—— 1980 'Mass migration in later prehistoric Europe', in P. Sörbom (ed) *Transport Technology and Social Change,* 33–42, Stockholm

CHAPOTAT, G. 1962 'Le char processionel de la Côte-St-André', *Gallia,* 20, 3–48

CHERDYNTSEV, V. V. *et al.* 1968 'Geological Institute Radiocarbon Dates I', *Radiocarbon* 10: 2, 419–25

CHERNYKH, E. N. 1976a 'Metallurgische Bereiche des 4–2 Jahrt. v. Chr. in der Ud SSR', *Les Débuts de la Metallurgie* (U.I.S.P.P. Nice), 177–208, Nice

—— 1976b 'Metallurgische Bereiche der jungeren und späten Bronzezeit in der Ud SSR', *Jahresb. Inst. Vorgesch. Univ. Frankfurt AM,* 130–49

CHILDE, V. G. 1950 'Axe and adze, bow and sling: contrasts in early Neolithic Europe', *Jahrb. Schweiz. Gesellsch. Urgesch.* 40, 156–62

—— 1951 'The First Waggons and Carts—from the Tigris to the Severn', *Proc. Prehist. Soc.* 17, 177–94

—— 1954a 'The Diffusion of Wheeled Vehicles', *Ethnograph.-Arch. Forschungen* 2, 1–17

—— 1954b 'Rotary Motion', in Singer *et al., Hist. Technology* 1, 187–215, Oxford

—— 1957 *The Dawn of European Civilization,* (6th ed.), London

CLARK, G. 1975 *The earlier stone age settlement of Scandinavia,* Cambridge

CLARK, R. M. 1975 'A calibration curve for radiocarbon dates', *Antiq.* 49, 251–66

CLARKE, D. L. 1972 'A provisional model of an Iron Age society and its settlement system', in D. L. Clarke (ed) *Models in Archaeology,* 801–69, London

COLES, J. M. 1960 'Scottish late Bronze Age metalwork: typology, distributions and chronology', *Proc. Soc. Antiq. Scot.* 93, 16–134

—— 1962 'European Bronze Age shields', *Proc. Prehist. Soc.* 28, 156–90

—— 1973 *Archaeology by experiment,* London

COLES, J. M. *et al.* 1978 'The use and character of wood in prehistoric Britain and Ireland', *Proc. Prehist. Soc.* 44, 1–45

COLES, J. M. & HARDING, A. F. 1979 *The Bronze Age in Europe,* London

COLLIS, J. 1975 *Defended sites of the late La Tène,* (B.A.R. Supp. Ser. 2), Oxford

—— 1976 'Town and market in Iron Age Europe', in B. Cunliffe & T. Rowley, *Oppida in barbarian Europe,* (B.A.R. Supp. Ser. 11) 3–23, Oxford

COOMBS, D. 1975 'Bronze Age weapon hoards in Britain', *Arch. Atlantica,* 1, i, 49–81

CORDIER, G. 1968 'Un nouveau tumulus à char hallstattien: Sublaines, (Indre-et-Loire)', *Ogam* 20, 5–12

—— 1975 'Les tumulus hallstattiens de Sublaines (Indre-et-Loire)', *L'Anthrop.* 79, 451–82: 579–628

CORNAGGIA CASTIGLIONI, O. & CALEGARI, G. 1978 'Le ruote preistoriche Italiane a disco ligneo', *Revist. Arch. dell' Antica Prov. e Diocesi di Como* 160, 5–50

COWEN, J. D. 1961 'The late Bronze Age chronology of Central Europe: some reflections', *Antiq.* 35, 40–44

CRANSTONE, B. A. L. 1969 'Animal husbandry: the evidence from ethnography', in P. J. Ucko & G. W. Dimbleby *The domestication and exploitation of plants and animals* 247–63, London

CROUWEL, J. H. 1978 'Aegean Bronze Age chariots and their Near Eastern Background', *Bull. Inst. Class. Studies* 25, 174–75

CSALOG, J. 1943 'Hallstattzeitliche Wagenurne aus Kanya, Komitat Tolna, Ungarn', *Arch. Értesítő,* 3rd S. 4, 41–49

CURLE, J. 1911 *A Roman frontier post: the fort at Newstead . . . ,* Glasgow

CUADRADO, E. 1955 'El carro iberico', *III Congreso Nac. de Arque. Galicia 1953* (Zaragozo), 116–35

DÉCHELETTE, J. 1924 *Manuel d'archéologie . . . II. 1. Age du bronze* (2nd ed.) Paris

—— 1927 *Manuel d'archéologie . . . IV Second Age du Fer ou époque de La Tène* (2nd ed.) Paris

DEFFONTAINES, P. 1938 'Sur la répartition géographique des voitures à deux roues et à quatre roues', *Cong. Internat. Folklore: Paris 1937,* 117–21

DEHN, W. 1963 'Frühe Drehscheibenkeramik nördlich der Alpen', *Alt-Thüringen* 6, 372–82

—— 1965 'Die Bronzeschüssel aus dem Hohmichele, Grab VI . . .', *Fundber. aus Schwaben,* NS 17, 126–34

—— 1966 'Eine Böhmische Zierscheibe der

Frühlatènezeit in Berlin', *Sbornik Narod. Muz. v Praze* A. 20, 137–48

—— 1969 'Keltische Röhrenkannen der älteren Latènezeit', *Pamatky Arch.* 60, 125–33

1970 'Ein keltisches Häuptlingsgrab aus Hallstatt', in Angeli *et al.* 1970, 72–81

—— 1971 'Hohmichele Grab 6—Hradenin Grab 28—Vace (Watsch) Helmgrab', *Fundber. aus Schwaben*, NS 19, 82–88

—— 1980 'Hessische Steinkisten und frühes Metall', *Fundber. aus Hessen* 19/20, 163–76

—— & FREY, O-H, 1979 'Southern imports and the Hallstatt and early La Tène chronology of central Europe', in D. & F. R. Ridgway, *Italy before the Romans*, 489–511, London/New York/San Francisco

DELEBECQUE, E. 1951 *Le cheval dans l'Iliade*, Paris

DEREGOWSKI, J. 1973 'Illusion and culture', in R. L. Gregory & E. H. Gombrich (edd), *Illusion in Nature and Art*, 161–91, London

DJUKNIĆ, M. & JOVANOVIĆ, B. 1966 *Illyrian princely necropolis at Atenica*, Čačak

DOWNEY, R. *et al.* 1979 *The Hayling Island temple, Third Interim Report*, London

DRACK, W. 1958 'Wagengräber und Wagenbestandteile aus Hallstattgrabhügeln der Schweiz', *Zeitschr. f. Schweiz. Arch. u. Kunstgesch.* 18, 1–67

—— (ed) 1959 *L'Age du Bronze en Suisse*, Basle

—— 1960 'Spuren von urnenfelderzeitlichen Wagengräber aus der Schweiz', *Jahrb. Schweiz. Gess. Urgesch.* 48, 74–77

—— *et al.* 1971 *Archäologie der Schweiz III, Bronzezeit*, Basle

—— 1974 *Archäologie der Schweiz IV, Eisenzeit*, Basle

DRIEHAUS, J. 1960 *Die Altheimer Gruppe*, Mainz

—— 1965a '"Furstengräber" und Eisenerze zwischen Mittelrhein, Mosel und Saar', *Germania* 43, 32–49

—— 1965b 'Eine frühlatenezeitliche Reiterdarstellung aus Kärlich', *Bonner Jahrb.* 165, 57–71

—— 1966a 'Zur Datierung des Graberfeldes von Bell in Hunsrück', *Bonner Jahrb.* 166, 1–26

—— 1966b 'Zur Verbreitung der eisenzeitlichen Situlen im mittelrheinischen Gebiet', *Bonner Jahrb.* 166, 26–47

DRIVER, H. E. & MASSEY, W. C. 1957 'Comparative studies of North American Indians', *Trans. Amer. Phil. Soc.* NS 47, 2, 165–456

DUŠEK, M. 1960 'Patince-pohrebisko severopanónskej kultúry', in Chopovsky, B., *et al.*, *Pohrebiská zo staršej doby bronzovej na Slovensku*, 219–96, Bratislava

—— 1966 *Thrakisches Gräberfeld der Hallstattzeit in Chotín*, Bratislava

DUVAL, A. 1975 'Une tombe à char de la Tène III: Inglemare', *Arch. Atlantica* 1, 2, 147–63

DVOŘÁK, F. 1933 'Ein Skelettgrab von Bylany Typus aus Plaňany', *Pamatky Arch.* 39, 35–38

—— 1938 *Wagengräber der älteren Eisenzeit in Böhmen*, Prague

EBERT, M. (ed) 1924–32 *Reallexikon der Vorgeschichte*, 15 vols. Berlin

EHRICH, R. & PLESLOVÁ-ŠTIKOVÁ, E. 1968 *Homolka: an eneolithic site in Bohemia*, Prague

EKHOLM, G. F. 1946 'Wheeled toys in Mexico', *Amer. Antiq.* 11, 222–28

EOGAN, G. 1964 'The later Bronze Age in Ireland in the light of recent research', *Proc. Prehist. Soc.* 30, 268–351

ESAYAN, S. A. 1966 *Oruzhie i voennoe delo drevney Armenii*, Erevan

—— 1967 'Pogrebenie no. 14 Astkhilurskogo mogil'nik', *Hist. Phil. Journ. Acad. Sciences Armenian SSR.* 1967 221–26

—— 1976 *Drevnyaya kultura plemen severo-vostochnoy Armenii.* Erevan

—— 1978 *Die Kunst des Gravierens im alten Armenien nach Abbildungen auf Bronzegürteln des II–I Jahrtausends v. u. Z.*, II Internat. Sympos. Armenian Art, Erevan

EVANS, D. E. 1967 *Gaulish Personal Names*, Oxford

EVANS, J. D. 1971 *The prehistoric antiquities of the Maltese Islands: a survey*, London

EWART, J. C. 1911 'Animal remains', in J. Curle, *A Roman Frontier Post and its People*, 362–77, Glasgow

FAVRET, P-M, 1936 'Les nécropoles des Jogasses à Chouilly, (Marne)', *Préhistoire* 5, 24–119

FELGENHAUER, F. 1962 'Eine Hallstättische Wagendarstellung aus Rabensburg, N. Ö.', *Mitt. Anthrop. Gess. Wien* 92, (Hančar Festschrift), 93–113

FENTON, A. 1963 'Early and traditional cultivating implements in Scotland', *Proc. Soc. Ant. Scot.* 96, 264–317

—— 1972 'Early Yoke types in Britain', *Magyar Mezőgazdasági Múzeum Közleményei* 1971–72, 69–75

— 1973 'Transport with pack-horse and slide-car in Scotland', in A. Fenton *et al.* (edd), *Land Transport in Europe,* 121–71, Copenhagen

FENWICK, V. 1978 *The Graveney Boat: a tenth-century find from Kent,* (Brit. Arch. Rep. 53), National Maritime Mus., Greenwich

FILIP, J. 1962 *Celtic civilization and its heritage,* Prague

— 1966–68 *Enzyklopädisches Handbuch zur Ur- und Frühgeschichte Europas,* Prague

FINLEY, M. I. 1956 *The World of Odysseus,* London

— 1959 'Technology in the Ancient World', *Econ. Hist. Review,* 2nd S. 12, 120–25

FISCHER, F. 1959 *Der spätatènezeit Depot-Fund von Kappel (Kreis Saulgau),* Stuttgart

— 1973 'KEIMELIA: Bemerkungen zur kulturgeschichten Interpretation des sogenannten Südimports in der Späten Hallstatt- und frühen Latène-Kultur des westlichen Mitteleuropa', *Germania* 51, 436–59

FISCHER, U. 1973 'Zur Megalithik der Hercynischen Gebirgsschwelle', in G. Daniel & P. Kjaerum (edd) *Megalithic Graves and Ritual,* 51–62, Moesgård

— 1979 *Ein Grabhügel der Bronze- und Eisenzeit in Frankfurter Stadtwald,* (Schr. Frankfurter Mus. f. Vor. u. Frügesch. IV), Frankfurt

FLOUEST, J. L. & STEAD, I. M. 1979 *Iron Age cemeteries in Champagne,* London (British Museum)

FOLTINY, S. 1955 *Zur Chronologie der Bronzezeit des Karpatenbeckens,* Bonn

FORRER, R. 1921 'Un char de culte, à quatre roues et trone, decouvert dans un tumulus gaulois à Ohnenheim (Alsace)', *Cahiers d'Arch. et Hist. d'Alsace* 45–48, 1195–1229

— 1932 'Les chars cultuels prehistoriques et leurs survivances aux epoques historiques', *Préhistoire* 1, 19–123

FOSTER, J. 1980 *The Iron Age moulds from Gussage All Saints,* London (British Museum)

FOURDRINGIER, E. 1878 *Double sépulture gauloise de la Gorge-Meillet...,* Paris & Châlons-sur-Marne

FOWLER, P. J. 1981 'Later prehistory', in S. Piggott (ed) *The Agrarian History of England and Wales* I. i., 63–298, Cambridge

FOX, C. 1931 'Sleds, carts and waggons', *Antiq.* 5, 185–99

— 1946 *A find of the early Iron Age from Llyn Cerrig Bach, Anglesey,* Cardiff (Nat. Mus. Wales)

— 1950 'Two Celtic bronzes from Lough Gur, Limerick, Ireland', *Antiq. Journ.* 30, 190–92

— 1958 *Pattern and purpose: a survey of Early Celtic art in Britain,* Cardiff (Nat. Mus. of Wales)

— & HYDE, H. A. 1939 'A second cauldron and an iron sword from the Llyn Fawr hoard', Rhigos, Glamorganshire', *Antiq. Journ.* 19, 369–404

FRANKENSTEIN, S. & ROWLANDS, M. J. 1978 'The internal structure and regional context of Early Iron Age society in south-western Germany', *Univ. London. Inst. Arch. Bulletin* 15, 73–112

FRANKFORT, H. 1954 *The Art and Architecture of the Ancient Orient,* Harmondsworth

FREDSJÖ, A. 1971 *Rock-carvings: Kville hundred, Svenneby parish, county of Bohuslän,* Gothenburg

— 1975 *Rock-carvings: Kville hundred, Bottna parish, county of Bohuslän,* Gothenburg

FREY, O-H. 1962 'Die Situla von Kuffarn', *Veröff. a. d. Naturhist. Mus. Wien,* N.F.4

— 1968 'Eine neue Grabstele aus Padua', *Germania,* 46, 317–20

— 1969 *Die Entstehung der Situlenkunst,* Berlin

— 1971 'Das keltische Schwert von Moscano di Fabriano', *Hamburger Beiträge z. Arch.* I. 2., 173–79

— 1976 'The chariot tomb from Adria: some notes on Celtic horsemanship and chariotry', in J. V. S. Megaw (ed) *To Illustrate the Monuments,* 172–79, London & New York

FROHBERG, U. 1974 'Zur Konstruktion der Wagenräder aus dem späthallstattzeitlichen Grab von Offenbach-Rumpenheim', *Jahrb. RGZ Mainz* 21, 144–48

GABALOWNA, L. 1970 'Radiocarbon dating of charcoal from the TRB cemetery... at Sarnowo (Barrow 8)...', *Prace i Materialy Muz. Arch. Lodz.* (Arch. series) 17, 89.

GABROVEC, S. et al. 1961 *Mostra dell' arte delle situle dal Po al Danube,* Florence

GALHANO, F. 1973 *O carro de bois em Portugal,* Lisbon

GALKIN, L. L. 1977 'Sosud srubnoy kul'tury s synzhetiym risunkom iz Saratovskogo zavolzhya', *Sov. Arkh.* 1977. 3, 189–96

GALLAY, A. 1972a 'Recherches préhistoriques au Petit-Chasseur à Sion', *Helvetia Arch.* 10/11, 35–61

—— 1972b *Sion, Petit-Chasseur (Valais, Suisse): Programme d'élaboration,* Geneva

GALLAY, G. & SPINDLER, K. 1972 'Le Petit-Chasseur — chronologische und kulturelle Probleme', *Helvetia Arch.* 10/11, 62–89

GALLUS, S. 1934 *Die Figuralvierzierten Urnen vom Soproner Burgstall, Arch. Hung.* 13

—— & HORVATH, T. 1939 *Un peuple cavalier préscythique en Hongrie,* Budapest

GANDERT, O-F. 1964 'Zur Frage der Rinderanschirrung in Neolithikum', *Jahrb. R.-G. Zentralmus. Mainz* 11, 34–56

GARCIA & BELLIDO, A. 1969 'Los bronces Tartesicos', in Maluquer de Motes, J., *ed. Tartessos y sus problemas,* 163–71, Barcelona

GENING, V. F. 1977 'Mogil'nik Sintashta i problema rannikh Indoiranskitch plemen', *Sov. Arkh.* 1977. 4, 53–73

—— & ASHICHMINA, L. I. 1975 'Mogil'nik epokhi bronzy na R. Sintashta', *Arkh. Otkrit. 1974,* 144–47

GEORGIEV, G. I. & MERPERT, N. YA. 1966 'The Ezero mound in south-east Bulgaria', *Antiq.* 40, 33–37

GHIRSHMAN, R. 1939 *Fouilles de Sialk II,* Paris

GHISLANZONI, E. 1930 'Il carro di bronzo della Camorta', *Rivista Arch. Como* 99–101, 3–25

—— 1944 'Una nuova tomba di guerriero scoperta a Sesto Calende', in *Munera: Raccolti di scritti in onore Antonio Guissani,* 1–55, Como

GIMBUTAS, M. 1958 'Middle Ural sites and the chronology of northern Eurasia', *Proc. Prehist. Soc.* 24, 120–57

—— 1965 *Bronze Age cultures in Central and Eastern Europe,* The Hague

—— 1970 'Proto-Indo-European culture: the Kurgan culture during the fifth, fourth and third millennia BC', in Cardona, G. *et al.* (edd), *Indo-European and Indo-Europeans,* 155–97, Philadelphia

—— 1974 'An archaeologist's view of PIE', *Journ. Indo-Europ. Studies* 2, 3, 289–307

—— 1978 'The first wave of Eurasian steppe pastoralists into Copper Age Europe', *Journ. Indo-Europ. Studies* 5, 4, 277–338

GLOB, P. V. 1951 *Ard og plov i Nordens Oldtid,* Aarhus

—— 1969a *Helleristninger i Danmark,* Copenhagen

—— 1969b *The Bog People,* London

GOMBRICH, E. H. 1960 *Art and Illusion: a study in the psychology of pictorial representation,* New York

GOODMAN, W. L. 1964 *A History of Woodworking Tools,* London

GORDON, J. E. 1978 *Structures, or why things don't fall down,* Harmondsworth

GRÄSLUND, B. 1967 'The Herzsprung shield type and its origin', *Acta. Arch.* 38, 59–71

GREENE, D. 1972 'The chariot as described in Irish literature', in C. Thomas (ed) *The Iron Age in the Irish Sea province,* (CBA Res. Rep. 9) 59–73, London

GREENHALGH, P. A. L. 1973 *Early Greek Warfare,* Cambridge

GRIGSON, C. 1966 'The animal remains . . .', in P. Ashbee, 'The Fussells Lodge Long Barrow Excavations 1957', *Arch.* 100, 63–73

GRINSELL, L. V. 1942 'The Kivik cairn, Scania', *Antiq.* 16, 160–74

GÜNTHER, A. 1934 'Gallisches Wagengräber im Gebiet des Neuwieder Beckens', *Germania* 18, 8–14

HABERLAND, W. 1965 'Tierfiguren mit Rädern aus El Salvador', *Baessler Archiv.* NF 13, 309–16

HADDON, A. C. 1898 *The Study of Man,* London & New York

HAFFNER, A. 1969 'Das Treverer–Graberfeld mit Wagenbestattungen von Hoppstädten-Wiersbach, Kreis Birkenfeld', *Trierer Zeitschr.* 32, 71–127

—— 1976 *Die westliche Hunsrück-Eifel-Kultur,* (R. G. Forschungen Bd. 36.)

HAGEN, A. 1967 *Norway,* London & New York

HALBERT, L. 1955 'Bronze Age finds from Lyngsjö', *Medd. Lunds Univ. Hist. Mus. 211–23*

HALL, A. R. 1961 'More on medieval pivoted axles', *Technology & Culture* 2, 17–21

HAMPEL, J. 1890 *Alterthümer der Bronzezeit in Ungarn,* Budapest

HANČAR, F. 1937 *Urgeschichte Kaukasiens,* Vienna

—— 1955 *Das Pferd in prähistorischer und früher historischer Zeit,* Vienna

HARBISON, P. 1969 'The chariot of Celtic funerary tradition', in *Marburger Beiträge zur Arch. der Kelten* (Fundber. aus Hessen Beiheft I), 34–58

—— 1971 'The old Irish "chariot" ', *Antiq.* 45, 171–77

HARDING, A. 1971 'The earliest glass in Europe', *Arch. Rozh.* 23, 188–200

—— 1975 'Mycenaean Greece and Europe: the evidence of bronze tools and implements', *Proc. Prehist. Soc.* 41, 183–202

—— 1980 'Radiocarbon calibration and the chronology of the European Bronze Age', *Arch. Rozh.* 32, 178–86

—— & WARREN, S. E. 1973 'Early Bronze Age faience beads in Central Europe', *Antiq.* 47, 64–66

HÄRKE, G. H. 1979 *Settlement types and patterns in the West Hallstatt province,* (Brit. Arch. Rep. Int. 57), Oxford

HARRISON, R. J. 1980 *The Beaker Folk: Copper Age archaeology in Western Europe,* London & New York

HARTUCHE, N. 1967 'Un car de lupta descoperit in regiunea Dobrogea', *Apulum* 6, 231–57

HASE, F.-W. VON 1969 *Die Trensen der Früheisenzeit in Italien,* (Prähist. Bronzefunde 16. 1), Munich

HASSELROT, P. & OHLMARKS, A. 1966 *Hällristningar,* Stockholm

HAUDRICOURT, A. G. 1948 'Contribution à la géographie et à l'ethnologie de la voiture', *Rev. de Géorg. Hum. et d'Ethnol.* 1948, 54–64

—— & DELAMARRE, M. J.-B. 1955 *L'homme et la charrue à travers le monde,* Paris

HÄUSLER, A. 1969a 'Die östlichen Beziehungen der schnurkeramischen Becherkulturen', *Veröffentl. Landesmus. Halle* 24, 255–74

—— 1969b [Review of Van der Waals 1964] *Zeitschr. für Anthrop. Z. Archäol.* 3. 141, 142–44

—— 1974 *Die Gräber der älteren Ockergrabkultur zwischen Ural und Dnepr,* Berlin

—— 1976 *Die Gräber der älteren Ockergrabkultur zwischen Dnepr und Karparten,* Berlin

—— 1981 'Zur ältesten Geschichte von Rad und Wagen im nordpontischen Raum', *Ethnog. Arch. Zeitschrift* 224, 581–647

HAWKES, C. 1968 'New thoughts on the Belgae', *Antiq.* 42, 6–16

—— 1974 'Bronze Age Hungary: a review of recent work', *Proc. Prehist. Soc.* 40, 113–17

—— 1977 'Britain and Julius Caesar', *Proc. Brit. Acad.* 63, 125–92

—— & SMITH, M. A. 1957 'On some buckets and cauldrons of the Bronze and Early Iron Age', *Antiq. Journ.* 37, 131–98

HAWORTH, R. 1971 'The horse harness of the Irish Early Iron Age', *Ulster Journ. Arch.* 34, 26–49

HAYEN, H. 1972 'Vier Scheibenräder aus Glum', *Die Kunde* NF 23, 62–86

—— 1973 'Räder und Wagenteile aus nordwestdeutschen Mooren', *Nachrichten aus Niedersachens Urgesch.* 42, 129–76

—— 1978a 'Ein bronzezeitliche Speichenrad', *Arch. Mitt. Nordwestdeutschland* I, 13–14

—— 1978b 'Ausgrabungen am Bohlenweg VI (Pr)', *Jahrb. f. das Oldenburger Münsterland* 81–94

HENCKEN, H. O'N. 1968 *Tarquinia, Villanovans and Early Etruscans,* Cambridge, Mass.

—— 1971 *The earliest European helmets,* Cambridge, Mass.

HENNELL, T. 1934 *Change in the farm,* Cambridge

HODDINOTT, R. F. 1975 *Bulgaria in Antiquity,* London

—— 1981 *The Thracians,* London & New York

HODGEN, M. T. 1939 'Domesday Water Mills', *Antiq.* 13, 261–79

HODSON, F. R. 1964 'La Tène chronology, continental and British', *Univ. Lond. Inst. Arch. Bulletin,* 4, 123–41

HOLSTE, F. 1953 *Die bronzezeitlichen Vollgriffschwerter Bayerns,* Munich

HUNDT, H.-J. & ANKNER, D. 1969 'Die Bronzeräder von Hassloch', *Mitt. Hist. Vereins der Pfalz* 67, (Festschrift 100 — Jahre Hist. Mus. Pfalz), 14–34

HÜTTEL, H-G. 1977 'Altbronzezeitliche Pferdefrensen', *Jahresber. Inst. f. Vorgesch. Univ. Frankfurt AM,* 65–86

IJZEREEF, G. F. 1974 'A medieval jaw-sledge from Dordrecht', *Bericht ROB* 24, 181–84

IVANOV, T. 1972 'Über die Kontinuät der Thrakischen Kultur in der Thrakischen Gebieten während der Römerherrschaft', *Thracia* 159–78

JACKSON, K. M. 1964 *The oldest Irish tradition: a window on the Iron Age,* Cambridge

JACOB-FRIESEN, G. 1969 'Skjerne und Egemose: Wagenteile südlicher Provenienz in Skandinavischen Funden', *Acta Arch.* 40, 122–58

JACOB-FRIESEN, K. H. 1927 'Der Bronzeräderfund von Stade', *Prähist. Zeitschr.* 18, 154–86

JACOBI, G. 1974 *Werkzeug und Gerät aus dem*

Oppidum von Manching (Ausgrab. Manching 5), Wiesbaden

JACOBI, L. 1897 *Das Romerkastell Saalburg,* Hamburg von der Hohe

JACOBSTHAL, P. 1944 *Early Celtic Art,* Oxford

— & LANGSDORFF, A. 1929 *Die Bronzeschnabelkannen,* Berlin

JANKUHN, H. 1957 *Denkmäler der Vorzeit zwischen Nord. und Ostsee,* Schleswig

— 1969 *Deutsche Agrargeschichte I: Vor- und Frühgeschichte,* Stuttgart

JAPARIDZE, O. M. 1960 *Archaeological Excavations at Trialeti 1957–58,* (Georgian with Russian and German summaries). Tbilisi

JAVAKISHVILI, A. I. & GLONTI, L. F. 1962 *Urbnisi I: Archaeological Excavations during 1954–61 at Kvatskhelebi (Tvlepia Kokhi),* (Georgian with Russian summary). Tbilisi

JAZDZEWSKI, K. 1965 *Poland,* London & New York

JELÍNKOVÁ, Z. 1959 'Kostěne součásti koňskeho postroje v Čechach a na Moravě', *Acta Univ. Carolinae,* Phil. et Hist. 3, 1959, 183–93

JENKINS, J. G. 1961 *The English Farm Wagon,* Reading

— 1962 *Agricultural transport in Wales,* Cardiff (Nat. Mus. Wales)

JESSEN, K. 1948 'Om naturalförholdene ved Trelleborg og forbruget af trae ved borgens opførelse', in P. Nørlund, *Trelleborg* (Nord. Fortidsmind. 4, i), 173–79, Copenhagen

JOACHIM, H-E. 1969 'Unbekannte Wagengräber der Mittel bis Spätlatènezeit aus dem Rheinland', *Marburger Beiträge zur Arch. der Kelten (Fundber. aus Hessen Beiheft I),* 84–111

— 1973 'Ein reich ausgestattes Wagengrab der Spätlatènezeit aus Neuwied...', *Bonner Jahrb.* 173, 1–52

— 1978 'Vom keltischen Wagen', *Rheinisches Landesmuseum Bonn: Berichte aus der Arbeit des Mus.* 4, 49–51

— 1979 'Latènezeitliche Siedlungsreste in Mechernich-Antweiler, Kr. Euskirchen', *Bonner Jahrb.* 179, 443–69

JOFFROY, R. 1954 *Le Trésor de Vix,* (Mon. & Mém. Piot, 18.1) Paris

— 1958 *Les sépultures à char du premier âge du fer en France,* Paris

— 1960 'Le bassin et le trépied de Sainte-Colombe (Côte-d'Or)', *Mon. et Mém. Piot* 51, 1–23, Paris

— 1973 'La tombe à char de Berru (Marne)', *Bull. Antiquités Nationales* 5, 45–57

— & BRETZ-MAHLER, D. 1959 'Les tombes à char de la Tène dans l'est de la France', *Gallia* 17, i, 5–35

JOPE, E. M. 1955 'Chariotry and paired-draught in Ireland during the Early Iron Age...', *Ulster Journ. Arch.* 18, 37–44

— 1956 'Vehicles and harness', in C. Singer *et al. A History of Technology* II, 537–62, Oxford

— 1962 'Iron Age brooches in Ireland: a summary', *Ulster Journ. Arch.* 24–25, 25–38

KADYRBAEV, M. K. & MARYASHEV, A. N. 1977 *Naskal'nye izobrasheniya khebta Karatau,* Alma-Ata

KALICZ, N. 1968 *Die Frühbronzezeit in nordost Ungarn,* Budapest

— 1976 'Novaya nakhodka modeli povozki epokhi eneolita iz okrestnostey Budapeshta', *Sov. Arkh.* 1976. 2, 106–17

KARAGEORGHIS, V. 1967 *Excavations in the Necropolis of Salamis I,* Nicosia

— 1969 *Salamis in Cyprus,* London & New York

— 1973 *Excavations in the Necropolis of Salamis III,* Nicosia

KARAKHANIAN, G. H. & SAFIAN, P. G. 1970 *The Rock-carvings of Syunik* (Armenian; Russian and English summaries), Erevan

KASTELIC, J. 1956 *The situla of Vače,* Belgrade

— 1964 *Situlenkunst,* Vienna/Munich

KEEGAN, T. 1978 *The Heavy Horse: its harness and harness decoration,* London

KELLER, J. 1965 *Das keltische Fürstengrab von Reinheim,* Mainz

KHACHATRYAN, T. S. 1963 *Material'naya kul'tura drevnego Artika,* Erevan

KIMMIG, W. 1938 'Vorgeschichtliche Denkmäler und Funde an der Ausoniusstrasse', *Trierer Zeitschr.* 13, 21–79

— 1950 'Ein Wagengrab der frühen Latenezeit von Laumersheim (Rheinpfalz)', *Germania* 28, 38–50

— 1964a 'Seevölkebewegung und Urnenfelderkultur', in R. v. Uslar & K. J. Knarr (edd) *Studien als Alteuropa,* 220–83. Cologne/Graz

— 1964b 'Bronzesitulen aus dem Rheinischen Gebirge...', *Bericht. Röm. Germ. Kom.* 43–44, 31–106

— 1974 'Zum Fragment eine Este-Gefässes von der Heuneburg an der oberen Donau',

Hamburg. Beiträge 2. Arch. 4, 33–102

—— & Rest, W. 1953 'Ein Fürstengrab der späten Hallstattzeit von Kappel am Rhein', *Jahrb. R-G. Zentralmus. Mainz* 1, 179–216

Kirk, G. S. 1964 'The Homeric poems as history', *Cambridge Ancient History* (rev. ed) II, Chap. XXXIXb, Cambridge

—— 1965 *Homer and the epic,* Cambridge

Kitov, G. 1979 *Trakiyskite mogile kray Strelcha,* Sofia

Klein, R. G. 1973 *Ice-Age hunters of the Ukraine,* Chicago

Klejn, L. S. 1963 'Bronze Age Earthen Wheels Models from the Northern Shore of the Black Sea', *Arch. Értesítő* 90, 61–63

Klindt-Jensen, O. 1949 *Foreign Influences in Denmark's Early Iron Age,* Copenhagen

—— 1953 *Bronzekedelen fra Brå,* Aarhus

Klobe, O. 1934 'Die Schnabelkanne vom Dürrnberg bei Hallein, Salzburg', *Wiener Prähist. Zeitschr.* 21, 83–107

Kohl, G. & Quitta, H. 1970 'Berlin Radiocarbon Measurements IV', *Radiocarbon* 12, 2, 400–20

Kohler, E. L. 1980 'Cremations of the Middle Phrygian period at Gordion', in *From Athens to Gordion* (Univ. Mus. Pennsylvania), 65–89, Philadelphia

Kosay, H. Z. 1951 *Alaca-Höyük: Das Dorf Alaca-Höyük: Materalien zur Ethnographie und Volkskunde Anatolien,* Ankara

Kossack, G. 1953a 'Pferdegeschirr aus Gräben der älteren Hallstattzeit Bayerns', *Jahrb. R-G. Zentralmus. Mainz* 1, 111–78

—— 1953b 'Hallstattzeitliches Pferdegeschirr aus Flavia Solva', *Schild von Steier* 2, 49–62

—— 1958 'Kammergräber der Hallstattzeit bei Grosseibstadt', (Unterfranken) in Krämer, W. (ed), *Neue Ausgrabungen in Deutschland,* 121–26, Berlin

—— 1959 *Südbayern während der Hallstattzeit,* Berlin

—— 1970 *Gräberfelder der Hallstattzeit am Main und Fränkischer Saale,* Kalmünz

—— 1971 'The construction of the felloe in Iron Age spoked wheels', in J. Boardman *et al.* (edd), *The European community in later prehistory,* 143–63, London & New York

Kostrewski, J. 1936a 'A unique discovery: a fortified Polish village of about 600 bc', *Illus. Lond. News* 8 Aug. 1936, 243–44

—— 1936b 'Osada bagienna w Biskupinie, v poz. zninskim', *Przegląd Arch.* 5, 121–40

Kothe, H. 1953 'Verbreitung und Alter der Stangenschleite', *Eth.-Arch. Forschungen* 1

Koutecky, D. 1966 'Das Bylaner Fürstengrab aus Rvenice bei Postoloprty', *Arch. Rozh.* 18, 12–21

—— 1968 'Grossgräber, ihre Konstruktion, Grabritus, und soziale struktur der Bevölkerung der Bylaner Kultur', *Pamatky Arch.* 59, 400–87

Kowalczyk, J. 1970 'The Funnel Beaker culture', in T. Wislanski (ed), *The Neolithic in Poland,* 144–77, Wroclaw

Krämer, W. 1952 'Neue Beobachtungen zum Grabrauch der mittleren Hallstattzeit in Südbayern', *Bayer. Vorgesch.* 18/19, 152–89

—— 1964 'Latènezeitliche Trensenanhänger in Omegaform', *Germania 42,* 250–57

—— & Schubert, F. 1979 'Zwei Achsnagel aus Manching: Zeugnisse keltischer Kunst der Mittellatènezeit', *Jahrb. Deutsch. Arch. Inst.* 94, 366–89

Kravets, V. P. 1951 'Glinanye tripol'skie model'ki sanochek i chelna v kollektsyach L'vovskogo istoricheskogo muzeya', *Kratkie Soobshch.* 39, 127–31

Kromer, K. 1959 *Das Gräberfeld von Hallstatt,* Florence

—— 1963 *Hallstatt: Katalog zur Ausstellung,* Vienna

Kruk, J. & Milisauskas, S. 1977 'Radiocarbon-Datierung aus Bronocice...', *Arch. Korrespond.* 7, 149–56

—— 1981 'Chronology of Funnel-Beaker, Baden-like, and Lublin-Volynian settlements at Bronocice, Poland', *Germania* 59, 1–19

Kruta, V. 1978 'Celtes de Cispadane et Transalpins aux IVe et IIIe siècles avant notre ère: données archéologiques', *Studi Etruschi* 46, 149–74

—— 1979 'Duchcov-Münsingen: nature et diffusion d'une phase Laténienne', in P-M. Duval & Kruta, *Les mouvements celtiques du Ve au Ier siècle avant notre ère,* Paris

Kubach, W. 1977 'Zum Beginn der bronzezeitlichen Hügelgräberkultur in Süddeutschland', *Jahresb. Inst. f. Vorgesch. Univ. Frankfurt AM* 119–63

Kuftin, B. A. 1941 *Arkheologicheskie raskopki v Trialeti,* Tbilisi

—— 1948 *Arckheologicheskie raskopki 1947 g. v chalkinskom raione,* Tbilisi

KUNWALD, G. 1970a 'Der Moorfund im Rappendam auf Seeland', *Prähist. Zeitschr.* 16, 42–88

—— 1970b 'Der Moorfund im Rappendam, Seeland, Danemark', in H. Jankuhn (ed), *Vorgeschichtliche Heiligtümer und Opferplätze in Mittel, und Nordeuropa* 100–18, Göttingen

KUZMINA, E. E. 1971 *Earliest evidence of horse domestication and spread of wheeled vehicles...* (Rapports & Comm.... de l'URSS, VIII Cong. Internat. Sci. Prehist. & Protohist.) Moscow

—— 1980 'Eshche raz o diskovidiykh psalikh evraziyskikh stepy', *Kratkie Soobshch.* 161, 8–21

DE LAET, S. J. & GLASBERGEN, W. 1959 *De voorgeschiedenis der Lage Landen,* Groningen

LANE, R. H. 1935 'Waggons and their ancestors', *Antiq.* 9, 140–50

LANG, D. M. 1966 *The Georgians,* London & New York

LANTING, J. N. & VAN DER WAALS, J. D. 1972 'British beakers as seen from the Continent', *Helinium,* 12, 20–46

LANTING, J. N. *et al.* 1973 'C^{14} chronology and the Beaker problem', *Helinium* 13, 38–58

LEMOINE, R. 1906 'Sépulture à char découverte... à Châlons-sur-Marne', *Mem. Soc. d'Agric., Commerce, Sciences et Arts du Département de la Marne 1904–05,* 123–50

LERCHE, G. 1968 'The radiocarbon-dated Danish ploughing implements', *Tools and Tillage* I, i, 56–58

LESKOV, A. M. 1964 'Drevneyshie rogovye psalii iz Trakhtemirova', *Sov. Arkh.* 1964 1, 299–303

LICHARDUS, J. 1980 'Zur Funktion der Geweihspitzen des Typus Ostdorf', *Germania* 58, 1–24

LITTAUER, M. A. 1968a 'The function of the yoke saddle in ancient harnessing', *Antiq.* 42, 27–31

—— 1968b 'A 19th and 20th Dynasty motif on Attic black-figured vases?', *Amer. Journ. Arch.* 72, 150–52

—— 1969 'Bits and pieces', *Antiq.* 43, 289–300

—— 1976a 'New light on the Assyrian chariot', *Orientalia* 45, Fasc. 1–2, 217–26

—— 1976b 'Reconstruction questioned', *Archaeology* 29, 212

—— 1977a 'Rock-carvings of chariots in Trans-

caucasia, Central Asia and Outer Mongolia', *Proc. Prehist. Soc.* 43, 243–62

—— 1977b 'Review of Greenhalgh 1973', *Class. Phil.* 72, 363–65

—— & BAHN, P. 1980 *Horse sense, or nonsense? Antiq.* 54, 139–42

—— & CROUWEL, J. 1973 'Evidence for horse bits from Shaft-Grave IV at Mycenae?', *Prähist. Zeitschr.* 48, 207–13

—— 1974 'Terracotta models as evidence for vehicles with tilts in the ancient Near East', *Proc. Prehist. Soc.* 40, 20–36

—— 1977a 'The origin and diffusion of the cross-bar wheel?', *Antiq.* 51, 95–105

—— 1977b 'Chariots with Y-poles in the ancient Near East', *Arch. Anzeiger* 1–8

—— 1979a *Wheeled vehicles and ridden animals in the ancient Near East.* Leiden/Cologne

—— 1979b 'An Egyptian wheel in Brooklyn', *Journ. Egypt. Arch.* 65, 107–20

LLOYD, S. 1978 *The archaeology of Mesopotamia,* London & New York

LOMBORG, E. 1959 'Donauländische Kulturbeziehungen und die relative Chronologie der frühen nordischen Bronzezeit', *Acta Arch.* 30, 51–146

LORIMER, H. 1903 'The country cart in ancient Greece', *Journ. Hell. Studies* 23, 132–51

LU, G-D. *et al.* 1959 'The wheelwright's art in ancient China: I. the invention of "dishing"', *Physis* I. 2, 103–26

LUCAS, A. T. 1952 'A Block-wheel Car from Co. Tipperary', *Journ. Royal Soc. Ant. Ireland* 82, 135–44

—— 1953 'Block-wheel Car from Slievenamon, Co. Tipperary', *Journ. Royal Soc. Ant. Ireland* 83, 100

—— 1972 'Prehistoric Block-wheels from Doogarymore, Co. Roscommon and Timahoe East, Co. Kildare', *Journ. Royal Soc. Ant. Ireland,* 102, 19–48

LUCKE, W. & FREY, O-H. 1962 *Die situla in Providence (Rhode Island),* Berlin

LUMLEY, H. DE, *et al.* 1976 *Vallée des Merveilles,* (Livret-Guide de l'excursion C 1.) Nice

MACDONALD, G. & PARK, A. 1906 'The Roman forts on the Bar Hill...', *Proc. Soc. Ant. Scot.* 40, 403–546

MCGRAIL, S. 1978 *Logboats of England and Wales,* (British Arch. Rep. 51) National Maritime Museum, Greenwich

MACGREGOR, M. 1962 'The Early Iron Age metalwork hoard from Stanwick, N. R. Yorks.', *Proc. Prehist. Soc.* 28, 17–57

—— 1976 *Early Celtic Art in North Britain,* Leicester

MACINTOSH, J. 1974 'Representations of furniture on the frieze plaques from Poggio Civitate (Murlo)', *Mitt. Deutsch. Arch. Inst. Rome* 81, 15–40

McLEOD, W. 1970 *Composite bows from the tomb of Tutankhamūn,* Oxford

MACQUEEN, J. G. 1975 *The Hittites and their contemporaries in Asia Minor,* London & New York

MADSEN, T. 1979 'Earthen long barrows and timber structures: aspect of the early neolithic mortuary practice of Denmark', *Proc. Prehist. Soc.* 45, 301–20

MAIER, F. 1958 'Zur Herstellungstechnik und Zierweise der späthallstatt-zeitlichen Gürtelbleche Südwestdeutschlands', *Bericht Rom-Germ. Komm.* 39, 131–249

MAIER, R. A. 1961 'Neolithische Tierknochen-Idole und Tierknochen-Anhänger Europas', *Bericht Rom-Germ. Komm.* 42, 171–305

MALMROS, C. & TAUBER, H. 1975 'Kulstof-14 dateringer af dansk enkeltgravskultur', *Aarbøger* 78–95

MARGABANDHU, C. 1973 'Technology of transport vehicles in early India', in D. P. Agrawal & A. Ghosh (edd), *Radiocarbon and Indian Archaeology,* 182–89, Bombay

MARIËN, M-E. 1958 *Trouvailles du Champ d'urnes et des Tombelles hallstattiennes de Court-Saint-Étienne,* Brussels

—— 1961 *La période de la Tène en Belgique: le groupe de la Haine,* Brussels

DE MARINIS, R. 1969 'Rivisione de vecchi scavi nella necropoli della Ca'Morta', *Revista Arch. Como* 150–51, 99–200

MARSTRANDER, S. 1963 *Østfolds jordbruksristninger,* Oslo

MARTIROSYAN, A. A. 1964 *Armeniya v epokhy bronzy i rannego sheleza,* Erevan

MAXWELL-HYSLOP, R. 1956 'Notes on some distinctive types of bronzes from Populonia, Etruria', *Proc. Prehist. Soc.* 22, 126–42

—— 1971 *Western Asiatic Jewellery c. 3000–612 BC,* London

MAYRHOFER, M. 1966 *Die Indo-Arier im alten Vorderasien,* Wiesbaden

MEDUNOVA-BENEŠOVÁ, A. 1972 *Jevišovice-Starý Zámek. Schicht B. Katalog des Funde,* Brno

MEGAW, J. V. S. 1970 *Art of the European Iron Age,* Bath

MELLAART, J. 1979 'Egyptian and Near Eastern chronology: a dilemma?', *Antiq.* 53, 6–18

VON MERHART, G. 1956 'Über blecherne Zierbuckel (Faleren)', *Jahrb. R. G. Z. Mus. Mainz* 3, 28–116

MERPERT, N. YA. 1961 'L'Enéolithique de la zone steppique de la partie européenne de l'URSS', in J. Böhm and S. de Laet (edd), *L'Europe à la Fin de l'Age de la Pierre,* 176–92, Prague

—— 1974 *Drevneyshie skotovod'i Volzhko-Ural'skogo mezhdurech'ya*

MIKAELIAN, G. H. *1968 Cyclopean fortresses in the basin of Lake Sevan,* (Armenian: Russian and English summaries), Erevan

MILISAUSKAS, S. & KRUK, J. 1977 'Archaeological excavations at the Funnel Beaker (TRB) site of Bronocice', *Arch. Polona* 18, 205–28

—— 1978 'Bronocice: a neolithic settlement in southeastern Poland', *Archaeology* 31, no. 6, 43–52

MINNS, E. H. 1913 *Scythians and Greeks,* Cambridge; repr. New York

MNATSAKANYAN, A. O. 1957 'Raskopki kurganov na poberezh'e oz. Sevan v 1956 g.' *Sov. Arkh.* 1957, 2., 146–53

—— 1960a *Bronze Age Culture on Lake Sevan Coast in Armenia,* (Papers pres. by USSR Deleg. to XXV Int. Cong. Orientalists), Moscow

—— 1960b 'Drevnie povozki iz kurganov bronzovogo veka na poberezh'e oz. Sevan' *Sov. Arkh.* 1960, 2., 139–52

MODDERMAN, P. J. R. 1964a 'The neolithic burial vault at Stein', *Analecta Praehist. Leidensia,* 1, 3–16

—— 1964b 'The chieftain's grave of Oss reconsidered', *Bull. ver. bevord. der kennis antieke beschaving* 39, 57–62

MODRIJAN, W. 1950 'Die figurale Bleiplastik von Frög', *Carinthia I,* 140, 91–120

MONTELIUS, O. 1922 *Swedish Antiquities: I. The Stone Age and the Bronze Age,* Stockholm

MOOREY, P. R. S. 1971 *Catalogue of the ancient Persian bronzes in the Ashmolean Museum,* Oxford

MOREAU, F. 1877 *Collection Caranda ... Album*

des principaux objets, St Quentin

MOREL, L. 1876 *Album des cimitières de la Marne,* Châlons-sur-Marne

—— 1877 *La Champagne Souterraine,* Châlons-sur-Marne

MORGAN, J. DE 1889 *Mission Scientifique en Caucase I,* Paris

MOSZOLICS, A. 1953 'Mors en bois de cerf sur le territoire du bassin des Carpathes', *Acta Arch. Hung.* 3, 69–109

—— 1956 'Azhuriye povozochiye nakladki pozdibronzovoy epokhi', *Acta Arch. Hung.* 7, 1–14

—— 1960 'Die Herkunftsfrage der ältesten Hirschgeweitrensen', *Acta Arch. Hung.* 12, 125–35

—— 1965 'Goldfunde des Depotfundhorizontes von Hajdúsámson', *Bericht R-G. Komm.* 46–47, 1–76

—— 1969 'La stratigraphie, base de la chronologie de l'âge du bronze de la Hongrie', *Origini* 3, 275–94

MOVSHA, T. G. 1965 'Arkhaicheskie kruglye psalii s shipami', *Materialy* 130, 201–4

MÜLLER, S. 1903 'Solbilledet fra Trundholm', *Nord. Fortidsmin.* I. 6, 303–25

—— 1907 'Et vognhjul', *Aarbøger* 2 R., 22, 75–9

—— 1920 'Nye fund og former, Stenaldern', *Aarbøger* 3 R., 10, 90–91

MÜLLER-KARPE, H. 1956 'Das Urnenfeldzeitliche Wagengrab von Hart a.d. Alz', *Bayerische Vorgeschichtsblätter* 21, 46–75

—— 1959 *Beiträge zur Chronologie der Urnenfelderzeit nördlich und südlich der Alpen,* Berlin

—— 1961 *Die Vollgriffschwerter der Urnenfelderzeit aus Bayern,* Munich

—— 1962a 'Die Metalbeigaben der früheisenzeitlichen Kerameikos-Gräber', *Jahrb. Deutsch. Arch. Inst.* 77, 59–129

—— 1962b 'Die späthallstattzeitliche Tierfibel von Kastlhof...', *Schriften. v. Bayer. Landesgesch.* 62, 101–5

MUNN-RANKIN, M. 1980 'Mesopotamian chronology: a reply to James Mellaart', *Antiq.* 54, 128–29

MUNRO, R. 1890 *The Lake Dwellings of Europe,* London

MURRAY, J. 1970 *The first European agriculture,* Edinburgh

MUSTY, J. & MACCORMICK, A. G. 1973 'An Early Iron Age wheel from Holme Pierrepont,

Notts.', *Antiq. Journ.* 53, 275–77

NASH, D. 1976 'Reconstructing Poseidonios's Celtic ethnography: some considerations', *Britannia* 7, 111–26

NAUE, J. 1887 *Die Hügelgräber zwischen Ammer- und Staffelsee,* Stuttgart

DE NAVARRO, J. M. 1972 *The finds from the site of La Tène: I. Scabbards and the swords found in them,* London

NEEDHAM, J. 1970 *Clerks and craftsmen in China and the west,* Cambridge

NEEDHAM, J. & LU, G. 1960 'Efficient Equine Harness: the Chinese Inventions', *Physis* 2, 121–62

NEEDHAM, S. & LONGLEY, D. 1980 'Runnymede Bridge, Egham: a Late Bronze Age riverside settlement', in J. Barrett & R. Bradley (edd), *Settlement and Society in the British Later Bronze Age* (Brit. Arch. Reports 83), 397–436, Oxford

NÉMEJCOVA-PAVÚKOVÁ, V. 1973 'Zu Ursprung und Chronologie der Boleraz-Gruppe', in Kalousek F. & Budinsky-Krička, V. (edd), *Symposium über die... Badener Kultur,* 297–316, Bratislava

NEUSTUPNÝ, E. & J. 1961 *Czechoslovakia before the Slavs,* London & New York

NEUSTUPNY, J. 1963 'The Bell Beaker culture in Bohemia and Moravia', *A Pedro Bosch Gimpera...,* 331–44. Mexico

—— et al. 1960 *Pravěk Čechoslovenska,* Prague

NICHOLSON, S. M. 1980 *Catalogue of the prehistoric metalwork in Merseyside County Museums,* Liverpool

NIKOLOV, D. 1961 'Trakiyski kolesnitsi kray Stara Zagora', *Arkheologiya* 3, 8–18

NOBIS, G. 1971 *Von Wildpferd zum Hauspferd* (Fundamenta Reihe B, Bd.6), Cologne

NOËTTES, L. DES 1931 *L'Attelage et le Cheval de Selle à travers les âges,* Paris

NORTHDURFTER, J. 1979 *Die Eisenfunde von Sanzeno im Nonsburg,* Mainz

NOVOTNY, B. 1972 'Výskum vo Vel'kej Lomnici, okr. Poprad', *Arch. Rozh.* 24, i, 10–17

O'CONNOR, B. 1975 'Six prehistoric phalerae in the London Museum', *Antiq. Journ.* 55, 215–26

OLMSTED, G. S. 1979 *The Gundestrup cauldron,* Brussels (Collection Latomus 162)

ONDRÁCEK, J. 1961 'Beiträge zur Erkenntnis der Glockenbecherkultur in Mähren', *Pamatky*

Arch. 52, 149

PALLOTTINO, M. 1975 *The Etruscans,* (rev. ed.) London

PÁRDUCZ, M. 1952 'Le cimitière hallstattien de Szentes-Vekerzug', *Acta Arch. Hung.* 2, 143–69

—— 1954 'Le cimitière hallstattien de Szentes-Vekerzug II', *Acta. Arch. Hung.* 4, 25–89

—— 1955 'Le cimitière hallstattien de Szentes-Vekerzug III', *Acta. Arch. Hung.* 6, 1–18

PARET, O. 1935 *Das Fürstengrab der Hallstattzeit von Bad Cannstatt,* (Supplement to *Fundber. aus Schwaben* NF 8). Stuttgart

PAULI, L. 1980 *Die Kelten in Mitteleuropa: Salzburger Landesausstellung 1980,* Salzburg

PEGGE, S. 1795 'Observations on the chariots of the antient *Britains*', *Arch.* 7, 211–13

PENNINGER, E. 1972 *Der Dürrnberg bei Hallein* I, Munich

PERONI, R. *et al.* 1975 *Studi sulla chronologia della civiltà di Este e Golasecca,* Florence

PESCHECK, C. 1972 'Ein reicher Grabfund mit Kesselwagen aus Unterfranken', *Germania* 50, 29–56

PESKE, L. 1976 'Osteologic finds from Praha-Michle', *Arch. Rozh.* 28, 159

PETERSEN, H. 1888 *Vognfunde i Dejbjerg Praestergaadsmose ved Ringjøbing 1881–1883,* Copenhagen

PETRIE, W. M. F. 1917 *Tools and Weapons,* London

PHILIPSON, J. 1882 *Harness: as it has been, as it is, and as it should be,* Newcastle-upon-Tyne & London (reprint 1971)

PIGGOTT, C. M. 1953 'An Iron Age barrow in the New Forest', *Antiq. Journ.* 33, 14–21

PIGGOTT, S. 1949a 'A wheel of Iron Age type from Co. Durham', *Proc. Prehist. Soc.* 15, 191

—— 1949b 'An Iron Age yoke from Northern Ireland', *Proc. Prehist. Soc.* 15, 192–93

—— 1950 *Prehistoric India,* Harmondsworth

—— 1952 'Celtic chariots on Roman coins', *Antiq.* 26, 87–88

—— 1953 'A Late Bronze Age hoard from Peebleshire', *Proc. Soc. Ant. Scot.* 87, 175–86

—— 1957 'A Tripartite Disc Wheel from Blair Drummond, Perthshire', *Proc. Soc. Ant. Scot.* 90, 238–41

—— 1959 'A Late Bronze Age wine trade?', *Antiq.* 33, 122–23

—— 1964 'Iron, Cimmerians and Aeschylus', *Antiq.* 38, 300–3

—— 1965 *Ancient Europe,* Edinburgh

—— 1966 'Mycenae and barbarian Europe: an outline survey', *Sbornik Narod. Muz. v Praze* Series A. 20, 117–25

—— 1968 'The earliest wheeled vehicles and the Caucasian evidence', *Proc. Prehist. Soc.* 34, 266–318

—— 1969a 'Early Iron Age "Horn-Caps" and yokes', *Antiq. Journ.* 49, 378–81

—— 1969b 'Review of V. Karageorghis 1967', *Antiq.* 43, 160–61

—— 1974 'Chariots in the Caucasus and in China', *Antiq.* 48, 16–24

—— 1975 'Bronze Age chariot-burials in the Urals', *Antiq.* 49, 289–90

—— 1976 'Summing up of the Colloquy', in P. M. Duval & C. Hawkes (edd) *Celtic art in Ancient Europe: five protohistoric centuries,* 281–89, London & New York

—— 1978a 'Chinese chariotry: an outsider's view', in P. Denwood (ed) *The Arts of the Eurasian steppelands* (Colloquies on Art and Archaeology in Asia no. 7), 32–51, London

—— 1978b '"Royal Tombs" reconsidered', *Prace i Materialy Muz. Arch. i Ethno. W Lodz* 25, 293–301

—— 1978c 'Nemeton, Temenos, Bothros: sanctuaries of the ancient Celts', *I Celti e la loro cultura nell'epoca pre-Romana e Romana nella Britannia* (Accad. Naz. dei Lincei) 37–54, Rome

—— 1979 '"The First Wagons and Carts": twenty-five years later', *Univ. Lond. Inst. Arch. Bulletin* 16, 3–17

PIJOAN, J. 1953 *Summa Artis VI: El arte prehistorico Europeo,* Madrid

PITTIONI, R. 1954 *Urgeschichte des Österreichischen Raum,* Vienna

PLEINER, R. 1959 'Bylaner Fürstengräber in Lovosice', *Arch. Rozh.* 11, 653–60

—— 1980 'Early iron metallurgy in Europe', in T. A. Wertime & J. D. Muhly, (edd), *The Coming of the Age of Iron,* 375–415, New Haven & London

—— & MOUCHA, V. 1966 'Sépultures princières de Lovosice (Bohème)', in J. Filip (ed) *Investigations archéologiques en Tchecoslovakie,* 178–79, Prague

PODBORSKY, V. 1967 'Der Štramberker Dolch mit kreuzförmigen Griff...', *Arch. Rozh.* 19,

194–220

Popova, T. B. 1963 *Dol'meny stanitsy Novosvobod-noy,* Moscow

Porada, E. 1965 'The relative chronology of Mesopotamia Part I. Seals and Trade (6000–1600 BC)', in R. W. Ehrich (ed), *Chronologies in Old World Archaeology* 133–200, Chicago & London

von Post, L. *et al.* 1939 *Ein eisenzeitliches Rad aus dem Filaren-See in Södermanland, Schweden,* Stockholm

Potratz, J. A. H. 1966 *Die Pferdetrensen des alten Orient,* Rome

Powell, T. G. E. 1950 'A late Bronze Age hoard from Welby, Leicestershire', *Arch. Journ.* 105, 27–40

—— 1960 'Megalithic and other art: centre and west', *Antiq.* 34, 180–90

—— 1963a 'The inception of the final Bronze Age in Middle Europe', *Proc. Prehist. Soc.* 29, 214–34

—— 1963b 'Some implications of chariotry', in I. Ll. Foster & L. Alcock (edd), *Culture and Environment,* 153–69, London

—— 1971a 'The introduction of horse-riding to temperate Europe: a contributory note', *Proc. Prehist. Soc.* 37, 1–14

—— 1971b 'From Urartu to Gundestrup: the agency of Thracian metal-work', in Boardman, J. *et al.* (edd) *The European Community in later prehistory* 183–210, London

—— 1976a 'South-western Peninsular chariot stelae', in J. V. S. Megaw (ed) *To Illustrate the Monuments* 164–9, London & New York

—— 1976b 'The inception of the Iron Age in temperate Europe', *Proc. Prehist. Soc.* 42, 1–14

—— 1980 *The Celts,* (2nd ed.) London & New York

—— *et al.* 1971 'Excavations in Zone VII peat at Storrs Moss', Lancashire, England, 1965–67, *Proc. Prehist. Soc.* 37, 112–37

Protsch, R. & Berger, R. 1973 'Earliest radiocarbon dates for domesticated animals', *Science* 179, 235–39

Pryor, F. 1980 *A catalogue of British and Irish prehistoric bronzes in the Royal Ontario Museum,* Toronto

Quitta, H. 1967 'The C14 chronology of the central and southeast European neolithic', *Antiq.* 41, 263–70

Rackham, O. 1976 *Trees and woodland in the British landscape,* London

—— 1977 'Neolithic woodland management in the Somerset Levels', in Coles *et al. Somerset Levels Papers.* 3, 3, 65–71

Raddatz, K. 1967 *Das Wagengrab der jüngeren vorrömischen Eisenzeit von Husby, Kreis Flensburg,* Neumünster

Radnóti, A. 1958 'Zur Frage der Beschläge von Brünn-Malmeritz (Brno-Maloměřice)', *Germania* 36, 28–35

Raftery, J. 1944 'The Turoe stone and the Rath of Feerwore', *Journ. Royal Soc. Ant. Ireland* 74, 23–52

Randall-MacIver, D. 1927 *The Iron Age in Italy,* Oxford

Rathje, A. 1979 'Oriental imports in Etruria in the eighth and seventh centuries BC: their origins and implications', in D. & F. R. Ridgway (edd), *Italy Before the Romans,* 145–83, London

Rausing, G. 1967 *The Bow: some notes on its origin and development,* Lund

Reed, R. 1972 *Ancient skins, parchments and leathers,* London & New York

Renfrew, C. 1973 *Before civilization,* London

Rest, W. 1948 'Das Grabhügelfeld von Bell im Hunsrück', *Bonner Jahrb.* 148, 133–89

—— & Röder, J. 1941 'Neue Wagengräber bei Kärlich, Landkreis Koblenz', *Bonner Jahrb.,* 146, 288–99

Richardson, N. J. & Piggott, S. n.d. 'Hesiod's ἄμαξα: text and technology', (*Journ. Hellenic Stud.* forthcoming)

Ridgway, D. 1979 ' "Editorial" to Part V', in D. & F. R. Ridgway (edd), *Italy before the Romans,* 415–18, London/New York/San Francisco

—— 1981 *The Etruscans,* (Univ. Edinburgh Dept. Arch. Occ. Paper 6) Edinburgh

—— & Ridgway, F. R. 1976 'From Ischia to Scotland: better configurations in old world protohistory', in J. V. S. Megaw (ed) *To Illustrate the Monuments* 146–52, London & New York

Ridgway, F. R. 1979 'The Este and Golasecca Cultures: a chronological guide', in D. & F. R. Ridgway (edd), *Italy before the Romans,* 419–87, London/New York/San Francisco

Riek, G. & Hundt, H. J. 1962 *Der Hohmichele,* (R-G. Komm. Forsch. 25), Berlin

Rieth, A. 1940 'Zur Technik antiker und prähistorische Kunst: das Holzdrechseln', *IPEK* 13/14, 85–107

—— 1950 'Werkstattkreise und Herstellungstechnik der hallstattzeitlichen Tonnenarmbänder', *Zeitschr. Schweiz. Arch. und Kunstgesch.* 11, 1–16

—— 1955 'Antike Holzgefasse', *Arch. Anzieger,* 1–26

Rieth, A. & Langenbacher, K. n.d. *Die Entwicklung der Drehbank,* Stuttgart/Cologne

Rivet, A. L. F. 1979 'A note on scythed chariots', *Antiq.* 53, 130–32

—— 1980 'Celtic names and Roman places', *Britannia* 11, 1–19

—— & Smith, C. 1979 *The Placenames of Roman Britain,* London

Roiz, J. P. G. & García, E. M. O. 1978 *Excavationes en la necropolis de 'La Joya', Huelva,* (Excav. Arq. Espana 96)

Rollas, A. N. 1961 'Les nouvelles acquisitions et un modèle de voiture en terre cuite', *Ann. Arch. Mus. Istanbul* 10, 130–31

Rolle, R. 1979 *Totenkult der Skythen: I. Das Steppengebiet,* Berlin/New York

—— 1980 *Die Welt der skythen,* Luzern/Frankfurt

Rosellini, I. 1836 *I monumenti dell'egitto e della Nubia II (Monumenti civili) III; Tavole (Monumenti civili),* Pisa

Rostholm, H. 1977 'Neolitiske Skivehjul fra Kideris og Bjerregarde i Midtjylland', *Kuml* 185–222

Rowlett, R. M. 1968 'The Iron Age north of the Alps', *Science* 161, 123–34

Rudenko, S. I. 1970 *Frozen tombs of Siberia,* (trans. M. W. Thompson), London

Rumyantsev, E. A. 1961 'Restavatsiya i konservatsiya drevnikh derevyannyk povozok iz Zakavkaz'ya i Altaya' *Sov Arkh.* 1., 236–42

Ruoff, U. 1978 'Die schnürkeramikischen Räder von Zürich "Pressehaus"', *Arch. Korrespondenzblatt* 8, 275–83

Rynne, E. 1962 'Late Bronze Age rattle-pendants from Ireland', *Proc. Prehist. Soc.* 28, 383–85

Sandars, N. K. 1957 *Bronze Age cultures in France,* Cambridge

—— 1962 'Wheelwrights and smiths', *Celticum* 3, 403–8

—— 1968 *Prehistoric Art in Europe,* Harmondsworth

—— 1971 'From Bronze Age to Iron Age: a sequel to a sequel', in Boardman J. *et al.* (edd), *The European Community in later prehistory,* 3–29, London

—— 1976 'Orient and orientalizing: recent thoughts reviewed', in P. M. Duval & C. Hawkes (edd), *Celtic art in Ancient Europe: five protohistoric centuries,* 41–55, London & New York

—— 1978 *The Sea Peoples,* London & New York

Saronio, P. 1969 'Revisione e presentazione di alcune tombe dell'età del ferro della necropoli della Ca' Morta...', *Rivista Arch. Como* 150–51, 47–98

Sauer, C. O. 1972 *Seeds, Spades, Hearths and Herds,* Cambridge, Mass.

Sauter, M. R. 1976 *Switzerland,* London & New York

Savory, H. N. 1958 'The late Bronze Age in Wales: some new discoveries and new interpretations', *Arch. Camb.* 1958, 3–63

—— 1971a *Excavations at Dinorben, 1965–69* (Nat. Mus. Wales), Cardiff

—— 1971b 'A Welsh Bronze Age hill fort', *Antiq.* 45, 251–61

—— 1976 *Guide Catalogue of the Early Iron Age collections* (Nat. Mus. Wales), Cardiff

Schaaf, U. 1969 'Versuch einer regionalen Gliederung frühlatènezeitlicher Fürstengräber', *Marburger Beiträge zur Arch. der Kelten* (Fundber. aus Hessen Beiheft I), 187–202

—— 1973 'Frühlatènezeitliche Grabfunde mit Helmen vom Typ Berru', *Jahrb. R.G.Z. Mus. Mainz* 20, 81–106

Schiek, S. 1954 'Das Hallstattgrab von Vilsingen', in W. Kimmig (ed) *Festschrift für Peter Goessler,* 150–67, Stuttgart

—— 1962 'Ein Brandgrab der Frühen Urnenfelderkultur von Mengen', Kr. Saulgau, *Germania* 40, 130–41

Schindler, R. 1970 'Das Wagengrab von Gransdorf (Kreis Wittlich)', *Trierer Zeitschr.* 33, 19–34

Schmid, W. 1934 *Der Kultwagen von Strettweg,* Leipzig

Schmidt, K. H. 1967 'Keltisches Wortgut in Lateinischen', *Glotta* 44, 151–74

Schneider, G. 1952 'Das vorgeschichtliche Wagenrad von Aulendorf', *Vorzeit am Bodensee,* 13–17

Schroller, H. 1933 *Die Stein- und Kupferzeit Sie-*

benbürgens, Berlin

SCHÜLE, W. 1969a *Die Meseta-Kulturen der Iberischen Halbinsel* (Madrider Forschungen, 3), Berlin

—— 1969b 'Glockenbecher und Hauspferde', in J. Boessneck (ed) *Archäologie und Biologie,* 88–93 Wiesbaden

SEEWALD, O. 1939 *Der Vogelwagen von Glasinac* (Praehistorica 4), Leipzig

SEHESTED, F. 1878 *Fortidsminder og Oldsager frå Egnen an Broholm,* Copenhagen

SEMYONSTOV, A. A. *et al.* 1972 'Radiocarbon dates of the Institute of Archaeology III', *Radiocarbon* 14.2., 336–67

SERRA RAFOLS, J. 1948 'Carrito iberico de bronce, del museo de Granollers', *Archivo Esp. Arque.* 21, 378–91

SHEFTON, B. B. 1979 *Die 'Rhodischen' Bronzekannen,* Mainz

SHEPPARD, T. 1941 'The Parc-y-meirch hoard, St George parish, Denbighshire', *Arch. Camb.* 96, 1–10

SHERRATT, A. 1981 'Plough and pastoralism: aspect of the Secondary Products Revolution', in I. Hodder *et al.* (edd) *Patterns of the Past: Studies in honour of David Clarke,* 261–305, Cambridge

SHORT, R. V. 1975 'The evolution of the horse', *Journ. Reprod. Fert.* Supplement 23, 1–6

SHRAMKO, B. A. 1971 'Der Hakenpflug der Bronzezeit in der Ukraine', *Tools & Tillage,* I. 4

SINITSYN, I. V. 1948 'Pamyatniki predskifskoy epokhi v stepyakh nizhnego Povolzh'ya', *Sov. Arkh.* 1948, 148–60

—— & ERDINEV, E. 1966 *Novye arkheologicheskie pamyatniki na territorii Kalmyshkoy ASSR,* Elista

SMIRNOV, K. F. 1961 'Arkheologicheskie dannye o drevnikh usadnikakh Povolshsko-Uralskikh stepy', *Sov. Arkh.* 1961. 1, 46–72

—— & KUZMINA, E. E. *1977 Proiskhoshdenie Indoirantsev v svete noveyshchikh arkheologicheskie otkrytiy,* Moscow

SNODGRASS, A. M. 1964 *Early Greek armour and weapons,* Edinburgh

—— 1965 'Barbarian Europe and Early Iron Age Greece', *Proc. Prehist. Soc.* 31, 229–40

—— *1967 Arms and armour of the Greeks,* London & New York

—— 1971a *The Dark Age of Greece,* Edinburgh

—— 1971b 'The First European body-armour', in J. Boardman *et al.* (edd), *The European Community in later prehistory,* 33–50, London

—— 1980 'Iron and early metallurgy in the Mediterranean', in T. A. Wertime & J. D. Muhly (edd), *The Coming of the Age of Iron,* 335–74, New Haven & London

—— 1973 'Bronze "Phalara": a review', *Hamburger Beiträge zur Arch.* 3.i., 41–50

SOUDSKA, E. 1976 'Hrob 196 z Manětína-Hrádku a další hroby s dvoukolovými vozy v Čechách', *Arch. Rozh.* 28, 625–54

SPINDLER, K. 1971 *Magdalenenberg I,* Villingen

SPRUYTTE, J. 1977 *Études experimentales sur l'attelage,* Paris

—— 1980 'Le véhicule a un essieu à deux brancards, ou à deux timons', *Almogaren* 9–10, 53–76

SPURNÝ, V. 1961 'Neue Forschungen über die Anfange der Lausitzer Kultur in Mahren', in A. Točik (ed), *Kommission für das Äneolithikum… Nitra 1958,* 125–37, Bratislava

STARY, P. F. 1979a 'Foreign influences in Etruscan arms and armour: 8th to 3rd centuries BC', *Proc. Prehist. Soc.* 45, 179–206

—— 1979b 'Feuerböcke und Bratspiesse aus eisenzeitlichen Gräbern der Appenin-Halbinsel', *Kleine Schriften Vorgesch. Seminar Marburg* 5, 40–61

—— 1979c 'Keltische Waffen auf der Appenin-Halbinsel', *Germania* 57, 99–110

—— 1980 'Zur Bedeutung und Funktion zweirädiger Wagen während der Eisenzeit in Mittelitalien', *Hamburger Beiträge zur Arch.* 7, 7–21

STEAD, I. M. 1965a *The La Tène cultures of Eastern Yorkshire,* York

—— 1965b 'The Celtic chariot', *Antiq.* 39, 259–65

—— 1967 'A La Tène III burial at Welwyn Garden City', *Archaeologia* 101, 1–62

—— 1979 *The Arras Culture,* York

STEINER, P. 1929 'Das erste Wagengräbnis der frühen Eisenzeit in der Eifel', *Trierer Zeitschr.* 4, 145–47

STEVENSON, R. B. K. 1966 'Metal-work and some other objects in Scotland…', in A. L. F. Rivet (ed), *The Iron Age in Northern Britain,* 17–44, Edinburgh

STJERNQUIST, B. 1958 'Ornamentation métallique sur vases d'argile', *Medd. Lunds Univ. Hist. Mus.* 107–69

— 1967 *Ciste a cordoni,* (*Acta Arch. Lund.* 4° no. 6), Bonn/Lund

STRAHM, C. 1969 'Die späten Kulturen', in W. Drack (ed), *Archäologie der Schweiz,* II, 97–116, Basle

STRUVE, K. W. 1973 'Hölzerne Scheibenräder aus einem Moor bei Alt-Bennebek, Kr. Schleswig', *Offa* 30, 205–18

STURT, G. 1923 *The Wheelwright's Shop,* Cambridge

SUESS, H. & STRAHM, C. 1970 'The neolithic of Auvernier, Switzerland', *Antiq.* 44, 91–99

SULIMIRSKI, T. 1945 'Scythian antiquities in central Europe', *Antiq. Journ.* 25, 1–11

— 1961 'Die Skythen in Mittel- und Westeuropa', *Bericht v. Internat. Kong. Vor. und Frügesch. Hamburg 1958,* 793–99, Berlin

— 1970 *Prehistoric Russia: an outline,* London

TAFFANEL, O. & J. 1958 *Le premier Âge du Fer Languedocien. II* (Inst. Int. d'études Ligures, Monograph III), Bordighera/Montpellier

— 1962 'Deux tombes de cavaliers du Ie. Âge du Fer à Maillac', *Gallia* 20, 3–32

TARR, L. 1969 *The History of the Carriage,* London

TAUBER, H. 1966 'Copenhagen Radiocarbon dates VII', *Radiocarbon* 8, 213–34

— 1972 'Radiocarbon chronology of the Danish Mesolithic and Neolithic', *Antiq.* 46, 106–10

TELEGIN, D. J. 1971 *Über einen der ältesten Pferdezuchtherde in Europa,* (Rapports et comm… de l'URSS VIII Cong. Internat. Sci. Prehist. et Protohist.) Moscow

— 1977 'Ob absolyutiom vozraste yamnoy kul'tury i nekotorye voprosy khronologii eneolita yuga Ukrainy', *Sov. Arkh.* 1977, 2, 5–19

TERENOZHKIN, A. I. 1980 'Die Kimmerier und ihre Kultur', in Angeli *et al.* 1980, 20–29

THILL, G. 1972 'Frühlatènezeitlicher Fürstengrabhügel bei Altrier', *Hémecht* 24, 487–501

THOMPSON, G. B. 1958 *Primitive land transport of Ulster,* Belfast

THRANE, H. 1958 'The rattle pendants from the Parc-y-Meirch hoard, Wales', *Proc. Prehist. Soc.* 24, 221–27

— 1962a 'The earliest bronze vessels in Denmark's Bronze Age', *Acta Arch.* 33, 109–63

— 1962b 'Hjulgraven fra Storehoj ved Tobol i Ribe Amt', *Kuml,* 80–122

— 1963 'De førsle broncebidsler i mellem- og nordeuropa' *Aarbøger,* 50–99

TIERNEY, J. J. 1960 'The Celtic ethnography of Posidonius', *Proc. Royal Irish Acad.* 60 (C), 180–275

TIHELKA, K. 1954 'Nejstarší hliněné napsdobeniny čtyřramenných kol na území ČSR', *Pamatky Arch.* 45, 219–24

— 1961 'The question of datation of the four-spoked clay wheel models…', *Arch. Rozh.* 13, 580–85

TOČIK, A. 1961 'Stratigraphie auf der befestigten Ansiedlung in Malé Kosihy, Bez. Štúrovo', in A. Točik (ed), *Kommission Nitra 1958,* 17–42, Bratislava

— 1964 *Befestigte bronzezeitliche Ansiedlung in Veselé,* Nitra

TODD, M. 1975 *The Northern Barbarians: 100 BC-AD 300,* London

TÓTH, E. H. 1976 'The equestrian grave of Izsák-Balápuszta from the period of the Magyar conquest', *Cumania* 4, 141–73

TRATMAN, E. K. 1925 'Second report on Kings' Weston Hill, Bristol', *Proc. Spelaeo. Soc. Univ, Bristol* 2:3, 238–43

TREUE, W. 1965 *Achse, Rad und Wagen,* Munich

TRINGHAM, R. 1971 *Hunters, fishers and farmers of Eastern Europe 6000–3000 BC,* London

UCKO, P. J. & ROSENFELD, A. 1967 *Paleolithic Cave Art,* London

UENZE, O. 1956 *Die ersten Bauern (Jungsteinzeit),* (Vorgesch. Nordhessen 2) Marburg/Lahn

— 1958 'Neue Zeichensteine aus dem Kammergrab von Züschen', in W. Krämer (ed) *Neue Ausgrabungen in Deutschland,* Berlin

ULRICH, K. 1973 'Ein späthallstattzeitliches Wagengrab von Offenbach-Rumpenheim', *Arch. Korresp.* 3, 313–35

VENEDIKOV, I. 1957 'Trakiyskata yuzda', *Izvestiya Arch. Inst. Sophia* 21, 153–201

— 1960 *Trakiyskata kolesnitsa,* Sofia

VENTRIS, M. & CHADWICK, J. 1959 *Documents in Mycenaean Greek,* Cambridge

VERWERS, G. J. n.d. *Het vorstengraf van Meerlo,* Bonnefanten Museum, Limburg

VIGNERON, P. 1968 *Le cheval dans l'antiquité Gréco-Romaine,* Nancy

VIZDAL, J. 1972 'Erste bildliche Darstellung eines zweirädigen Wagens vom Ende der mittleren Bronzezeit in der Slovakien', *Slovenska Arch.* 20, 223–31

VOGEL, J. C. & WATERBOLK, H. T. 1963 'Groningen radiocarbon dates IV', *Radiocarbon* 5, 163–202

—— 1972 'Groningen radiocarbon dates X', *Radiocarbon* 14, 6–110

VOGT, E. 1947 'Basketry and woven fabrics of the European Stone and Bronze Ages', *CIBA Review* 54, 1938–1964

VOUGA, P. 1923 *La Tène*, Leipzig

VULPE, A. 1967 *Necropola hallstattiană de la Fergile*, Bucharest

WAALS, J. D. VAN DER 1964 *Prehistoric Disc Wheels in the Netherlands*, Groningen

WAILES, B. 1976 'Dun Ailinne: an interim report', in D. W. Harding (ed), *Hillforts: later prehistoric earthworks in Britain and Ireland*, 319–38 London/New York

WAINWRIGHT, G. J. 1979 *Gussage All Saints: an Iron Age settlement in Dorset*, London

—— & SPRATLING, G. 1973 'The Iron Age settlement of Gussage All Saints', *Antiq.* 47, 109–30

WANKEL, H. 1882 'Bericht über die Ausgrabung der Býčí Skala-Höhle', in Angeli *et al.* 1970, 99–138

WARD-JACKSON, C. H. & HARVEY, D. E. 1972 *The English Gypsy Caravan*, Newton Abbot

WARD PERKINS, J. B. 1940 'Two early linchpins...' *Antiq. Journ.* 20, 358–67

—— 1941 'An Iron Age linchpin..., from Cornwall', *Antiq. Journ.* 21, 64–67

WELLS, H. B. 1974 'The position of the large bronze saws of Minoan Crete in the history of tool making', *Expedition* 16.4, 2–8

WELLS, P. S. 1980 *Culture contact and culture change: Early Iron Age central Europe and the Mediterranean world*, Cambridge

WESTERN, A. C. 1973 'A wheel-hub from the tomb of Amenophis III', *Journ. Egypt. Arch.* 49, 91–94

WHITE, K. D; 1967 *Agricultural implements of the Roman world*, Cambridge

WHITE, L. JR. 1962 *Medieval Technology and Social Change*, Oxford

—— 1970 'The origins of the coach', *Proc. Amer. Phil. Soc.* 114, 423–31

—— 1978 *Medieval religion and technology*, Berkeley/Los Angeles/London

WIESNER, J. 1968 *Fahren und Reiten*, (Arch. Homerica I. F.) Göttingen

VAN WIJNGAARDEN-BAKKER, L. H. 1974 'The animal remains from the Beaker settlement at Newgrange...', *Proc. Royal Irish Acad.* 74, (C), 313–83

—— 1975 'Horses in the Dutch Neolithic', in A. T. Clason (ed), *Archaeozoological Studies*, 341–44, Amsterdam

WILD, J. P. 1970 'Borrowed names for borrowed things?', *Antiq.* 44, 125–30

WITT, T. 1969 'Egerhjul og Vogne', *Kuml*, 111–48

WOYTOWITSCH, E. 1978 *Die Wagen der Bronze- und frühen Eisenzeit in Italien*, (*PBF* XVII. 1), Munich

YADIN, Y. 1963 *The art of warfare in Biblical lands*, London

ZACCAGNINI, C. 1977 'Pferde und Streitwagen in Nuzi, Bemerkungen zur Technologie', *Jahresber. Inst. für Vorgesch. Univ. Frankfurt A. M.*, 21–38

ZBENOVICH, V. G. 1973 'Chronology and cultural relations of the Usatovo group in the USSR', in Kalousek F. & Budinský-Krička, V. (edd), *Symposium über... der Badener Kultur*, 513–24, Bratislava

ZEUNER, F. E. 1963 *A History of Domesticated Animals*, London

ZHORZHIKASHCHVILI, L. & GOGADZE, E. M. 1974 *Pamyatniki Trialeti epokhi ranney i sredney bronzy: katalog Trialetskikh materialoz*, Tbilisi

ZÜRN, H. 1972 'Fundchronik Jahre 1965–67', *Bayerische Vorgesch.* 37, 170

Index of places and subjects

Note: In this index 'radio-carbon date from' is abbreviated to 'rcd'. References to illustration pages are shown in italic type.

Index of places

Index of subjects

models, 16; baked clay, 29; chariot, 70, 136; lead, 151; metal, 29; pottery, 37, 40–1, 45–6, *55*, 56, 58; sledge, 37; vehicle, 15, 31, 45–6; wheel, 13, 40–1, 46, 48–9, 58

mortice, 21, 24, 25, 27, 74, 95, 98; circular, 58; dovetail, 25, 50, 51–2, 85–6; internal tubular, 25, 31, 68, 72, 198; square, 25, 51, 52

'Mother Goddess', 83

moulds, 224

Mycenaean: chariot representations, 112; civilization, 187; contacts, 65–6, 104, 112; spiked discs, 100

nails: bronze, 160; close-set, 167, 193; domed, 167; fastening, 168, 216; iron, 163, 183, 186; stud, 170; types of, for iron hoop tyres, 167–8, 169, 170; wide-spaced, 168, 185, 193

nave, 18–19, 27, 28, 97, 154, 160–5, 213–15; annular, 194; biconical, 163–4; bronze—bands, 161; bronze sheathing, 112–13, 122, 125, 127, 168, 183; cast ribbed tube, 124, 133, 134; classification of, 163–5; coloured, 57; composite and clamped, 213; conical, ribbed iron, 160; crimped iron fillet in, 213, 229; cylindrical, 114, 164, 165, 213; cyma moulded, 213; double-reel, 111, 115; integral, 24, 25, 49, 50, 52, 57, 68; inserted tubular, 24, 25, 26, 50–1, 85–6, 98, 103, 107, 198; iron, 128, 165, 183, 185; iron band, 158, 160, 183, 186, 204, 205, *213;* iron sheathing, 140, 148, 149, 158, 160; long, decorated, 160; metal sheathing, 159–60, 162–5; ogee mouldings, 164–5, 222–3, 227; pairs of nave bands, 213, 223, 225, 226, 228, 236; perforation, 111; and spoke, 27; tubular, 114; turned, 95, 162; wooden with iron bushes, 165, 186

Neolithic, 20, 35, 39, 40; European, 63, 83, 240; late, 54, 88; middle, 44

Nerthus, 227

North American Indians, 36, 63, 243; adoption of horse riding by, 63

oak, 19, 26, 28, 49, 50, 56, 73, 124, 157, 160, 166, 172, 199, 218; age and size of, 19–20, 125; cork-, 193

ochre, red, 54, 61

Ochre- Grave culture, 54

Old Testament, 9

onagers, 103

'orientalizing phase', 128–9, 130, 132, 137, 169, 178, 187, 192, 207

ornament: animal, 72–3; crimped fillet, 213–14, 229; Greek key pattern, 212; geometric, 73, 101, 115, 116; 'leaf-crowns', 184; Mycenaean, 104; ogee moulding, 164–5, 170, 183; openwork, compass-drawn circles, 220; openwork triangles, 112, 113, 127; ring and dot, 101, 102, 161, 163, 183; 'sea-serpent', 197; spiral motifs, 73, *77;* spiral and wave, 102, 104

ovicaprids, 34

ox, 10, 27, 90, 243; castrated, 26, 34–5, 239; -drawn, 17, 36, 37, 38, 43, 48, 59, 60, 64, 66, 75, 95, 106, 236; models of, 38, 42; paired, 47, 48, 64, 84, 192; protomes, 46; representations of, 35, 42–4, 53–4, 79, 80, 108, 119; skeletons, 72; skull, 71, 72, 75, 76; speed of, 89–90, 241

Palladius, 231

palstave, 115

peat bog, 31, 36, 49, 85, 108, 197, 216, 225, 226

Pegge, Samuel, 10

pendant, bronze, 102

Persians, 9, 228

petorritum, 231

Philargyrius, 232

Philip of Macedon, 209, 210

Phoenicians, 128, 130–1, 132, 187, 193, 194

pictographs, 12, 38, 39

pig, 14, 40; cremated, 226; joints in burial, 207

pigeon, 207

pigments/colouring materials, 123–4

Piliny culture, 93–4, 109, 114, 159

pin: gold, 60; hammer-headed, 56, 61; silver, ·70; wheel-headed, 102

pine, 26, 56, 69

Pinus sp., see pine

pitch, 214

Pit-Grave culture, 39–40, 45, 54–61, 67, 75, 78, 142, 241; dates of vehicle burial, 56; end of, 90; phases of, 55–6

plank: felloe, 124–5; floor, 46, 68, 69, 73, 74, 75–6, 147, 154; morticed, 177; sawn, 18; size of, 19–20; split, 18, 27, 49, 50, 65, 148, 198; spoked wheel cut from, 112

platform: planked, 76, 154, 208; wheeled, 125, 150–1

plaustrum, see *ploxenum,* 9

Pliny, 231

plough, 14, 34–6, 38; cultivation, 65; marks, 35–6; ox-drawn, 36; representations of, 35, 53; share, 35; traction-, 35, 36, 64, 88, 239

ploxenum, 231

Polada culture, 94

pollen analysis, 30, 34

Polybius, 230, 234

Pomponius Mela, 9, 230, 233

pony, 89, 90; bronze cap for, 223; draught, 10

poplar, 19, 28

populus sp. see poplar

Posidonius, 208, 211, 230, 234, 236

potter's wheel, 17

pottery, 31, 42, 71, 72; Attic black-figured, 140, *179,* 235; Attic painted, 139, 165, 170, 179, 188; Attic red-figured, 184, 196, 228; cinerary urn on wheels, 93, 110; collared flask, 44; cross-foot bowl, 56; cup on wheels, 46–7; early Neolithic, 44; four-wheeled wagon, 83–5, *83;* Greek geometric funerary, 188, 189; Greek cups, 201; handle with yoked beasts, 38, 42; Iberian painted, 194; incised drawings on, 29, 41–2, 91, 92, 94, 116, 149–50; Iron Age, 177; Kara-Araxes type, 59; Late Corded Beakers, 44; Mesoamerican toys 15; model wheels, 82, 91, 92–3; 'Pomeranian Face-Urn', 150, 153, 156, 228; tin inlay on, 116; vehicle models, 37, 41, 45–6, *55,* 56; wheel-turned, 163

priests, 122

Propertius, 231, 232

prothesis, 140

Prunus, sp. see cherry-wood

Pu-Abi, 75

pulses (lentils, peas), 207

Pyrus, see fruitwood

Quercus aegilops, 172; sp. see oak

Quintillian, 231

radio-carbon dates: Catacomb Graves, 56; hill-fort, 177; planking, 18, 197; plough marks, 33, 36; Pit-Graves, 40, 55–6; shield, 197; Srédny Stog culture, 87; 'tell', 39; Timber-Graves, 90; trackways, 49, 65; vehicle graves, 52; wagon, 68, 70; wheels, 33, 49, 51, 56, 112, 193, 198–9; wooden sledge, 36

radio-carbon dating, 13, 26, 31, 32, 58, 62–3, 65–6, 106, 108, 198, 240; calibration of, 32, 33, 63, 66

raeda, 231

rafts, 14

razor, 177; bronze, 171

red deer, 99, 101, 126

Remedello culture, 35, 54

Remi, 197, 209; chariots on coins, 209

ring, finger: bronze, 177; gold, 205, 226

roadways, see also trackway, 65

rock-carvings, 29, *37,* 49, 53–4, 76, *80, 81,* 94, 106, 107–8, 109, 116–19, 148; Italy, 106, 152; Karatau, 82; Monte Bego, 35; Scandinavian, 35, 79, 98, 106, 151, 152, 156; Syunik, 17, 37, 76, 78–82; Val Camonica, 35, 53–4, 107–8

rock-paintings: Saharan, 90; Spanish, 76, 94, 98

Roman: Conquest, 137, 197, 231, 233; later-period, 242; occupation of Britain, 223, 231–2; period, 90, 212; vehicle words, 230–33

roofing, 54, 142, 154, 202, 203–4

rope, 126; guide, 82, 95, 118, 119; mouthpiece, 101; nose, 39, 75

Royal Ascot, 243

salix sp. see willow

Sammes, Aylett, 10, 233

sanctuaries, 193

Sargon I of Akkad, 62

saw, 17, 18, 25

scabbard, engravings on, 187, 212, 220

Scheffer, John, 9–10, 239

scissors, 26

sculptures: Assyrian reliefs, *169,* 169, 170, 172, 180; bronze yoke ornaments, 75, 96–7; Pergamon reliefs, 197

scythe, 233–4

Scythian culture, 9, 99, 185, 186, 187, 228–9

seat, 69; with carvings, 72, 73

Seine-Oise-Marne (SOM) culture, 89

Senecca, 232

Sennacherib, 170

Senones, 197

Servius, 232

settlements, 83, 87–8, 123, 138–9, 201, 214, 222, 224; defended, 195; hill-top, 139; lakeside, 51, 112, 127; nobleman's seat (*Adelsitze*), *139; oppidum,* 201; princely residences (*Fürstensitze*), 139, 140; 'tell', 34, 39, 44

sheep, 14, 34, 40; cremated, 226

shield: shield-boss, 226; on Iberian stelae, 131–2, 193; Irish circular wooden, 19; moulds for leather, 19; oval, 150, 196, 210, 228; rcd for, 197; representation of, 197, 228; V-notch type 132; wood, 197, 228

shipbuilding, 18; representations of, 108, 118

'shoe-last celt', 35

'shrine', 83

sickle: bronze, 177; iron, 162, 177

side (of vehicle): double inverted 'U' hoop, 210, 211; wickerwork, 72, 73, 76, 209

Silius Italicus, 233

silver, 70, 225; metalwork, 75, 225, 228

Single-Grave culture, 50

situla: art, 132, 149, 178–9; bronze, 122, 124, 125, 135, 183, 201; cross-handle-attachment, 151; distribution of, 178; silver, 70; vehicles on, 149, 178–82, 193

sled, see also sledge, 12

sledge, 10, 14, 16, 36–9, 60, 64, 239; -car, 38; children's, 237; ox-drawn, 75; pictographs of, 38, 61, 63; pottery models of, *37;* representations of, 37, 78, 79, 193; rollers, 39, 239; runners, 16, 36, 37, 239; -throne, 38; triangular/boat-shaped, 37; on wheels, 239

slide-car, 10, 12, 14, 16, 36–9, 60, 76, 193; Irish

Sources of Illustrations

The nine maps and a number of line illustrations have been redrawn by H. A. Shelley from material provided by the author and from other sources which are indicated by a reference to the *Bibliography*.

2 courtesy of American Museum of Natural History, New York; *4* Lucas, 1952; *6* Kravets, 1951; *7 (1, 4, 7, 8)*, *39* Karakhanian and Safian; *7 (2, 6)* Breuil, 1933; *7 (3, 5)* Bicknell, 1913; *9* Arkeolojii Müzesi, Ankara; *10, 11* courtesy Professor S. Milisauskas; *12, 17, 25, 28, 29, 30, 31, 32, 33, 34, 35, 36, 37* Piggot, 1968; *13* Staatliche Kunstsammlung, Kassel; *14* courtesy Dr Němejcova-Pavúková; *15, 16* Magyar Nemzeti Múzeum, Budapest; *18, 56, 89* Piggott, 1965; *19* Ruoff, 1978; *20* Büro für Archäologie, Zürich; *21, 60, 93* copyright 1982 by Centro Camuno di Studi Preistorici, 25044 Capo di Monte, Italy; *22a, 79* Almagro, 1966; *22 b & c* Van Berg-Osterrieth, 1972; *40* courtesy Professor J. V. S. Megaw; *42* Staatliches Museum für Naturkunde und Vorgeschichte, Abt. Moorforschung, Oldenburg; *43, 53* Woytowitsch, 1978; *45* photo Larry Burrows; *46* Hermitage Museum, Leningrad; *48* courtesy Dr L. V. Galkin; *50* courtesy Dr J. Vizdal; *51* Piggott, 1974; *59* Federseemuseum, Buchau; *61* Národni Muzej, Belgrade; *62, 65, 66* Hampel, 1890; *63, 73* Museum für Vor- und Frühgeschichte, Berlin; *64* Moszolics, 1956; *67, 95* Národni Muzeum, Prague; *68, 119, 141* by permission of the Danish National Museum, Copenhagen; *69* Cordier, 1975; *70* courtesy Professor J. M. Coles; *71* courtesy Dr C. Peschek; *72, 99, 102, 130* Naturhistorisches Museum,

Vienna; *74* photo R. L. Wilkins; *76* photo Rihse (Jankuhn, 1957); *77* Balkwill, 1973; *78a* The Metropolitan Museum of Art, Purchase 1956, Walter C. Baker Gift; *78b, 105, 116* British Museum, London; *80* Savory, 1976; *81* Nationalmuseum, Stockholm; *83* Esayan, 1967; *86, 87* Kossack, 1970; *88* Fischer, 1979; *90* Joffroy, 1958; *92* La Baume, 1928; *94* courtesy Dr. J. Biel and Landesdenkmalamt, Baden-Württemberg; *96* Felgenhauer, 1962; *100* courtesy Dr J. A. Brongers and Rijksdienst voor het Oudheidkundig Bodemonderzoek, Amersfoort; *101* Drack, 1958; *103* Joffroy 1954; *104, 133* Kossack 1971; *106, 110* Mariën, 1958; *107* courtesy of Dr R. Pleiner and Dr V. Moucha, photo Archaeological Institute, Prague; *108* courtesy Dr U. Fischer and Museum für Vor- und Frühgeschichte, Frankfurt-am-Main; *111 a & b* courtesy Professor O. H. Frey (Lucke & Frey, 1962); *112a* Kastelic, 1956; *113* Museo Civico, Como; *114* Djuknic and Jovanic, 1966; *115* courtesy Piggott; *116b* National Museum, Athens; *116c* courtesy Museum of Fine Arts, Boston; *117* Woytowitsch, 1978; *119* Joachim, 1979; *121* Morel, 1876; *122* Fourdringier, 1878; *123* Lemoine, 1906; *125, 131* courtesy Dr A. Cahen-Delhaye; *127a* Bibliothèque Nationale, Paris; *127 b–d* photo R. L. Wilkins; *128* Museo Civico, Padua; *132* courtesy Dr H. J. Hundt and Bad Nauheim; *132* Kramer and Schubert, 1979; *135* Stead, 1979, drawn by J. Thorn; *136* Macdonald and Park, 1906; *137* Jacobsthal, 1944. Photo Dr Günther; *138* Piggott, 1969a; *139* photo Department of the Environment, Crown Copyright reserved; *142* The Aldine Romance of Invention, Travel and Adventure Library, No. 8.